Intracellular Staining
in
Neurobiology

Intracellular Staining
in
Neurobiology

Edited by

Stanley B. Kater

and

Charles Nicholson

Springer-Verlag Berlin Heidelberg GmbH

1973

Library of Congress Cataloging in Publication Data

Kater, Stanley B
　　Intracellular staining in neurobiology.

　　1. Stains and staining (Microscopy)　2. Neurobiology—Technique.　3. Re-
active dyes. I. Nicholson, Charles, Joint author. II. Title. [DNLM: 1.
Histological techniques.　2. Stains and staining.　QH237 K19i 1973]
QH237.K33　　578'.64　　73-77837

ⓒ 1973 by Springer-Verlag Berlin Heidelberg
Originally published by Springer-Verlag Berlin Heidelberg New York in 1973
Softcover reprint of the hardcover 1st edition 1973

ISBN 978-3-642-87125-2　　　ISBN 978-3-642-87123-8 (eBook)
DOI 10.1007/978-3-642-87123-8

To the memory of

Donald Maynard

Preface

The purpose of this book is to provide basic reference material to facilitate the further development and application of the intracellular staining techniques which originated with the introduction of Procion dye as a neurocellular stain. We had several specific objectives in compiling these chapters. First, we recognized that both the techniques and results of intracellular staining have been derived from a variety of preparations and published in diverse journals. Consequently, we tried to make this information more readily available by incorporating several reviews by original authors. This should provide the reader with a knowledge of the kinds of neurobiological problems for which intracellular staining has already been employed. A second objective was to facilitate extension of these methods to a wider variety of problems. To this end there are chapters dealing with Procion dye as a marker for ultrastructural investigations, the use of cobalt as a staining compound and the implementation of computer techniques for quantitative studies of neuronal relationships. As background for further extension of these methods this volume contains material on the history of the discovery of Procion dyes as intracellular stains, the chemistry of these dyes, and fundamentals of fluorescence microscopy. Our final specific objective was to present details of technical procedures that were not readily available in the literature. In addition to asking contributors to stress this aspect and providing a final technical chapter, a two-day symposium was held at the University of Iowa in October of 1972. This meeting acted as a forum for discussion of technical details; much of this, often animated, discussion has been included in the book. It is hoped that, by virtue of the breadth of expertise represented by those in attendance, this volume gives a true, "state of the art", picture of intracellular staining technology.

The organization of this volume has proceeded smoothly, due, in large part, to the cooperation of our contributors, each of whom met stringent deadlines and wrote chapters which we believe make this book a relatively cohesive entity rather than simply a collection of papers. Our publishers, Springer-Verlag, New York, cooperated fully in realizing our desires for high quality and rapid publication. We also wish to particularly acknowledge Dr. Konrad Springer who took a personal interest in this endeavor. The whole venture was, of course, dependent upon generous financial support from several sources. We are indebted to the Alfred P. Sloan Foundation, which not only financed much of the symposium and editing expenses but also paid for the production of the color plates of this volume. Another major financial contributor was the University of Iowa Interdisciplinary Committee for Neurobiology (NSF-GU 2591). Other valuable support came from the University of Iowa Neurobehavioral Sciences Program (5 T01 MH 10641–07), the University of Iowa Graduate College Learned Publications Fund, and the Department of Physiology and Biophysics and the Department of Zoology.

On an individual basis, we wish to thank Dr. Rodolfo Llinás who provided a lot of the initiative for this undertaking. Dr. Jerry Kollros likewise fully supported our efforts and provided much timely advice. Both Dr. Llinás and Dr. Kollros assisted with preliminary readings of this volume. Dy Mistick bore the brunt of the copy editing with professional patience and unfailing good humor. We owe special thanks to Chris Kaneko who master-minded the practical details of the symposium. Joleen Brue provided a most human interface with the computer used to store this manuscript. To these and the many other individuals who facilitated this effort, we offer our thanks.

Iowa City, Iowa   Stanley B. Kater

August 15, 1973   Charles Nicholson

List of Contributors

John N. Barrett, Department of Neurobiology, Harvard Medical School, Boston, Massachusetts.

M. V. L. Bennett, Department of Anatomy, Kennedy Center for Research, Albert Einstein School of Medicine, Bronx, New York.

D. J. Bihary, Department of Computer Science, Carnegie-Mellon University, Pittsburgh, Pennsylvania.

Melvin J. Cohen, Department of Biology, Yale University, New Haven, Connecticut.

William J. Davis, Division Natural Sciences I, University of California, Santa Cruz, California.

Yoko Hashimoto, Tokyo Women's Medical College, 10 Kaeada-cho, Shinjuku-Ku, Tokyo 106, Japan.

E. Jankowska, Department of Physiology, University of Göteborg, Göteborg, Sweden.

Akimichi Kaneko, Department of Physiology, School of Medicine, Keio University, Shinaneomachi, Shinjuku-Ku, Tokyo, Japan.

C. R. S. Kaneko, Department of Zoology, University of Iowa, Iowa City, Iowa.

Stanley B. Kater, Department of Zoology, University of Iowa, Iowa City, Iowa.

James Kelly, Department of Neurobiology, Harvard Medical School, Boston, Massachusetts.

Donald Kennedy, Department of Biology, Stanford University, Palo Alto, California.

E. A. Kravitz, Department of Neurobiology, Harvard Medical School, Boston, Massachusetts.

S. Lindström, Department of Physiology, University of Göteborg, Göteborg, Sweden.

Rodolfo Llinás, Division of Neurobiology, Department of Physiology and Biophysics, University of Iowa, Iowa City, Iowa.

U. J. McMahan, Department of Neurobiology, Harvard Medical School, Boston, Massachusetts.

William H. Miller, Department of Ophthalmology and Visual Sciences, School of Medicine, Yale University, New Haven, Connecticut.

Brian Mulloney, Department of Biology, University of California San Diego, La Jolla, California.

R. K. Murphey, Department of Zoology, University of Iowa, Iowa City, Iowa.

Charles Nicholson, Division of Neurobiology, Department of Physiology and Biophysics, University of Iowa, Iowa City, Iowa.

R. B. Ohlander, Department of Computer Science, Carnegie-Mellon University, Pittsburgh, Pennsylvania.

Robert N. Pitman, Department of Physiology and Biochemistry, University of Southampton, Southampton, England.

D. Purves, University College London, London, England.

D. R. Reddy, Department of Computer Science, Carnegie-Mellon University, Pittsburgh, Pennsylvania.

Takehiko Saito, Department of Ophthalmology and Visual Sciences, School of Medicine, Yale University, New Haven, Connecticut.

Allen I. Selverston, Department of Biology, University of California San Diego, La Jolla, California.

C. V. Stead, Research Department, Organics Division, Imperial Chemical Industries Limited, Blackley, Manchester, England.

A. O. W. Stretton, Department of Zoology, University of Wisconsin, Madison, Wisconsin.

Tsuneo Tomita, Department of Ophthalmology and Visual Sciences, School of Medicine, Yale University, New Haven, Connecticut.

Charles D. Tweedle, Department of Zoology, Michigan State University, East Lansing, Michigan.

David Van Essen, Department of Neurobiology, Harvard Medical School, Boston, Massachusetts.

L. S. Van Orden III, Department of Pharmacology, University of Iowa, Iowa City, Iowa.

Table of Contents

Intracellular Staining
in
Neurobiology

CHAPTER 1

The Development of Intracellular Staining

Charles Nicholson and Stanley B. Kater

If a neurobiologist were asked to select the two most indispensible experimental techniques in his field, he would probably choose the histological method of impregnation named after Golgi and an electrophysiological recording system using the glass micro-pipette invented by Ling and Gerard. It is quite true that our understanding of the nervous system owes a great debt to many other techniques, including multitudinous histological stains based on normal and degenerating material, ablation methods, gross electrical stimulation and recording, sophisticated chemical assays, and electron micros-copy. Nevertheless it was the Golgi method and the micropipette which furnished the major advances in the understanding of basic neural mechanisms. In 1968 a technique was announced (Kravitz et al., 1968; Stretton and Kravitz, 1968) which combined features of both the Golgi method and the micropipette and led many to forecast a new era of progress in neurobiology. The present book surveys the results which have stemmed from the new technique; it documents unexpected findings and applications, and, above all, brings together the experience gained by the many people who have used the new methods. This chapter will begin with a brief review of the staining techniques which were available prior to 1968. The chapter will then indicate the scope of research that has been carried out with the new intracellular staining methods, introduce subsequent chapters of the book, and comment on future trends.

Earlier Methods for Correlating the Structure and Function of Neurons

The Classical Histological Procedures

Prior to 1968 there were only two histological methods for the selective and complete staining of single neurons. These were the Golgi and Methylene Blue methods.

The merit of the Golgi method, discovered by Camillo Golgi in 1873, lies in its potential ability to stain a nerve cell in its entirety without staining other elements in the vicinity. Such a stain is highly informative because (1) the pattern of processes of a neuron is often a characteristic of one type of cell, thus a whole-cell stain enables classification; (2) the trajectories of dendrites and axons are suggestive of cellular connec-tions, which may be confirmed by the additional evidence of other stains, electron

1

microscopy, and electrical recording; (3) the specific geometry of a cell may have functional significance.

The Golgi method has the drawback of being capricious. It requires enormous patience and considerable experience with the method to stain systematically all types of neurons in even a restricted region of a nervous system. Fortunately, the famous contemporary of Golgi, Ramón y Cajal, possessed the motivation, skill, and insight to exploit the Golgi method to the full, and his monumental contributions are a continuing source of invaluable data (Ramón y Cajal, 1909–1911). Ramón y Cajal worked mainly on the vertebrate nervous system but even he was not able to induce every type of cell to stain. When the Golgi method was extended to invertebrates, refractoriness to staining was found to be even more common and the apparent lack of organization within the neuropil further complicated the analysis of these systems. Attempts to understand the mechanism of the Golgi method have met with little success, although the chemical nature of the precipitate is now known in at least one case (Fregerslev et al., 1971). The current status of the method has been reviewed recently in the book by Nauta and Ebbesson (1970).

The only other classical method of staining nerve cells in their entirety is the Methylene Blue technique (Ehrlich, 1886). Like the Golgi method, the Methylene Blue stain is highly capricious, although with invertebrates it usually gives better results than does Golgi impregnation (Bullock and Horridge, 1965).

Information that is otherwise unobtainable by either the Golgi or Methylene Blue methods can be provided by experimental degeneration techniques (e.g., one of the Nauta methods; see Nauta and Ebbesson, 1970). These methods allow one to trace long axons by virtue of their characteristics during degeneration. However, they provide only information about axonal trajectories and not complete neuronal geometry.

By the early part of the twentieth century, much insight into the morphological characteristics of neuronal ensembles had been obtained through the use of the Golgi and Methylene Blue methods. For example, Ramón y Cajal was able to draw a circuit diagram of the afferents, efferents, and the five main classes of neurons in the cerebellar cortex. Furthermore he indicated the probable direction of information flow. Recent data, obtained using the electron microscope and electrical recording, have substantiated all Ramón y Cajal's results. Ramón y Cajal found it difficult to surmise the possible functional interactions between the elements of his cerebellar circuit, and this type of problem was common among anatomists. In essence, the resolution of the anatomical data during the early part of this century exceeded the available physiological information.

The Electrical Recording Micropipette

The situation changed dramatically in 1949 when Ling and Gerard produced a recording electrode with a sufficiently fine tip to enable the electrode to be pushed through the membrane of a muscle fiber without changing the electrical activity of the cell (Ling and Gerard, 1949). The major discovery of Ling and Gerard was that a hard glass capillary tube could be pulled out to a submicron tip without sealing. When filled with a suitable electrolyte, such a micropipette could be employed as a recording electrode. The simplicity of this device belies its enormous utility. Today micropipettes are the dominant tool for electrophysiological recording in the nervous system. Prior to the glass micropipette, sharp insulated metal electrodes had been used, but their mechanical and electrical properties precluded their use for intracellular recording in most cells of interest. Detailed information about micropipettes is given by Frank and Becker (1964) and in the book by Lavallée et al. (1969). Basically, glass micropipettes enabled the recording of intracellular, as well as extracellular, electrical activities of nerve cells and thus opened up a vast field of investigation.

After its invention, the micropipette was used extensively in the nervous system of vertebrates (e.g., Eccles, 1957). With very few exceptions the neurons of the central nervous system are invisible in the living preparation. Thus the successful penetration of a nerve cell depended on monitoring the electrical activity recorded by the electrode. In regions of the nervous system where only a few cells were located in a well defined geometrical arrangement, or where tests such as antidromic stimulation could be employed, it was often feasible to use deductive reasoning to correlate the picture revealed by the Golgi method with the intracellular electrical activity. Thus it became possible to associate a particular class of electrical activity with a particular type of cell and finally to deduce the action of one cell type upon another. A dramatic discovery originating from this type of analysis was that many cells inhibited, rather than excited, their neighbors (Anderson et al., 1964a, b; Eccles, 1969). This result could not have been deduced from anatomical data alone.

Work on invertebrates appeared at first sight to be easier, both because the number of cells in any region of the nervous system was vastly less than that in the vertebrate and because the cell bodies were often visible in the living preparation. Thus one could hope to penetrate and record from a high percentage of cells in a given ganglion of an invertebrate. Much progress was made, but gradually two problems became apparent. First, the anatomy and connectivity of most of the neuropil were unknown, thus one had little idea which processes were joined to a given soma. Second, much of the electrical activity of the cell took place within the complex matrix of the neuropil, while the soma in some animals was electrically inexcitable and played little role in the cellular electrical transactions. Recording from the invisible elements of the neuropil often proved as technically difficult as recording from the unseen elements of the vertebrate nervous system. Thus by the mid 1960's electrophysiology had caught up with anatomy and, in some areas at least, had utilized all the information available at the light and electron microscopic levels. Two crucial problems hampered further progress. First, there was no reliable method for routinely staining many nerve cells in their entirety, particularly in the invertebrate systems. It might, literally, take years to stain a member of a given class of cells with the Golgi or Methylene Blue method, especially when the class was very small. Second, many of the most intriguing questions required the correlation of electrical and anatomical characteristics of a cell, particularly in the central nervous system of vertebrates where the abundance of cell types in a given area might make it impossible to correlate electrical activity with cell types revealed by the Golgi method. A further and potentially more interesting question was whether differences in individual electrical activity could be related to differences in the individual morphology of neurons.

Origins of Staining Methods using Microelectrodes

One way to achieve the correlation of anatomy and physiology would be to fill the micropipette with an electrically conductive solution which could also be made visible at the microscopic level. The electrode could be used to record electrical activity and then to infiltrate the solution into the cell in order to visualize it. This concept is not new and has its origins in extracellular methods designed to produce a localized tissue stain around the tip of a recording electrode for the purpose of locating the recording site. A brief review may be helpful in illustrating the reasons for the superiority of the new methods and in giving some clues to further avenues of research.

Iron Stains

The first attempts at localized tissue staining arose out of the need to pinpoint the placement of the tip of gross recording metal electrodes within the tissue. By passing a relatively large current through the electrode a lesion would be produced which was visible on subsequent histological examination. This technique often extensively damaged the tissue and was superseded by a staining method based on the Prussian Blue reaction for ferric iron (Perls, 1867). In this method a steel electrode was used, and, after recording, a current was passed through it to deposit iron in the tissue. Subsequently the tissue was bathed in a solution containing ferrocyanide, forming an insoluble precipitate of Prussian Blue (ferric ferrocyanide) around the position of the tip (Hess, 1932; Adrian and Moruzzi, 1939; Marshall, 1940; Scheibel and Scheibel, 1956; Green, 1958). The method has also been used with steel microelectrodes for limited intracellular marking in the large Mauthner cell of the goldfish (Furshpan and Furukawa, 1962); these authors used a bathing solution containing both ferro– and ferricyanide so that the latter constituent would react with any ferrous iron to yield a precipitate of Turnbull's Blue (ferro ferricyanide). The blue precipitates formed with these methods were not unduly washed out by subsequent histological processing of the tissue: an essential requirement for any marking technique.

With the advent of the micropipette the Prussian Blue method was readily available by simply filling the pipette with one of the reactants involved and expelling it into the tissue by iontophoretic current. Thus Rayport (1957) filled pipettes with ferric ammonium sulfate and, after passing some of the ferric ion into the tissue, bathed it in potassium ferrocyanide, while Bultitude (1958) filled the pipette with sodium ferrocyanide and immersed the tissue in ferric chloride; both these workers thus obtained a localized precipitate of Prussian Blue. Other investigators filled pipettes with potassium ferricyanide and bathed the tissue in ferrous chloride, ferrous sulfate, or a mixture of ferrous and ferric ammonium sulfate to obtain Turnbulls's Blue (Scheere and Mumbach, 1960; Gouras, 1960; Talbott et al., 1967). In all cases the micropipette continued to record apparently normal electrical activity from the neural tissue.

All the methods so far cited were used for extracellular staining. The first documented application of Prussian Blue for intracellular staining appears to have been that by Kerkut and Walker (1962), who used an electrode filled with potassium ferrocyanide and a bathing solution of ferric chloride. Kerkut and Walker noted that the stain remained localized to the region of injection and proximal axon, but they suggested that extended iontophoresis might cause migration to distant regions of the cell; a later paper (Kerkut et al., 1970) indicated that these expectations were not realized. Behrens and Wulff (1965) achieved localized intracellular staining using a micropipette filled with potassium ferrocyanide and, in this case, cobaltous chloride in the bathing solution. Stretton and Kravitz (Chapter 2) also report that intracellular ferrocyanide ions do not give good staining of cellular processes.

Other Metal Stains

Many other metal ions were used as the electrolyte in micropipettes and tried as extracellular markers, including silver nitrate (Motokawa et al., 1957; Oikawa et al., 1959), lithium carmine (Mitarai 1958, 1960), gold trichloride (von Baumgarten et al., 1960) and complex copper ions (Galifret and Szabo, 1960; Frank and Becker, 1964). More recently an attempt was made to use copper as an intracellular stain in motoneurons (Lux and Globus, 1968). The copper was precipitated with diethyl-dithiocarbamate, and the soma and some proximal dendrites were stained. Comparing the results of these numerous markers it was a common finding that the techniques were unreliable and the degree of staining was difficult to control. When used intracellularly only the site of injection and proximal major processes were stained. It was found that with metal ions the electrodes often blocked (increased their resistance to the point of becoming unusable) as current was passed. Galifret and Szabo (1960) suggested that this was due to the rapid formation of a metal ion-protein complex at the electrode tip, and they therefore suggested filling electrodes with complex ions which, they believed, would bind less rapidly to proteins. Electrode blocking may also be caused by local heating due to the relatively large current used; in some cases this heating, and not iontophoresis, may be the primary mechanism for expelling electrolyte from the micropipette (Oikawa et al., 1959).

Dyes

Since the use of metallic ions led to many problems, dyes became popular as extracellular and intracellular markers; dyes would, of course, be directly visible without the necessity for employing a chemical reaction. For extracellular staining, MacNichol and Svaetichin (1958) tried Crystal Violet. Later the dyes Fast Green (Thomas and Wilson, 1965) and Alcian Blue (Lee et al., 1969) were proposed as good extracellular markers. Bortoff (1964) used Trypan Blue, and Behrens and Wulff (1965) used both Analine Blue and a mixture of Orange G and Fast Green for intracellular staining. Subsequently Niagara Blue [Niagara Blue 6B, Chicago Blue 6B and Pontamine Sky Blue 6B are synonymous (Lille, 1969)] (Potter et al., 1966; Kaneko and Hashimoto, 1967; Harris et al., 1970) Methyl Blue and Fast Green (Thomas and Wilson, 1966; Kelly et al., 1967; Holubář et al., 1967; Talbott et al., 1967; Erulkar et al., 1968; Kato et al., 1968; Rovainen and Birnberger, 1971) became popular intracellular markers. These dyes appear to stain somewhat more reliably than metals and all withstand routine histological processing to some extent. The electron microscopy of cells filled with Methyl Blue or Fast Green was described by Kato et al. (1968), and Stretton and Kravitz (Chapter 2) present new data on the staining limitations of Fast Green.

Cell Marking with Radioactive Amino Acid

Before concluding this section, mention should be made of another important and recent intracellular marking method based on autoradiography. This method was introduced by Globus et al. (1968), and other reports have been published by Lux et al. (1970a, b). These workers iontophoretically injected tritiated glycine into cat motoneurons and then performed autoradiography on the fixed and sectioned material. The radioactively labeled amino acid was transported throughout the cell processes of the motoneuron and autoradiography yielded a picture of the cellular morphology. Other important results included information on transport rates, increased labeling with antidromic stimulation (Lux et al., 1970a), and some labeling of adjacent cells, possibly via axon collaterals of the injected cell (Globus et al., 1968). This is obviously an important technique, particularly in regard to questions of cytoplasmic transport and protein synthesis. As a tool for cell marking the method is limited by the period required to produce the autoradiograph (5–6 weeks) and the technical complexity of the procedure. Globus et al. (1968) claimed a resolution of about 6 μm. Nevertheless, such methods may be of great value in studying basic mechanisms of cell marking.

Conclusions

The fact that the stains in use before 1968 often failed to delineate any of the neuron beyond the immediate region of electrode penetration (usually the soma or large proximal processes) made it difficult to distinguish a filled cell from artifact, particularly in the vertebrate nervous system. Even when identification could be accomplished, it was often impossible to recognize the anatomical class to which the cell belonged because so little was visible. One may therefore conclude that some of the requirements for a usable intracellular stain would be the following: (1) solubility in water; (2) no insoluble precipitate formed with common intracellular salts; (3) slow binding to proteins so that stain can migrate a long distance from the site of injection; (4) sufficiently small hydrated radius (for ionic compounds) to permit rapid movement from the electrode as well as in the cell; (5) sufficiently large hydrated radius to prevent the stain crossing the cell membrane; (6) visibility in its own right or after binding to a cellular constituent or after a specific chemical reaction which forms an insoluble compound; (7) the visible form should withstand some form of routine histological processing; (8) toxicity low enough to avoid modification of cellular electrical activity, by electrode leakage, during the recording phase; (9) stain which is deposited outside the cell should be removable, either by natural processes within the tissue or during histological processing.

Procion Dyes and Cobalt

Introduction of Procion

Procion dye belongs to a comparatively new class of reactive textile dyes. Its relationship to other textile dyes and its detailed chemistry are described in Chapter 3.

The successful development of a method for staining individual, living nerve cells with Procion dye was first revealed in 1968 (Kravitz et al., 1968). A more detailed publication (Stretton and Kravitz, 1968) indicated that a long search had been carried out at the Harvard laboratories in order to find a suitable dye. The details of this search, together with data on 63 Procion dyes and the anatomical information obtained with Procion, are described by Stretton and Kravitz in Chapter 2. Shortly after the above mentioned publications in 1968, another paper detailing results with the new Procion technique was published by Remler et al. (1968). Taken together, these papers clearly indicated that the near perfect intracellular stain had been found (Anonymous, 1968).

Stretton and Kravitz iontophoretically injected Procion yellow from a micropipette into living nerve cells of the lobster abdominal ganglion. The dye did not noticeably affect the electrophysiological state of the neuron. On subsequent histological fixation and examination under fluorescent microscopy, it was seen that virtually the entire cell, including the fine processes, had filled with the dye. Stretton and Kravitz were interested in the important anatomical question of whether or not a given cell had a similar geometry and made similar connections in different members of a particular species. The resolution of this question demanded the ability to stain the given cell at will, and the results of Stretton and Kravitz showed that this could be accomplished. Visualization of the pattern of cell processes then provided an essential first step in the analysis of intercellular connectivity.

Remler et al. (1968) used pressure to inject Procion yellow into the lateral giant fibers of the crayfish. These investigators sought the location of the cell body and its physiological relationship to the giant fibers. The Procion technique showed that the cell body lay contralateral to the fiber, and electrophysiological recording demonstrated that the soma was electrically inexcitable, thus explaining why it had never been located by conventional physiological methods. Further aspects of the crayfish motor system, including the demonstration that the visualization of giant fiber contacts using Procion led to successful predictions about connections, were described in several communications (Selverston and Kennedy, 1969; Kennedy et al., 1969; Kennedy, 1971).

After these studies were published, the Procion technique was applied to a large number of invertebrates. The volume of work may be indicated by listing publications in the three invertebrate phyla where work has been done. Annelids: Baylor and Nicholls (1969); Baylor and O'Bryan (1971); Lent (1973); McMahan and Purves (1972); Mulloney (1970a, b); Nicholls and Purves (1970); Purves and McMahan (1972); Stuart (1969, 1970); (see also Chapter 5). Molluscs: Gillette and Pomeranz (1973); Kater et al. (1971); Kater and Rowell (1973); Kerkut et al. (1970); Sakharov and Salánki (1971); Shaw (1972); (see also Chapter 10). Arthropods: (a) Crustaceans: Anderson and Smith (1971); Davis (1970); Jones et al. (1971); Kennedy (1971); Kennedy et al. (1969); Larimer et al. (1971); Paul (1972); Payton et al. (1969); Sandeman (1969, 1971); Selverston and Kennedy (1969); Stretton and Kravitz (1968); Remler et al. (1968); Zucker et al. (1971); (see also Chapters 2 and 17). (b) Insects: Alawi and Pak (1971); Alawi et al., (1972); Autrum et al. (1970); Bentley (1970); Iles (1972); Iles and Mulloney (1971); Järvilehto and Zettler (1970, 1971); Kerkut et al. (1969); Milburn and Bentley (1971); Mote and Goldsmith (1971); Zettler and Järvilehto (1971) (see also Chapters 7 and 9).

It also quickly became clear that the Procion method was a powerful tool for solving problems in the vertebrate nervous system and it has now been applied in all the major classes, as can be seen from the following list.

Fishes: Hayward (1972); Kaneko (1970, 1971); Kriebel et al. (1969); Terävainen and Rovainen (1971). Amphibians and Reptilians: Laties and Liebman (1970); McMahan and Kuffler (1971); McMahan and Spitzer (1971); Llinás and Nicholson (1971); Matsumoto and Naka (1972). Mammals: Barrett and Crill (1971); Barrett and Graubard (1970); Dennis and Gershenfeld (1969); Jankowska and Lindström (1970a, b, 1971, 1972); Kellerth (1973); Van Essen and Kelly (1973); Van Keulen (1971), and Steinberg and Schmidt (1970). Further details will be found in Chapters 7, 11, 12, 13, 14, 15, and 18.

Some Results Achieved with Procion Dyes

Despite the brief period since its introduction, some major achievements may be credited to the use of Procion dyes. Comparing results with invertebrates and vertebrates, one can see a difference in the type of progress which has been made.

Prior to the introduction of Procion, some areas of invertebrate nervous systems were already amenable to quite detailed analysis. The role of Procion has been, for the most part, to greatly reduce the time necessary for the analysis of invertebrate systems and to enable the neuropil to be studied systematically. These developments are having far reaching effects since they make possible the examination of invertebrate systems in sufficient numbers and completeness to enable general trends in neuronal organization to be discerned.

As detailed above, the first application of Procion was by Stretton and Kravitz to the visualization of cells having known function in the lobster. These authors report their studies in detail in Chapter 2. The integrative properties implicit in neuronal geometry can be studied using dye injection combined with electrical recording (Sandeman, 1969; Milburn and Bentley, 1971). Murphey reports similar studies in the cricket in Chapter 9. Further sophistication of this approach is now feasible since Purves and McMahan (Chapter 5) demonstrate that Procion-stained leech neurons can be examined at the ultrastructural level to reveal synaptic contacts. In demonstrating neuronal geometry, Procion frequently indicates possible connections between nerve cells, and this aids the analysis of network properties in invertebrates (Davis, 1970; Sandeman, 1971; Larimer et al., 1971). Selverston illustrates this application of dye injection to the lobster in Chapter 17. Intracellular stains have been applied to a fairly restricted selection of invertebrate preparations but the technique may be applied to all animals, and C. Kaneko and Kater show the potential value of the method in the snail (Chapter 10).

In the vertebrates the role of Procion has been rather different. The major advance brought about by the new technique has been the correlation of electrical activity of an unseen cell with its morphology, as revealed by Procion dye injection and subsequent histological study. This has led to several important results.

With the aid of Procion dye, most known cells in the retina have been recorded from electrically and stained (Kaneko, 1970, 1971; Matsumoto and Naka, 1972; see also Hashimoto et al., Chapter 12). This area is reviewed by Kaneko in Chapter 11. The retina is probably the area where Procion is presently having the most impact in vertebrate studies (Anonymous, 1970; Baylor and O'Bryan, 1971; Hodgkin, 1972). More centrally in the visual system, Van Essen and Kelly (Chapter 13) have stained cells in the visual cortex, designated as simple, complex, and hypercomplex by Hubel and Wiesel (1962).

Other areas of the central nervous system have been stained with Procion dyes. Llinás reviews some of this work in Chapter 15, while Jankowska and Lindström report on spinal interneurons, including the Renshaw cell, in Chapter 14 (Jankowska and Lindström, 1971; Van Keulen, 1971). Barrett (Chapter 18) reports studies of the membrane properties of motoneurons.

Procion has also contributed uniquely to several other types of problems. Payton et al. (1969) demonstrated that Procion crosses electrotonic synapses in the crayfish septate axon; thus Procion contributes a new tool for studying intercellular junctions (Bennett, Chapter 8). Dennis and Gershenfeld (1969) used Procion to identify "silent cells" as glial cells. Finally, Procion has been combined with histochemical techniques (Laties, 1972; Sivitz et al., 1973) and used as a section stain (Payton, 1970).

Evolution of the Procion Technique

Stretton and Kravitz (1968 and Chapter 2) selected Procion brown H3RS, brilliant red H3BNS, navy blue H3RS, and yellow M4RS as the best dyes for neurobiological applications. In Chapter 3, Stead confirms that Procion orange, red and yellow M dyes are likely to be the most suitable for neurobiological purposes. Subsequently Procion yellow M4RS or M4RAN has been used in the vast majority of all studies. Procion navy blue H3R has been used by Järvilhehto and Zettler (1970) and Autrum et al. (1970); however, this dye quenches fluorescence. Several investigators have used two dyes of different colors to indicate relationships between contiguous structures. Kaneko (1971 and Chapter 11) used Procion yellow M4RS and Brilliant red H3BNS. Mote and Goldsmith (1971) also used Procion yellow and red, and Shaw (1972) used mixtures of Procion yellow M4RAN and navy blue H3RS, and yellow and brilliant red H3BNS.

The most common method of dye infusion is by iontophoresis of the negatively charged Procion ion from the micropipette. Pressure has also been used by several workers. An important development was iontophoresis through the cut end of an axon (Iles and Mulloney, 1971; see also Chapters 7, 10, and 15). This latter technique obviates the necessity for penetration of the cell with a micropipette but usually precludes electrical recording from the cell. Techniques are discussed in detail in Chapter 20.

Limitations of Procion

It is generally agreed that Procion is superior to any other dye which has been used for intracellular staining. However, Procion has limitations which will be reviewed here and discussed in detail in other chapters.

Infusion of the Dye

Iontophoresis is the most frequently used method. Several workers have noted that Procion does not always pass out of the pipette (Barrett and Graubard, 1970; Barrett,

Chapter 18; Kaneko, 1970; Stretton and Kravitz, Chapter 2; Van Essen and Kelly, Chapter 13; authors' personal observations).

Pressure injection from a micropipette gives good results but usually requires a pipette tip size of 1 μm or more (Curtis, 1964). Again there is considerable variation in the ability of electrodes to eject dye (see Chapter 20).

Axonal iontophoresis does not always fill cells, although many variations have been tried to make the technique more reliable (see Chapters 7 and 20).

Physiological and Anatomical Effects of the Dye

The majority of publications have not reported any noticeable modifications in the electrophysiology of cells injected with Procion dye. This feature is one of the major assets of the stain. A minority have found physiological change. A prolongation of action potentials and, in some cases, reduction and blockage were reported by Barrett (Chapter 18), and Barrett and Graubard (1970), Jankowska and Lindström (1970a), Kerkut et al. (1970), and Van Keulen (1971). No conclusive evidence for any effect on synaptic potential has been reported. A reduction in resting potential was reported by Jankowska and Lindström (1970a). New data indicating an absence of intracellular effect of Procion yellow, when used in low concentrations, are reported by Llinás (Chapter 15). In contrast to these findings of the comparatively minor effects of intracellular injection of Procion dye, Payton et al. (1969) and Payton (1970) found in preliminary experiments that external application of Procion yellow produced a reversible block of action potentials, similar to the action of a local anesthetic. (See also discussion following Chapter 8.)

Purves and McMahan (1972 and Chapter 5) have studied the ultrastructure of cells injected with Procion yellow and report little modification of the basic structures, although a number of vacuoles appear in the endoplasmic reticulum.

The above results are not in complete agreement. It is possible that different results are obtained in different cells. In some cases the effects attributed to the properties of the dye may actually be caused by electrode damage, by osmotic effects of the injected dye solution, or even by the passage of large currents across the membrane. It should be noted that Purves and McMahan (1972) found that some cells showed severe damage at the ultrastructural level after penetration with a micropipette.

Dye Filling of Cell Processes

In general, cell processes fill well and their appearance rivals the results obtainable by the best Golgi and Methylene Blue histology. Considerable variation may be encountered, however, in filling the same cell on different occasions. The mechanisms responsible for stain migration in neurons are unclear at the present time. Electrophoresis (either by extrinsic or intrinsic currents), hydrostatic pressure, osmotic pressure, passive diffusion, or even axoplasmic transport may all play a role in moving stain within the cell. Further research in this area would be worthwhile. Cellular processes less than 1 μm in diameter have been filled with Procion (Stretton and Kravitz, 1968) and dendritic spines (0.2 μm in diameter) have been filled in Purkinje cells (see Chapter 15).

In general it has not proved possible to fill axons over distances of several centimeters by either microelectrode injection or axonal iontophoresis of Procion. Thus, at present, Procion cannot be used for extensive hodological mapping of the nervous system.

Binding of the Dye

It is known that in the conditions prevailing in a nerve cell, Procion dye will bind fairly slowly, thus permitting extensive travel in the cell (Stretton and Kravitz, 1968 and Chapter 2; see also Chapter 3). Stead has also suggested the use of low reactivity Procion dyes which can be subsequently activated by suitable compounds (see discussion, Chapter 3). It appears that at the typical intracellular pH, the binding will take place through an amino group in the cell (Chapter 3). It is also observed that the cell nucleus sometimes stains more densely than the surrounding cytoplasm. This is probably due to pH differences (see Chapter 15). Laties and Liebman (1970), on the basis of studies on retinal cells with a polarized microscope, concluded that Procion was isotropically bound to the membrane, with the chromophoric end of the molecule retaining a high degree of rotational mobility.

Fluorescence

The fluorescence spectrum of intracellular Procion yellow is illustrated in Chapters 2 and 4. As Stead points out in Chapter 3, the fluorescence of Procion inside cells is surprising since dyed textiles do not show this feature. The fluorescence is usually quite strong (in comparison with typical histochemical fluorescence) but it does exhibit photodecomposition and thus observation must be limited in duration (Kerkut et al., 1970; Van Orden, Chapter 4).

In recent years fluorescent probes have been used to study molecular mechanisms in nerve cell membranes (Tasaki et al., 1972). When Procion yellow is infused into the giant axon of the squid and the membrane excited, there is no change in the Procion fluorescence (I. Tasaki, personal communication). Thus Procion is not a fluorescent probe.

Introduction of Cobalt

In a surprising development, Pitman et al. (1972) reported that intracellular iontophoresis of cobaltous chloride into cockroach motoneurons, followed by treatment with ammonium sulfide, yielded results of comparable information content to those obtained with Procion. The ammonium sulfide produced an insoluble brown-black precipitate of cobaltous sulfide which withstood subsequent histological processing in nonacidic media.

Two advantages were claimed for the cobalt method over Procion. First, the cobalt precipitate was more visible under conventional light microscopy than Procion under fluorescence microscopy. Second, cobalt produced a precipitate which was sufficiently electron opaque to be detectable with electron microscopy, thus enabling stained neuronal elements to be located and studied at the ultrastructural level.

The above claims have been further substantiated in this book by Pitman et al. (Chapter 6). Cobalt may also be infused into cells by retrograde axonal iontophoresis (see Mittenthal and Wine, 1973; Chapters 6, 7, 10, and 15). It thus appears that cobalt is an important intracellular stain. This is unexpected in view of the comparative lack of success experienced with other metals prior to the introduction of Procion (see above). At the present time, it remains an important, but open, question as to whether the staining properties of cobalt are unique or whether other metals can give similar results under the right conditions. Some success has recently been reported with an "intracellular Golgi impregnation," using chromium iontophoretically injected from a micropipette (Alawi et al., 1972).

Cobalt and Procion Compared

Experience with cobalt is far more limited than with Procion, and the following statements may therefore be revised in the near future.

Cobalt may be injected by methods similar to those used for Procion. When injected iontophoretically cobalt appears to block the electrode far more readily than Procion and this blockage is frequently irreversible. Blockage was often reported during earlier experiments with metal iontophoresis (see above). Pitman et al. (Chapter 6) report that iontophoresis is superior to pressure for injection of cobalt in cockroach neurons. Retrograde axonal iontophoresis works adequately (Chapters 7 and 15).

Pitman et al. (1972 and Chapter 6) report that apparently normal action potentials could be recorded with a cobalt-filled micropipette. C. Kaneko and Kater (Chapter 10) have reported the rapid loss of action potentials and membrane potential after nerve cell penetration in the mollusc. It is known that externally applied cobalt affects calcium currents in membranes (Hagiwara and Takahashi, 1967; Geduldig and Junge, 1968; Wald, 1972). Cobalt applied externally at the nodes of Ranvier causes a prolongation of the action potential (Tasaki, 1959); however, the action potential is not prolonged when cobalt is applied to the squid axon (I. Tasaki, personal communication). When cobalt is internally applied to motoneurons it causes a long lasting depolarizing shift of the equilibrium potential for inhibitory postsynaptic potentials (Lux and Schubert, 1969). Thus it is possible that cobalt may adversely affect neuron physiology, and this may limit its value in certain applications, since there is always significant passive diffusion of the electrolyte from a micropipette (Geisler et al., 1972). Anatomically, the use of ammonium sulfide as a precipitate can lead to significant distortion of the cell at the ultrastructural level.

Cell processes fill well with cobalt, and Pitman et al. (1972) claim that processes less than 1 μm in diameter may be filled. The movement of the stain appears comparable to that achieved with Procion. Llinás (Chapter 15) has shown that nodes of Ranvier are revealed with cobalt.

The principal advantages of cobalt over Procion are its light and electron opacity when precipitated in cells and treated appropriately. However, it appears that the labor involved in locating stained processes for electron microscopy is about equal with both cobalt and Procion (see Chapters 5 and 6). In addition, Stead (Chapter 3) indicates that it may be possible to incorporate uranium into a Procion-like structure.

Quantitative Analysis

The new intracellular staining techniques are enabling neurocellular geometry to be revealed almost at will. The problem now is to analyze the geometry in terms which lead to insight into neuronal function. It is becoming apparent that the computer will be a valuable tool in this area.

Quantitative histology is not new; stereology has reached a state of considerable refinement both in theoretical and practical aspects. (See, for example, Proceedings of the Third International Congress for Stereology, *Journal of Microscopy,* Vol. 95, parts 1 and 2, 1972.) This discipline primarily yields information on cell numbers, cross sectional areas, and tissue volumes. The information which is presently required concerns the pattern of cell processes, the sizes and shapes of dendritic and axonal branches, and the location of synapses. Cells are three-dimensional objects and the processes of neurons often form highly complex patterns, thus quantification of these elements is a considerable task.

At the simplest level, the geometry of a neuron can be revealed by projection drawings with a camera lucida or more elegantly by stereophotography of whole mount preparations (Chapters 7 and 8). For more detailed analysis, quantification and computer aided reconstruction is necessary.

Computer Aided Three-Dimensional Reconstruction

At the present time there are two approaches to computer aided three-dimensional reconstruction of biological structure. The first method consists of recording and digitizing the image of the whole object when viewed in transmitted light or electron beam. By repeating these measurements for several orientations of the object, enough information can be obtained to enable the object to be reconstructed. This method has reached considerable theoretical sophistication (Ramachandran and Lakshminarayan, 1971; Klug and Crowther, 1972) but it seems best suited to simple structures and has not yet been applied to neuronal elements.

The second method of three-dimensional reconstruction begins with a sequence of serial sections through the objects of interest, thus yielding a set of planar profiles which are then digitized and reassembled into a three-dimensional structure by the computer. It is this method which is discussed by Reddy et al. (Chapter 16) and Selverston (Chapter 17) and which appears to offer a powerful tool for the study of neuronal geometry.

Reconstruction from serial sections necessitates the digitization of the boundaries of the nerve cell. The simplest way to achieve this is for the investigator to transmit the coordinates of the boundary to the computer with a suitable encoding device, e.g., a writing tablet (Levinthal and Ware, 1972; Lopresti et al., 1973; Macagno et al., 1973; Reddy et al., Chapter 16; Selverston, Chapter 17). At the next level of sophistication, the image of the cell boundaries may be scanned automatically by a suitable digital sensor system. The computer performs appropriate pattern recognition operations to select objects of importance (Ledley, 1972; Reddy et al., Chapter 16). This latter method is technically difficult, particularly in its pattern recognition phase. Reconstruction of the cell requires further interaction by the investigator. Finally, the display of the three-dimensional structure demands the use of modern computer graphic systems to produce stereo pairs and rotating images.

Computation of Membrane Properties

We are now beginning to see electrophysiological properties of cells predicted from their geometry, using the theoretical models of cellular cable properties (Rall, 1959, 1962, 1967, 1969), and compared with actual electrical measurement on that cell (Barrett and Crill, 1971; Murphey, Chapter 9). Barrett describes this approach, its possibilities and present limitations, in Chapter 18. There is little doubt that this approach can lead to new insights into the functional significance of anatomical structure.

Acknowledgments. This work was supported in part by U.S.P.H.S. grant NS09916 from N.I.N.D.S. (C. Nicholson) and by U.S.P.H.S. grant NS 09696 from N.I.N.D.S. (S. B. Kater).

References

Adrian, E. D., and G. Moruzzi: Impulses in the pyramidal tract. J. Physiol., Lond. *97*, 153–199 (1939).

Alawi, A. A., and W. L. Pak: On-transient of insect electroretinogram: its cellular origin. Science *172*, 1056–1057 (1971).

———, V. Jennings, J. Crossfield, and W. L. Pak: Phototransduction mutants of *Drosophila melanogaster.* In: The Visual System: Neurophysiology, Biophysics and their Clinical Applications. Ed. G. B. Arden. pp. 1–21. New York: Plenum Press, 1972.

Anderson, M. E. and D. S. Smith: Electrophysiological and structural studies on the heart muscle of the lobster *Homarus americanus.* Tissue and Cell *3*, 191–205 (1971).

Anderson, P., J. C. Eccles, and Y. Løyning: Location of postsynaptic inhibitory synapses on hippocampal pyramids. J. Neurophysiol. *27*, 592–607 (1964a).

——— ———: Pathway of postsynaptic inhibition in the hippocampus. J. Neurophysiol. *27*, 608–619 (1964b).

Anonymous: Fluorescent neurones. Nature, Lond. *220*, 332–333 (1968).

———: How to dye cells. Nature, Lond. *226*, 1003 (1970).

Autrum, H., F. Zettler, and M. Järvilehto: Postsynaptic potentials from a single monopolar neuron of the ganglion opticum I of the blowfly *Calliphora.* Z. vergl. Physiol. 70, 414–424 (1970).

Barrett, J. N., and W. E. Crill: Specific membrane resistivity of dye-injected cat motoneurons. Brain Res. *28*, 556–561 (1971).

———, and K. Graubard: Fluorescent staining of cat motoneurons *in vivo* with bevelled micropipettes. Brain Res. *18*, 565–568 (1970).

Baylor, D. A., and J. G. Nicholls: Chemical and electrical synaptic connexions between cutaneous mechanoreceptor neurones in the central nervous system of the leech. J. Physiol., Lond. *203*, 591–609 (1969).

———, and P. M. O'Bryan: Electrical signaling in vertebrate photoreceptors. Fedn Proc. Fedn Am. Socs exp. Biol. *30*, 79–83 (1971).

Behrens, M. E., and V. J. Wulff: Light-initiated responses of retinula and eccentric cells in the *Limulus* lateral eye. J. gen. Physiol. *48*, 1081–1093 (1965).

Bentley, D. R.: A topological map of the locust flight system motor neurons. J. Insect Physiol. *16*, 905–918 (1970).

Bortoff, A.: Localization of slow potential responses in the *Necturus* retina. Vision Res. *4*, 627–636 (1964).

Bullock, T. H., and G. A. Horridge: Structure and Function in the Nervous Systems of Invertebrates. San Francisco: W. H. Freeman and Co., 1965.

Bultitude, K. H.: A technique for marking the site of recording with capillary microelectrodes. Q. Jl microsc. Sci. *99*, 61–63 (1958).

Curtis, D. R.: Microelectrophoresis. In: Physical Techniques in Biological Research, Vol. VA. Ed. W. L. Nastuk. pp. 144–192. New York: Academic Press, 1964.

Davis, W. J.: Motoneuron morphology and synaptic contacts: determination by intracellular dye injection. Science *168*, 1358–1360 (1970).

Dennis, M. J., and H. Gerschenfeld: Some physiological properties of identified mammalian glial cells. J. Physiol., Lond. *203*, 211–222 (1969).

Eccles, J. C.: The Physiology of Nerve Cells. Baltimore: Johns Hopkins Press, 1957.

———: The Inhibitory Pathways of the Central Nervous System. Springfield, Ill., U.S.A.: Charles C. Thomas and Liverpool, England: Liverpool University Press, 1969.

Ehrlich, P.: Über die Methylenblaureaktion der lebenden Nervensubstanz. Biol. Centralbl. *6*, 214–224 (1886).

Erulkar, S. D., C. W. Nichols, M. B. Popp, and G. B. Koelle: Renshaw elements: localization and acetylcholinesterase content. J. Histochem. Cytochem. *16*, 128–135 (1968).

Frank, K., and M. C. Becker: Microelectrodes for recording and stimulation. In: Physical Techniques in Biological Research. Vol. VA. Ed. W. L. Nastuk. pp. 23–88. New York: Academic Press, 1964.

Fregerslev, S., T. W. Blackstad, K. Fredens, and M. J. Holm: Golgi potassium-dichromate silver-nitrate impregnation. Histochemie *25*, 63–71 (1971).

Furshpan, E. J., and T. Furukawa: Intracellular and extracellular responses of the neural region of the Mauthner cell of the goldfish. J. Neurophysiol. *25*, 732–771 (1962).

Galifret, Y., and T. H. Szabo: Locating capillary microelectrode tips within nervous tissue. Nature, Lond. *188*, 1033–1034 (1960).

Geduldig, D., and D. Junge: Sodium and calcium components of action potentials in the *Aplysia* giant neurone. J. Physiol., Lond. *199*, 347–365 (1968).

Geisler, C. D., E. N. Lightfoot, F P. Schmidt, and F. Sy: Diffusion effects of liquid-filled micropipette: a pseudobinary analysis of electrolyte leakage. IEEE Trans. Biomed. *19*, 372–375 (1972).

Gillette, R., and B. Pomeranz: A study of neuron morphology in *Aplysia californica* using Procion Yellow dye. Comp. Biochem. Physiol. *44A*, 1257–1260 (1973).

Globus, A., H. D. Lux, and P. Schubert: Soma dendritic spread of intracellularly injected tritiated glycine in cat spinal motoneurons. Brain Res. *11*, 440–445 (1968).

Gouras, P.: Graded potentials of bream retina. J. Physiol., Lond. *152*, 487–505 (1960).

Green, J. D.: A simple microelectrode for recording from the central nervous system. Nature, Lond. *182*, 962 (1958).

Hagiwara, S., and K. Takahashi: Surface density of calcium ions and calcium spikes in the barnacle muscle fiber membrane. J. gen. Physiol. *50*, 583–601 (1967).

Harris, G. G., L. S. Frishkopf, and A. Flock: Receptor potentials from hair cells of the lateral line. Science *167*, 76–79 (1970).

Hayward, J. N.: Physiological properties of dye-injected preoptic neurons in the hypo-thalamus of the goldfish (*Carassius auratus*). Anat. Rec. *172*, 454 (1972).

Hess, R.: Beiträge zur Physiologie des Hirnstammes. Leipzig: Thieme, 1932.

Hodgkin, A. L.: Address of the President Professor A. L. Hodgkin at the Anniversary Meeting, 30 November 1971. Proc. R. Soc., Lond. B. *180*, v–xx (1972).

Holubář, J., B. Hanke, and V. Malík: Intracellular recording from cortical pyramids and small interneurons as identified by subsequent staining with the recording microelectrode. Exp. Neurol. *19*, 257–264 (1967).

Hubel, D. H., and T. N. Wiesel: Receptive fields, binocular interaction and functional architecture in the cat's visual cortex. J. Physiol., Lond. *160*, 106–154 (1962).

Iles, J. F.: Structure and synaptic activation of the fast coxal depressor motor neurone of the cockroach, *Periplaneta americana*. J. exp. Biol. *56*, 647–656 (1972).

——, and B. Mulloney: Procion yellow staining of cockroach motor neurones without the use of microelectrodes. Brain Res. *30*, 397–400 (1971).

Jankowska, E., and S. Lindström: Morphological identification of physiologically defined neurones in the cat spinal cord. Brain Res. *20*, 323–326 (1970a).

—— ——: Intracellular staining of physiologically identified interneurones in the cat spinal cord. Acta physiol. scand. *79*, 4A–5A (1970b).

—— ——: Morphological identification of Renshaw cells. Acta physiol. scand. *81*, 428–430 (1971).

—— ——: Morphology of interneurones mediating Ia reciprocal inhibition of motoneurones in the spinal cord of the cat. J. Physiol., Lond. *226*, 805–824 (1972).

Järvilehto, M., and F. Zettler: Micro-localisation of lamina-located visual cell activities in the compound eye of the blowfly *Calliphora*. Z. vergl. Physiol. *69*, 134–138 (1970).

—— ——: Localized intracellular potentials from pre- and postsynaptic components in the external plexiform layer of an insect retina. Z. vergl. Physiol. *75*, 422–440 (1971).

Jones, C., J. Nolte, and J. E. Brown: The anatomy of the median ocellus of *Limulus*. Z. Zellforsch. mikrosk. Anat. *188*, 297–309 (1971).

Kaneko, A.: Physiological and morphological identification of horizontal, bipolar and amacrine cells in goldfish retina. J. Physiol., Lond. *207*, 623–633 (1970).

——: Electrical connexions between horizontal cells in the dogfish retina. J. Physiol., Lond. *213*, 95–105 (1971).

——, and H. Hashimoto: Recording site of the single cone response determined by an electrode marking technique. Vision Res. 7, 847–851 (1967).

Kater, S. B., C. B. Heyer, and J. P. Hegmann: Neuromuscular transmission in the gastropod mollusc *Helisoma trivolvis:* identification of motoneurons. J. Comp. Physiol. *74*, 127–139 (1971).

——, and C. H. F. Rowell: Integration of sensory and centrally programmed components in generation of cyclical feeding activity of *Helisoma trivolvis*. J. Neurophysiol. *36*, 142–155 (1973).

Kato, M., B. Fujimori, and Y. Hirata: An electron microscopic study of intracellularly stained neurons. Brain Res. *9*, 390–393 (1968).

Kellerth, J-O.: Intracellular staining of cat spinal motoneurons with Procion Yellow for ultrastructural studies. Brain Res. *50*, 415–418 (1973).

Kelly, J. S., K. Krnjevic, and G. K. W. Yim: Unresponsive cells in cerebral cortex. Brain Res. *6*, 767–769 (1967).

Kennedy, D.: Nerve cells and behavior. Am. Scient. *59*, 36–42 (1971).

——, A. I. Selverston, and M. P. Remler: Analysis of restricted neural networks. Science *164*, 1488–1496 (1969).

Kerkut, G. A., R. M. Pitman, and R. J. Walker: Iontophoretic application of acetylcholine and GABA onto insect central neurones. Comp. Biochem. Physiol. *31*, 611–633 (1969).

——, M. C. French, and R. J. Walker: The location of axonal pathways of identifiable neurones of *Helix aspersa* using the dye Procion yellow M-4R. Comp. Biochem. Physiol. *32*, 681–690 (1970).

——, and R. J. Walker: Marking individual nerve cells through electrophoresis of ferrocyanide from a microelectrode. Stain Technol. *37*, 217–219 (1962).

Klug, A., and R. A. Crowther: Three-dimensional image reconstruction from the viewpoint of information theory. Nature, Lond. *238*, 435–440 (1972).

Kravitz, E. A., A. O. W. Stretton, J. Alvarez, and E. J. Furshpan: Determination of neuronal geometry using an intracellular dye injection technique. Fedn Proc. Fedn Am. Socs exp. Biol. *27*, 749 (1968).

Kriebel, M. E., M. V. L. Bennett, S. G. Waxman, and G. D. Pappas: Oculomotor neurons in fish: electrotonic coupling and multiple sites of impulse initiation. Science *166*, 520–524 (1969).

Larimer, J. L., A. C. Eggleston, L. M. Masukawa, and D. Kennedy: The different connections and motor outputs of lateral and medial giant fibres in the crayfish. J. exp. Biol. 54, 391–402 (1971).

Laties, A. M.: Specific neurohistology comes of age—look back and a look forward. Inv. Ophth. 11, 555–584 (1972).

——, and P. A. Liebman: Cones of living amphibian eye: selective staining. Science 168, 1475–1477 (1970).

Lavallée, M., O. F. Schanne, and N. C. Hebert: Glass Microelectrodes. London: Wiley, 1969.

Ledley, R. S.: Analysis of cells. IEEE Transactions on Computers 21, 740–753 (1972).

Lee, B. B., G. Mandl, and J. P. B. Stean: Micro-electrode tip position marking in nervous tissue: a new dye method. Electroenceph. clin. Neurophysiol. 27, 610–613 (1969).

Lent, C. M.: Retzius' cells from segmental ganglia of four species of leeches: comparative neuronal geometry. Comp. Biochem. Physiol. 44A, 35–40 (1973).

Levinthal, C. and R. Ware: Three dimensional reconstruction from serial sections. Nature, Lond. 236, 207–210 (1972).

Lillie, R. D.: H. J. Conn's Biological Stains. Baltimore: William and Wilkin Co., 1969.

Ling, G. and R. W. Gerard: The normal membrane potential of frog sartorius muscle fibers. J. Cell comp. Physiol. 34, 383–396 (1949).

Llinás, R., and C. Nicholson: Electrophysiological properties of dendrites and somata in alligator Purkinje cells. J. Neurophysiol. 34, 532–551 (1971).

Lopresti, V., E. R. Macagno, and C. Levinthal: Structure and development of neuronal connections in isogenic organisms: cellular interaction in the development of the optic lamina of Daphnia. Proc. natn Acad. Sci. U.S.A. 70, 433–437 (1973).

Lux, H. D., and A. Globus: Effect on IPSPs of cat spinal motoneurones due to intra- and extracellular iontophoresis of CuSo4. Brain Res. 9, 377–380 (1968).

——, and P. Schubert: Postsynaptic inhibition: intracellular effects of various ions in spinal motoneurons. Science 166, 625–626 (1969).

—— ——, G. W. Kreutzberg, and A. Globus: Excitation and axonal flow: autoradiographic study on motoneurons intracellularly injected with a 3H-amino acid. Exp. Brain Res. 10, 197–204 (1970a).

—— —— ——: Direct matching of morphological and electrophysiological data in cat spinal motoneurons. In: Excitatory Synaptic Mechanisms. Ed. P. Anderson and J. K. S. Jansen. pp. 189–198. Oslo: Universitetsforlaget, 1970b.

Macagno, E. R., V. Lopresti, and C. Levinthal: Structure and development of neuronal connections in isogenic organisms: variations and similarities in the optic system of Daphnia magna. Proc. natn Acad. Sci. U.S.A. 70, 57–61 (1973).

MacNichol, E. F. Jr., and G. Svaetichin: Electrical responses from the isolated retinas of fishes. Am. J. Ophthal. 46, 26–40 (1958).

Marshall, W. H.: An application of the frozen sectioning technic for cutting serial sections thru the brain. Stain Technol. 15, 133–138 (1940).

Matsumoto, N., and K.-I Naka: Identification of intracellular responses in the frog retina. Brain Res. 42, 59–71 (1972).

McMahan, U. J., and S. W. Kuffler: Visual identification of synaptic boutons on living ganglion cells and of varicosities in postganglionic axons in the heart of the frog. Proc. R. Soc., Lond. B 177, 485–508 (1971).

——, and D. Purves: An electron-microscopic study of a physiologically identified motoneurone in the leech C.N.S. after injection of the fluorescent dye Procion yellow. J. Physiol., Lond. 222, 64P–66P (1972).

——, and N. C. Spitzer: Viewing neuromuscular junctions in live, unstained skeletal muscle. J. Physiol., Lond. 213, 35P–37P (1971).

Milburn, N., and D. R. Bentley: On the dendritic topology and activation of cockroach giant interneurons. J. Insect Physiol. 17, 607–623 (1971).

Mitarai, G.: The origin of the so-called cone potential. Proc. Japan Acad. 34, 299–304 (1958).

———: Determination of ultramicroelectrode tip position in the retina in relation to S potential. J. gen. Physiol. *43*, 95–99 (1960).

Mittenthal, J. E., and J. J. Wine: Connectivity patterns of crayfish giant interneurons: visualization of synaptic regions with cobalt dye. Science *179*, 182–184 (1973).

Mote, M. I., and T. H. Goldsmith: Compound eyes: localization of two color receptors in the same ommatidium. Science *171*, 1254–1255 (1971).

Motokawa, K., T. Oikawa, and K. Tasaki: Receptor potential of vertebrate retina. J. Neurophysiol. *20*, 186–199 (1957).

Mulloney, B.: The structure of the giant fibres of earthworms, as disclosed by Procion yellow injections. J. Physiol., Lond. *210*, 22P (1970a).

———: Structure of the giant fibers of earthworms. Science *168*, 994–996 (1970b).

Nauta, W. J. M., and S. O. E. Ebbesson: Contemporary Research Methods in Neuroanatomy. New York, Heidelberg, Berlin: Springer-Verlag, 1970.

Nicholls, J. G., and D. Purves: Monosynaptic chemical and electrical connexions between sensory and motor cells in the central nervous system of the leech. J. Physiol., Lond. *209*, 647–667 (1970).

Oikawa, T., T. Ogawa, and K. Motokawa: Origin of the so-called cone action potential. J. Neurophysiol. *22*, 102–111 (1959).

Paul, D. H.: Decremental conduction over "giant" afferent processes in an arthropod. Science *176*, 680–682 (1972).

Payton, B. W.: Histological staining properties of Procion yellow. J. Cell Biol. *45*, 659–662 (1970).

———, M. V. L. Bennett, and G. D. Pappas: Permeability and structure of junctional membranes at an electrotonic synapse. Science *166*, 1641–1643 (1969).

Perls, M.: Nachweis von Eisenoxid in gewissen Pigmenten. Virch. Arch. *39*, 42–48 (1867).

Pitman, R. M., C. D. Tweedle, and M. J. Cohen: Branching of central neurons: intracellular cobalt injection for light and electron microscopy. Science *176*, 412–414 (1972).

Potter, D. D., E. J. Furshpan, and E. S. Lennox: Connections between cells of the developing squid as revealed by electrophysiological methods. Proc. natn. Acad. Sci. U.S.A. *55*, 328–336 (1966).

Purves, D., and U. J. McMahan: The distribution of synapses on a physiologically identified motor neuron in the central nervous system of the leech: an electron microscopic study after the injection of the fluorescent dye Procion yellow. J. Cell Biol. *55*, 205–220 (1972).

Rall, W.: Branching dendritic trees and motoneuron membrane resistivity. Exp. Neurol. *1*, 491–527 (1959).

———: Theory of physiological properties of dendrites. Ann. N.Y. Acad. Sci. *96*, 1071–1092 (1962).

———: Distinguishing theoretical synaptic potentials computed for different soma-dendritic distributions of synaptic inputs. J. Neurophysiol. *30*, 1138–1168 (1967).

———: Time constants and electrotonic length of membrane cylinders and neurons. Biophys. J. *9*, 1482–1508 (1969).

Ramachandran, G. N., and A. V. Lakshminarayan: Three-dimensional reconstruction from radiographs and electron micrographs: application of correlation instead of fourier transform. Proc. natn. Acad. Sci. U.S.A. *68*, 2236–2240 (1971).

Ramón y Cajal, S.: Histologie du système Nerveux de l'homme et des Vertébrés. 2 vol. Paris: Maloine, 1909, 1911.

Rayport, M.: Anatomical identification of somatic sensory cortical neurons responding with short latencies to specific afferent volleys. Fedn Proc. Fedn Am. Socs exp Biol. *16*, 104 (1957).

Remler, M. P., A. I. Selverston, and D. Kennedy: Lateral giant fibers of crayfish: location of somata by dye injection. Science *162*, 281–283 (1968).

Rovainen, C. M., and K. L. Birnberger: Identification and properties of motoneurons to fin muscle of the sea lamprey. J. Neurophysiol. *34*, 974–981 (1971).

Sakharov, D. A., and J. Salánki: Study of neurosecretory cells of *Helix pomatia* by intracellular dye injection. Experientia 27, 655–656 (1971).

Sandeman, D. C.: Integrative properties of a reflex motoneuron in the brain of the crab *Carcinus maenas*. Z. vergl. Physiol. 64, 450–464 (1969).

———: The excitation and electrical coupling of four identified motoneurons in the brain of the Australian mud crab, *Scylla serrata*. Z. vergl. Physiol. 72, 111–130 (1971).

Scheere, B. T., and M. W. Mumbach: The locus of the electromotive force in frog skin. J. cell. comp. Physiol. 55, 259–266 (1960).

Scheibel, M. E., and A. B. Scheibel: Histological localization of microelectrode placement in brain by ferrocyanide and silver staining. Stain Technol. 31, 1–5 (1956).

Selverston, A. I., and D. Kennedy: Structure and function of identified nerve cells in the crayfish. Endeavour 28, 107–113 (1969).

Shaw, S. R.: Decremental conduction of the visual signal in barnacle lateral eye. J. Physiol., Lond. 220, 143–175 (1972).

Sivitz, M., R. G. Kallen, and A. M. Laties: Procion Yellow and Catecholamine derivatives: chemical relationships. J. Histochem. Cytochem. 21, 87–92 (1973).

Steinberg, R. H., and R. Schmidt: Identification of horizontal cells as S-potential generators in the cat retina by intracellular dye injection. Vision Res. 10, 817–820 (1970).

Stretton, A. O. W., and E. A. Kravitz: Neuronal geometry: determination with a technique of intracellular dye injection. Science 162, 132–134 (1968).

Stuart, A. E.: Excitatory and inhibitory motoneurons in the central nervous system of the leech. Science 165, 817–819 (1969).

———: Physiological and morphological properties of motoneurones in the central nervous system of the leech. J. Physiol., Lond. 209, 627–646 (1970).

Talbott, R. E., A. L. Towe, and T. T. Kennedy: Physiological and histological classification of cerebellar neurons in chloralose-anesthetized cats. Exp. Neurol. 19, 46–64 (1967).

Tasaki, I.: Correlation of two stable states of the nerve membrane in potassium-rich media. J. Physiol., Lond. 148, 306–331 (1959).

———, A. Watanabe, and M. Hallett: Fluorescence of squid axon membrane labelled with hydrophobic probes. J. Mem. Biol. 8, 109–132 (1972).

Teräväinen, N., and C. M. Rovainen: Fast and slow motoneurons to body muscle of the sea lamprey. J. Neurophysiol. 34, 990–998 (1971).

Thomas, R. C., and V. J. Wilson: Precise localization of Renshaw cells with a new marking technique. Nature 206, 211–213 (1965).

——— ———: Marking single neurons by staining with intracellular recording microelectrodes. Science 151, 1538–1539 (1966).

Van Essen, D., and J. Kelly: Correlation of cell shape and function in the visual cortex of the cat. Nature, Lond. 241, 403–405 (1973).

Van Keulen, L. C. M.: Morphology of Renshaw cells. Pflügers Arch. ges. Physiol. 328, 235–236 (1971).

von Baumgarten, R., E. Kanzoe, M. P. Kowpchen, and F. Timm: Beitrag zur Technik der extra- und intracellulären sowie der stereotaktischen Mikroableitung im Gehirn. Pflügers Arch. ges. Physiol. 271, 245–256 (1960).

Wald, F.: Ionic differences between somatic and axonal action potentials in snail giant neurons. J. Physiol., Lond. 220, 267–281 (1972).

Zettler, F., and M. Järvilehto: Decrement-free conduction of graded potentials along the axon of a monopolar neuron. Z. vergl. Physiol. 75, 402–421 (1971).

Zucker, R. S., D. Kennedy, and A. I. Selverston: Neuronal circuit mediating escape responses in crayfish. Science 173, 645–650 (1971).

CHAPTER 2

Intracellular Dye Injection: The Selection of Procion Yellow and its Application in Preliminary Studies of Neuronal Geometry in the Lobster Nervous System

A. O. W. Stretton and E. A. Kravitz

The purpose of this chapter is to provide a more complete description than has been available of the experiments that led to the choice of Procion yellow as an intracellular agent suitable for determining neuronal geometry (Stretton and Kravitz, 1968; Kravitz et al., 1968). We would also like to give proper credit to Drs. E. J. Furshpan and J. Alvarez for their generosity in providing us with the basic technology and underlying philosophy of the technique of intracellular dye injection. In addition, we will describe the preliminary results we obtained in applying this method to neurons of lobster central ganglia, in what was a most exciting and rewarding year of scientific collaboration for both of us.

The Problem and the Experimental Approach

Otsuka et al. (1965) first reported the mapping of positions of identified cell bodies in lobster abdominal ganglia. Their work, published in complete form two years later (Otsuka et al., 1967), demonstrated that the same cells, defined by the specific muscles they innervated, could be found in similar positions in ganglia from different animals (Fig. 1). Although the maps from different individuals or different abdominal ganglia from the same individual were similar, they were not identical: identified cells were found in roughly the same position relative to each other, but nearest neighbors were not always the same. Lobster ganglion neurons are monopolar cells with no synaptic contacts on cell bodies or on the initial portions of the process that leaves the cell body. Rather, synaptic contacts are made between branches of cells in the neuropil. It is easy to imagine the possibility that the variability in the position of cell bodies is unimportant, and that the structure of the neuropil, where the synaptic contacts are made, is much more highly ordered and perhaps completely invariant. On the other hand, the neuropil might be less ordered than the cell body position, maybe to such an extent that the structure appears random. A direct way of exploring these possibilities is to ask whether the same cell (cell serving the same function) in different animals has a recognizable or even unique geometric shape, or whether such a cell can assume a variety of shapes to perform the same function.

21

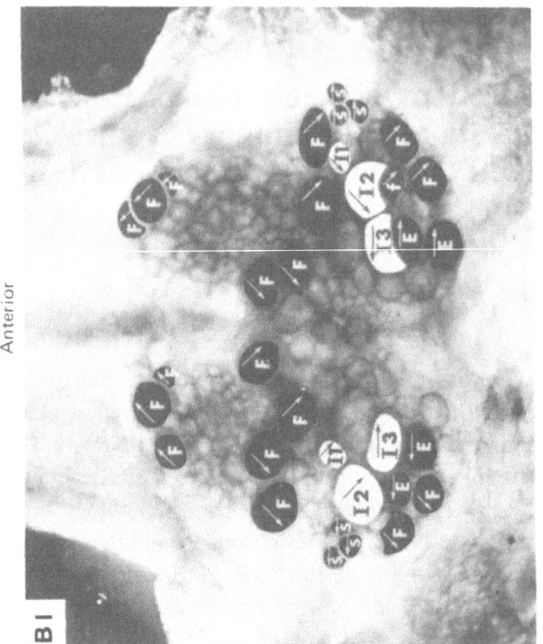

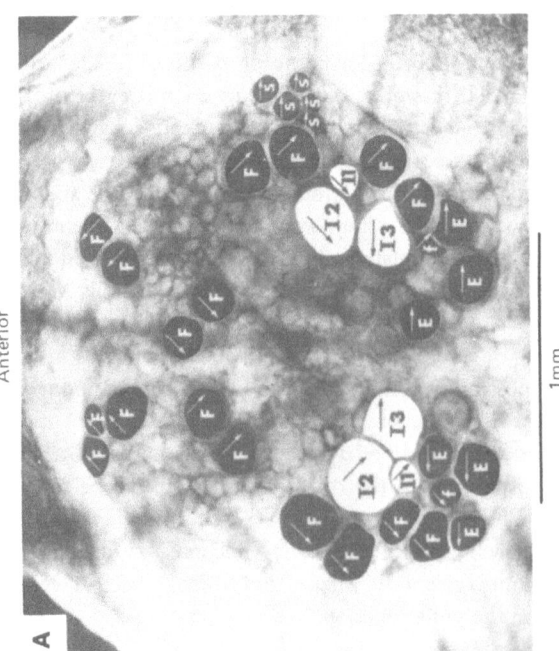

Fig. 1. Maps of the positions of physiologically identified cell bodies in abdominal ganglia from two different lobsters. Cells marked in black are excitatory and those in white are inhibitory. Cells to flexor muscles are marked F (fast) or f (slow), cells to extensors are marked E and to swimmerets, S. The arrows indicate whether axons emerge from the CNS on the same or opposite side of the ganglion from the cell bodies. Reprinted from M. Otsuka, E. A. Kravitz, and D. D. Potter, *J. Neurophysiol.* 30:725–752, 1967.

To attempt to solve these problems, we required a procedure that not only had the resolution of conventional neuroanatomical techniques (e.g., Methylene Blue and Golgi stains), which stain an entire cell (all its branches) and differentiate it clearly from its neighbors, but which also enabled us to choose the cell to be stained rather than relying on apparently chance selective staining.

We were fortunate that Alvarez and Furshpan in our department had developed an excellent method for attempting this (see also Thomas and Wilson, 1965). In their procedure, dyes were injected electrophoretically from a microelectrode into a cell and then allowed to spread through the cytoplasm. In developing the technique, Alvarez and Furshpan had solved many technical problems, such as methods for filling electrodes with dye solution and for cutting serial sections of plastic embedded material with a steel knife. They had also invested much effort in screening dyes to select those with suitable properties. The dye of choice should (1) be easily injected electrophoretically into cells, (2) spread freely through the cytoplasm of the cell, (3) remain within the injected cell during the histological manipulations involved in preparing the tissue for sectioning, and (4) be readily detectable at the low concentrations which may result at points of the cell distant from the site of injection. Many of these properties are hard to predict from the chemical structure of dyes, so in each case the whole procedure from injection to histology had to be completed. Several dyes were found which could be injected easily and which spread through the cytoplasm [e.g., Fluorescein, Chromotrope 2R, Light Green SF Yellowish, Fast Green FCF, and Chicago Blue 6B (synonomous with Niagara Blue 6B)]. The problem of keeping the dye localized in the injected cell, however, was much more severe; some dyes slowly leaked out of live cells, while others leaked out during the alcohol dehydration steps when the permeability barrier of the cell was broken down. Furshpan and Alvarez had finally selected two dyes for use, Fast Green FCF and Chicago Blue. With acid fixatives they observed little diffusion of dye from injected cells to their neighbors. Our studies with lobster ganglia began from this point. There is little doubt that the assistance provided by Alvarez and Furshpan saved us many months of labor and was sufficiently essential to our studies to suggest that we might never have found Procion yellow without their help.

The Selection of a Suitable Dye

Initial Results: Definition of a Dye of Choice

Our first results, using Fast Green in identified lobster cells, were disturbing. Reconstructions of cell shape from serial sections of a ganglion in which the large inhibitory cell I_2 was injected, showed that the cell body was attached to a large process which descended through the neuropil and left the ganglion without branching. Essentially the same results were found in other injected cells, although occasionally large branches containing dye were found. If the technique were displaying the true anatomy of the cell, we would have to conclude that cells with efferent contacts with muscles did not make fine branches in the neuropil. We were reluctant to accept this because the Methylene Blue staining of Retzius (1890) on the crayfish and of Allen (1894) on larval lobster showed that most cells in such ganglia have many elaborate branches.

In order to test stringently the adequacy of the technique, we injected Fast Green into lobster abdominal stretch receptor neurons. The dye diffused into the very fine branches of the cell, and the branching pattern seen was as rich as that shown in classical Methylene Blue preparations made by Alexandrowicz (1951). The dye survived fixation, but during dehydration in methanol the dye was leached out of the fine branches, although the cell body, axon, and some larger branches still retained dye. Fast Green is much easier to localize in tissue by observation of its red fluorescence than by conventional optical examination of green-stained profiles. Fluorescence microscopy of sections of these cells showed that the connective tissue around the fine branches was heavily stained. On re-examination by fluorescence microscopy of sections of ganglia with cells injected with Fast Green, we observed the same extensive spread of dye throughout the tissue. We attempted to arrest dye diffusion with a variety of techniques (e.g., by using metal salts, Al^{3+} and Pb^{2+}, at various stages in the histological procedure) but without success. Poor color localization also resulted when $Fe(CN)_6^{3-}$ was injected into a cell, allowed to diffuse, and the tissue subsequently treated with a $FeCl_3$ solution (cf., Kerkut and Walker, 1962; also see Chapter 1).

Our experience with Fast Green suggested that the most suitable dyes would be those with reactive groups which would form covalent bonds with macromolecules in the cell. With such dyes, after fixation there would be little chance of wash out or spread. It further appeared that fluorescent dyes were highly desirable for two reasons. First, fluorescence methods provided far greater sensitivity in the enhanced signal to noise ratio and color differentiation of dye-filled profiles. Second, there was a good chance of detecting more easily the very fine processes of a cell (those below the resolving power of the light microscope) due to scattering of the emitted light.

The Search for a Fluorescent Dye that would be Covalently Bonded to Tissues

Fluorescein isothiocyanate, which reacts with amino groups of proteins, was used with fair success, although very often the electrode tips became plugged with solid dye during the course of an injection. Another drawback was that the greenish fluorescence of the dye resembled the background fluorescence of the tissue. Rhodamine isothiocyanate did not come out of the electrode. Mercurichrome, which is highly fluorescent, and which has the attractive feature of containing a heavy metal (Hg) which might be useful in subsequent electron microscopy, did not spread through the cytoplasm, possibly because of binding to sulfur on proteins.

Among E. J. Furshpan's large dye collection were several Procion dyes. Procion dyes are used for dyeing textiles and paper (see Chapter 3) and are derivatives of cyanuric chloride (Fig. 2) known to form covalent bonds with carbohydrates and proteins (cf. Hess and Pearse, 1959). A sample of one of these (Procion brilliant orange 2RS) was injected into a lobster stretch receptor neuron. After carrying the tissue through our standard histological procedure (fixative 1, Table 2, dehydration in methanol, embedding in plastic, see below), we were delighted to find that the dye had apparently remained within the fine branches of the cell. Examination of serial sections of the receptor cell confirmed this. Other cells were injected with Procion brilliant orange but we had great difficulty seeing dye-filled profiles since the fluorescence was weak. We therefore sought other Procion dyes to find one which spread through the cytoplasm

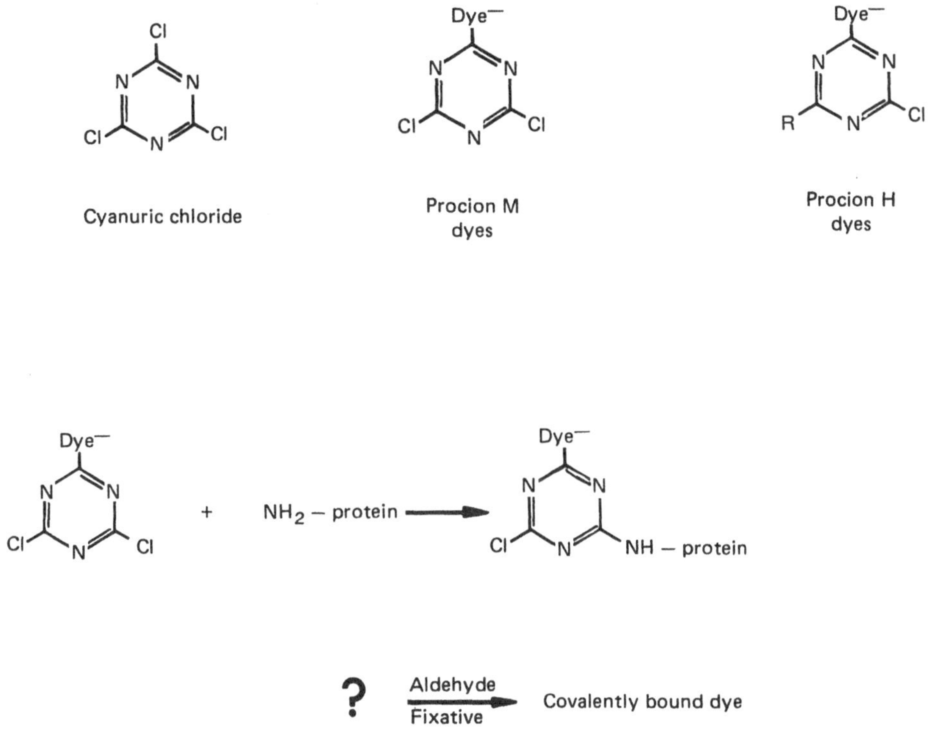

Cyanuric chloride

Procion M
dyes

Procion H
dyes

Dye⁻ + NH₂ — protein ⟶ Dye⁻ ... Cl ... NH — protein

? $\xrightarrow[\text{Fixative}]{\text{Aldehyde}}$ Covalently bound dye

Fig. 2. Generalized structural formulae and types of reactions of Procion dyes with tissue constituents. Cyanuric chloride is the parent compound of Procion M (mono-substituted) and H (disubstituted) dyes. The second substituent on H dyes can be a second dye species. Procion dyes are alkylating agents and can react with amino, sulfhydryl, and hydroxyl groups in tissues. In addition, fixation may play some role in binding the dyes to tissues.

and survived histology as well as Procion brilliant orange and which was more highly fluorescent. Through the courtesy of Mr. R. Thornton of ICI Organics, Inc. we obtained samples of 58 Procion dyes and five related dyes, classified as Procion Supra dyes. All of these substances were tested at least once by injection into neurons in central ganglia. They were categorized according to their ability to pass out of electrodes, spread through cell body cytoplasm, survive fixation, dehydration, and embedding, and penetrate into fine branches. The dyes tested are listed in Table 1. With dyes that came out of electrodes, we never saw the extensive diffusion out of the cell that we observed with previous dyes. As can be seen in Table 1, many of the 63 tested either did not spread extensively through the cell cytoplasm or did not come out of electrodes. Those which showed good migration properties were examined to see which were most fluorescent.

The Selection of Procion Yellow: Experimental Technique

Although many of the dyes we tested may be useful in some situations, the four dyes which were most suitable for our purposes were Procion brown H3RS, brilliant

Table 1. Procion dyes tested

Procion dye	Out of electrode	Diffusion	Fluorescence
1. Yellow M4RS	+++	+++	++, orange yellow
2. Yellow MRS	− (only tried once)		
3. Yellow MGRS	−		
4. Yellow HAS	+*	+	
5. Brilliant yellow M4GS	++*	++	
6. Brilliant yellow H5GS	−		
7. Brilliant yellow M6GS	−		
8. Brilliant yellow H3GS	−		yellow
9. Brilliant yellow H4GS	−		
10. Gold yellow HRS	++	+	
11. Blue M3GS	±		
12. Blue HBS	±	±	
13. Blue H3GS	+	+	weak, orange
14. Blue H5RS	+	+	
15. Brilliant blue MRS	±	±	
16. Brilliant blue HAS	−		
17. Brilliant blue H5GS	−		
18. Brilliant blue H7GS	−		
19. Brilliant blue HGRS	++	±	
20. Navy blue M3RS	++	+	
21. Navy blue H3RS	++	+	+, red
22. Navy blue H4RS	±	±	
23. Red MGS	+	+	
24. Brilliant red M2BS	++	+	+, red
25. Brilliant red M5BS	−		
26. Brilliant red M8BS	−		
27. Brilliant red H3BS	+	+	
28. Brilliant red H3BNS	++	+++	+, orange
29. Brilliant red H8BS	++	+	+, orange
30. Brilliant red H7BS	+	±	
31. Brilliant red H8BNS	−		
32. Scarlet MGS	++*	++	
33. Scarlet HRS	++	+	weak
34. Scarlet HRNS	++	++	+, orange
35. Scarlet H2GS	−		
36. Rubine MB80	±		
37. Rubine HBS	++*	++	
38. Brilliant orange MGS	++	+	+, orange
39. Brilliant orange M2RS	++	++	+, orange
40. Brilliant orange HGRS	−		
41. Brilliant orange H2RS	++	±	weak
42. Green M2BS	+	+	
43. Brilliant green HGS	−		
44. Brilliant green H4GS	−		
45. Olive green M3GSA	−		
46. Olive green H7GS	+	+	
47. Brown H3RS	++	++	+, red
48. Red brown M4RS	++	+	
49. Red brown H4RS	−		
50. Orange brown HGS	−		

Table 1. (Continued)

Procion dye	Out of electrode	Diffusion	Fluorescence
51. Orange brown H2GS	+	+	
52. Turquoise HGS	+	−	
53. Turquoise H2GS	−		
54. Brilliant purple H3RS	+	+	poor
55. Violet H2RS	+	+	+, red
56. Grey MGS	+	±	
57. Black HGSA	+ +	+	
58. Black H7NS	+ +	+	weak
59. Supra yellow H2RP	+	+	+, light yellow
60. Supra yellow H4GP	+	+	−
61. Supra black HBP	+	+	−
62. Supra turquoise H2GP	±	−	−
63. Supra red H4BP	+	+	+, red

* Did not come out of electrode on repeat trial.

red H3BNS, navy blue H3RS and yellow M4RS; Procion yellow M4RS is appreciably more fluorescent than the other three, and finally we used it exclusively. Its fluorescence (as free dye) is activated at 460 nm, and it emits at 550 nm (see Fig. 3). A 4% (w/v) solution of dye in distilled water was used to fill electrodes; electrode resistance ranged from 10–20 MΩ. In large cells of the lobster ganglia (ca. 150 μm diameter) the dye was routinely injected over a period of 0.5–1.0 hr, the preparation being held at 10°C. We used pulses of current (1×10^{-8} A, 0.5 sec duration at a frequency of 1 Hz) to pass dye into the cell, as it was found that prolonged passage of current caused blockage of the electrode and damage to cells. After injection, the time allowed for the dye to migrate (at 4°C) before fixation was important. The results were dramatically improved if the migration time was increased from 6 to 16 hr.

Various fixatives were used (see Table 2), the main difference being in the intensity of the tissue autofluorescence. While almost all of our early experiments were done with glutaraldehyde-acrolein at an acid pH as fixative, later experience suggested that, because of the lower background fluorescence, glutaraldehyde:formaldehyde or formaldehyde alone might be more desirable. One problem with formaldehyde at acid pH is that tissue sections are virtually invisible in the fluorescence microscope. It is therefore difficult to pick up the outlines of other cells or cell processes to use as landmarks in reconstruction of cell shape.

After fixation, the tissue was dehydrated in a graded methanol series and embedded in Epon 812 (46 ml Epon Resin 812, 31 ml dodecenyl succinic anhydride, 23 ml methyl nadic anhydride and 1.5 ml DMP-30). Serial 10 μm sections were cut with a steel knife, mounted on slides in Lustrex (Monsanto Chem. Co.), and examined for fluorescent dye profiles. Recently, Maraglas has also been used as an embedding agent (Erlandson, 1964). The optical clarity of Maraglas makes it particularly effective for viewing whole mounts and blocks cut easily with a steel knife. Photographic enlargements of phase-contrast images (on semitransparent paper) of the serial sections were made. A Zeiss projection microscope was used for this purpose and enlarged images

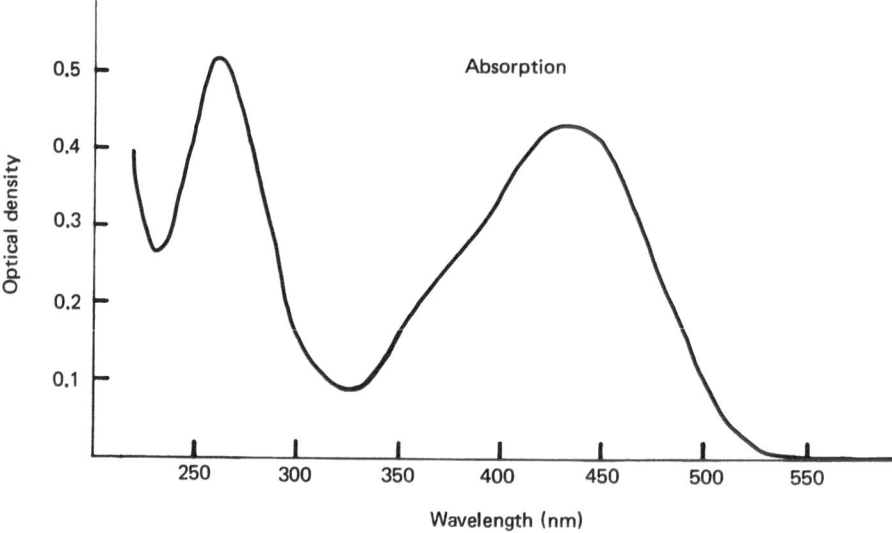

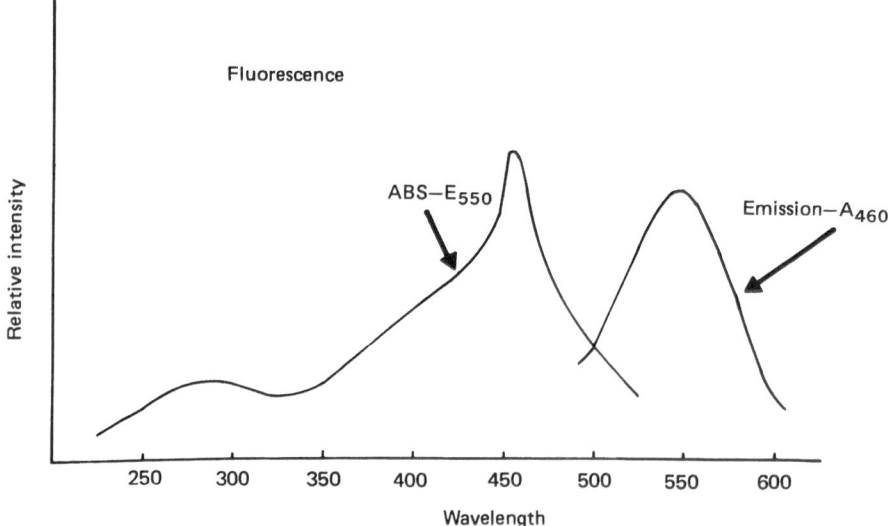

Fig. 3. Absorption and fluorescence spectra of Procion yellow M4RS. The dye concentration used was 0.02 mg/ml in phosphate buffer at pH 7.4. The fluorescence emission spectrum was measured at an activating wavelength of 460 nm and the fluorescence absorption spectrum at an emitting wavelength of 550 nm. Both spectra were measured in an Aminco-Bowman Spectrophotofluorometer. The absorption spectrum was measured in a Cary recording spectrophotometer.

Table 2. Composition of fixative and background fluorescence of tissues

Fixative	Composition							Result	
	25% glutaral-dehyde ml	acrolein ml	*formal-dehyde gm	**buffer ml	+2x sea water ml	++salts ml	H₂O ml	Dye localization	Background fluorescence intensity
Glutaraldehyde acrolein pH 4.0	2.6	0.2	—	5(A)	2.5	—	—	Good	Medium
Glutaraldehyde acrolein pH 7.2	2.6	0.2	—	5(P)	—	—	2.2	Good	High
Glutaraldehyde formaldehyde pH 4.0	0.1	—	0.4	5(A)	—	2.5	2.5	Good	Low
Glutaraldehyde formaldehyde pH 7.4	0.1	—	0.4	5(C)	—	2.5	2.5	Good	Medium
Formaldehyde — pH 4.0	—	—	0.4	5(A)	—	2.5	2.5	Good	Almost Invisible
Formaldehyde — pH 7.4	—	—	0.4	5(C)	—	2.5	2.5	Good	Low

* Solid paraformaldehyde was weighed out, heated to 90° in distilled water, and 1 drop 1N NaOH was added.

** Buffers were (A) = 0.1 M acetate pH 4, (P) = 0.2 M phosphate pH 7.2, (C) = 0.2 M collidine pH 7.4.

+ 2x sea water contains NaCl, 920 mM; KCl, 32 mM; CaCl₂, 32 mM; MgCl₂, 16 mM.

++ Salts contain 2.0% (w/v) CaCl₂, 12% (w/v) NaCl, 12% (w/v) Sucrose.

were projected directly on Fotorite paper in an enlarging easel. Prints were developed using a Kodak Ektamatic Processor. The positions of dye-containing profiles were marked on the phase-contrast enlargements by viewing sections in a fluorescence microscope (using a high intensity white light source, a thin BG-12 activation filter and a barrier filter that passed light above 500 nm). The profiles were then drawn on the photographs of the same sections. The photographs were oriented relative to each other on a light table and register marks were made. Projections of the profiles of the cell onto horizontal and vertical planes were constructed in the standard way from the oriented photographs and these projections will be illustrated below.

We have not carried out any definitive experiments on the chemistry of the binding of Procion yellow to cell constituents. The fact that the dye spreads easily through the cell implies that covalent bond formation is slow. It is also possible that aldehyde fixation plays a role in the binding (see Fig. 2).

Disadvantages of Procion Yellow as an Intracellular Marker

Two properties of Procion yellow make it a less than optimal intracellular marker. Its fluorescence is relatively weak, and it neither contains nor will it precipitate heavy metals in sufficient amounts to be visible with electron microscopy.

We tried to obtain a dye with better fluorescent properties by coupling cyanuric chloride to fluorescein amine and napthylamine di-sulfonic acid. The resulting dye was very fluorescent but did not spread well through the cytoplasm. Moreover, it was difficult to see the green-yellow fluorescence of the fluorescein derivative over the green tissue background. In any event, the synthesis of dye derivatives of cyanuric chloride is relatively easy, and one should be able to design highly fluorescent or electron opaque compounds for testing (see Chapter 3). Recently McMahan and Purves have developed techniques for fixation of Procion yellow-injected cells such that fluorescence microscopy and electron microscopy can be carried out on the same preparation (see Chapter 5).

Another approach to electron microscopic examination of cells is the recently described experimental technique of cobalt injection (Pitman et al., 1972a and Chapter 6).

Analysis of Lobster Neuronal Geometry

General Anatomy of the Lobster Abdomen: A Brief Description

The central nervous system (CNS) of the lobster is a ventral chain of segmental ganglia joined together by connectives. The ganglia contain cell bodies concerned mostly with innervating the body segment in which they are located; the connectives carry interganglionic fibers.

In the abdomen, two separate sets of muscles control movement. The fast, or phasic, muscles are responsible for the characteristic lobster tail flip used in the escape response. The slow, or tonic, muscles control abdominal posture. Each set of muscles consists of two antagonistic groups, flexors and extensors. An additional distinct set of muscles in each segment controls the swimmerets.

In the abdominal portion of the CNS, the first to fifth abdominal ganglia each have three roots on either side. The first root includes nerves controlling the swimmeret musculature together with many sensory fibers. The second root contains sensory fibers and nerve fibers innervating extensor muscles. The third root contains only axons to the flexor muscles. The third root leaves the CNS from the connective, about a quarter of the way between its ganglion and the next posterior one, while the first and second roots emerge directly from the ganglion.

Ganglia and connectives are surrounded by an outer connective tissue sheath. Within the ganglion, cell bodies lie ventrally and are surrounded by soft perineural tissue. We have counted about 750 neurons in several second abdominal ganglia. It is difficult to make exact determinations of the number of cells since there are many small cells adjacent to the neuropil which may be either neurons or glia.

Occasionally, in our examination of over 100 ganglia, we found one or two medium-sized cells on the opposite side of the neuropil from the usual position, i.e., on the dorsal surface, close to the giant fibers. On rare occasions we have seen large cells in this position. While none of these cells has been identified, they do send processes into the neuropil. We do not know whether these are cells displaced from their usual positions but making their normal contacts in the neuropil, or whether they are supernumary cells making adidtional contacts.

Our anatomical findings using the dye injection technique relate to three aspects of neuronal geometry: (1) the path taken by the axons of identified cells both within and outside of the neuropil region; (2) the reproducibility of branching pattern of the same cells in different preparations; and (3) a comparison of the branching patterns of different cells.

Course of the Axons of Identified Cells

Otsuka et al. (1967) had assigned the axons of identified cells to particular roots by antidromic stimulation of the root. We were able to confirm their results directly by following dye-injected axons into the appropriate roots. In the case of some of the flexor cells we could follow axons into the three main divisions of the third root: the superficial, middle, and deep branches. This provided valuable additional criteria for identifying cells by antidromic stimulation of these branches (see Table 3). The superficial branch of the third root in the lobster is different from that of the crayfish (Kennedy and Takeda, 1965) in that in addition to the six axons supplying the superficial flexor muscle, the lobster root also carries two fast flexor axons, M_7 and a branch of I_2. The superficial branch sends a further branch to the fast flexor musculature shortly before entering the tonic muscle.

In tracing the axons through the neuropil, our results, including those we obtained using Fast Green FCF where only the axons and very large branches were stained, gave striking indications that certain structures of the neuropil are regular and reproducible from animal to animal. For many cells the path through the neuropil and the relationship to other "landmarks" in the ganglion, such as tracts and bundles of axons, is so reproducible that subsequently they can be recognized in uninjected ganglia. This regularity is also displayed in the easily recognizable shapes, found in transverse sections, that are generated by the particular path taken by some cells. We found many

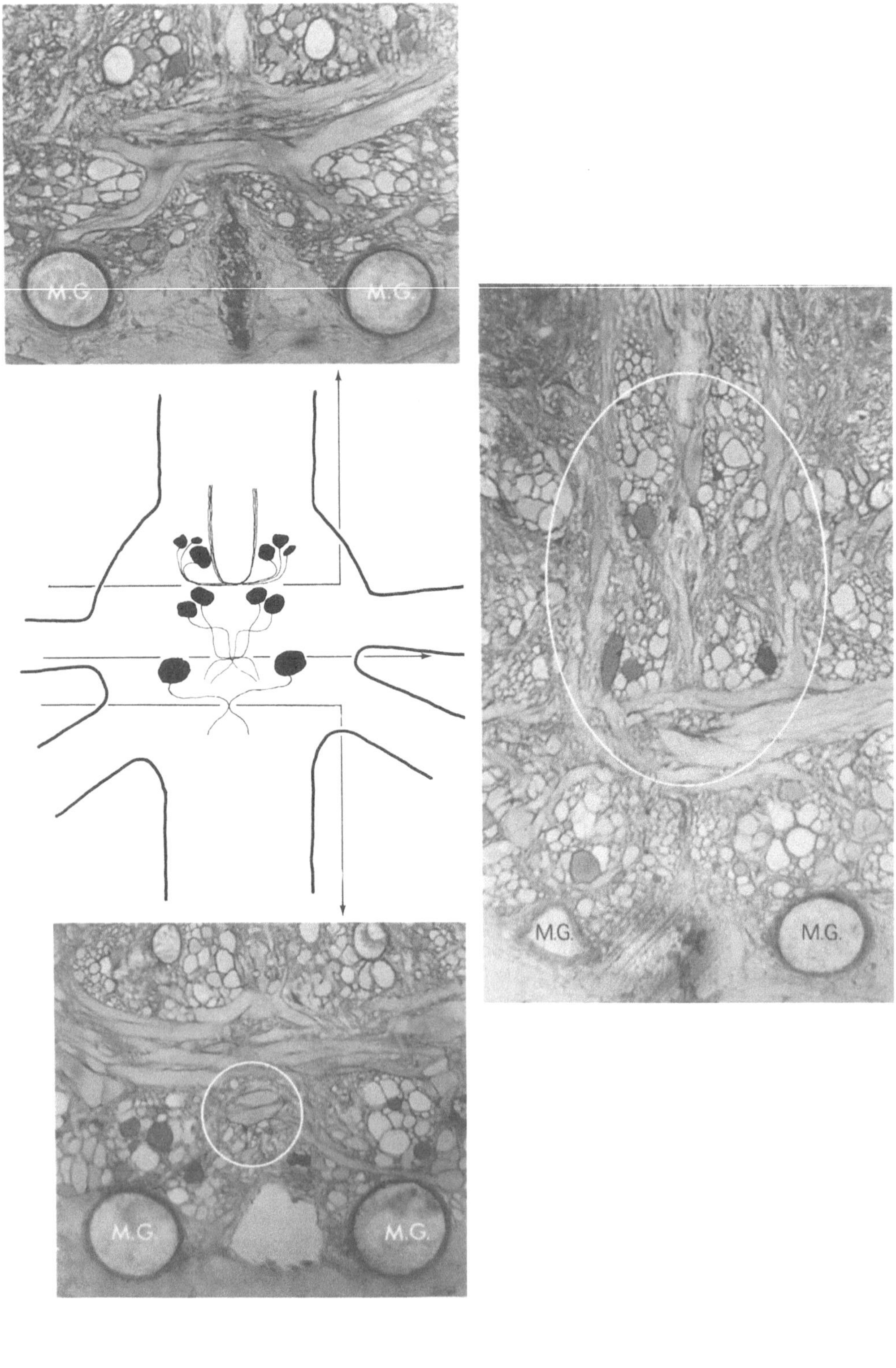

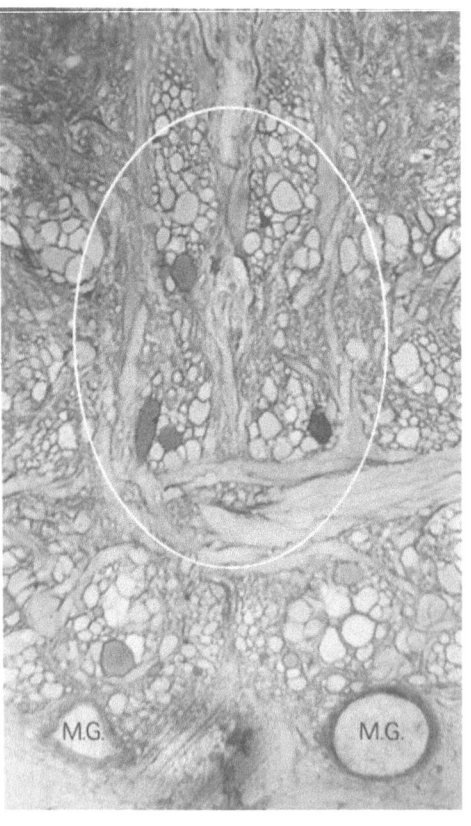

Table 3. Roots of emergence of axons of identified cells:
anatomical confirmation of physiological evidence

Cell	Root of emergence	Axon crosses midline	Position of cell body
M₁	*3 ant. M & D	Yes	Variable in cluster
M₂	3 ant. M & D	Yes	
M₃	3 ant. M	Yes	Always anterior & medial to M5
M₄	3 M & D	Yes	
M₅	3 M & D	Yes	
M₆	3 M	No	interchangeable
M₇	3 S	No	
M₈ series	1	No	
M₉	3 D	No	
M₁₀	3 D	No	
M₁₁	3 S	No	
M₁₂	2	No	
M₁₃	2	No	
M₁₄	2	No	
I₁	3 S	Yes	
I₂	3 M, D & S	Yes	
I₃	2	Yes	

* M, D, & S refer respectively to the middle, deep and superficial branches of
the third root. In the anatomical study the members of the anterior motor group
were not identified individually.

such features, and three examples are shown in Fig. 4. The first is the shallow, stretched-
out X-shape found where the six cells of the anterior motor group cross the midline.
This is the most anterior region where axons decussate. The second is an E-shaped
structure formed by the axons of M_4 and M_5 near the point where they cross the
midline. The third is the region where the I_2 axons are closely apposed in the midline
over a distance of 30–60 μm. These features have been found with great reliability
in all ganglia examined.

Reproducibility of the Branching Pattern of Cells

The horizontal plane projections (plans) of the branching patterns of cells are
shown in Figs. 5 through 7. The richness of cell shape is best illustrated in this projection

Fig. 4. Neuropil landmarks initially revealed by dye injection of functionally identified
cells. (1). The decussation and intertwining of the axons of the anterior motor group of
cells, M₁, M₂, and M₃ of both sides. This region is one of the first crossing points of axons
in the ganglion and is revealed by a broad band forming a shallow X-shaped structure (in
cross section) sending branches down to the medial giant fibers (M.G.). (2). The descent
of axons M₄ (lateral) and M₅ (medial) form an E-shaped structure approximately midway
through the neuropil. These axons intertwine with numerous unidentified axons in forming
this structure. (3). The crossing of the I₂ axons in the midline. A white circle surrounds the
double oval profiles of the two I₂ axons. The axons lie together at the crossing point over a
distance of about 30–60 μm.

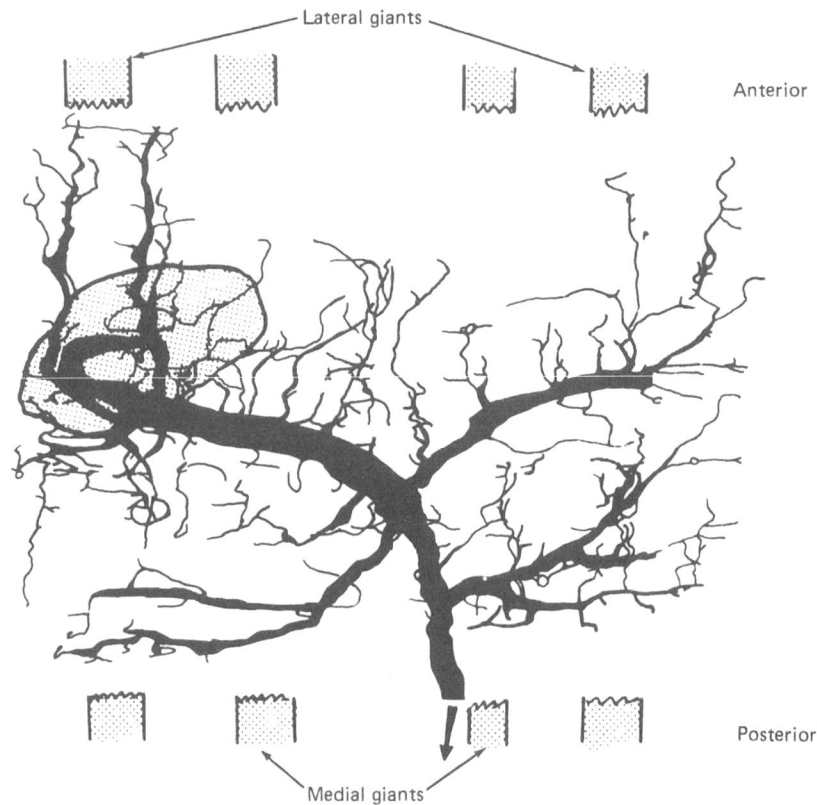

Fig. 5. Plan (horizontal projection) of cell I_2 in the lobster second abdominal ganglion. Reprinted from A. O. W. Stretton and E. A. Kravitz, *Science, 162:*132–134, Fig. 3A. (Copyright 1968 by the American Association for the Advancement of Science.)

since the branching pattern of most cells is restricted to a rather narrow horizontal section of the ganglion. Moreover, plan projections also allow a ready comparison of the shape of different cells. For some purposes, other projections are useful and they will be described below.

The cell I_2 (the inhibitory cell to the main mass of fast flexor muscles) was selected to test the reproducibility of branching patterns of the same cell in different animals. The pattern of I_2 branches is very complicated (Fig. 5). When a number of I_2s are examined, however, the overall shapes of the cells are similar. To aid the description of similarities and differences between branches of I_2s, simplified and reduced versions of six cell profiles are shown in Fig. 6; only a few points of difference will be considered. One striking dissimilarity between cells is in the relative points of origin of branches B and C. In five cases they join the main axon very close together; in the sixth cell, branch C is displaced considerably (150 μm) towards the cell body. Branch C was identified in this cell by its size and the final destination of its fine outer branches. Another clear difference between the cells is in the origins of the branches labeled A in Fig. 6. In some cells they arise independently from the main axon, whereas in others they are branches off a single limb.

The significance of differences such as these is not easy to assess. One possible source of variation was that the animals used were not isogenic; wild animals of unknown genetic background were studied. Moreover, we have had no control over the experiential background of the animals, nor did we search for overt behavioral differences between them. We do not know if changes in the shapes of particular cells are compensated for by changes in other functionally related cells so that the activity of a functional unit consisting of several cells might be preserved.

The consequences of branch displacements, like the ones that we found, would almost certainly differ depending on whether the branches concerned are input or output channels (or both) for a cell. For example, if a branch carries only output signals to other regions of the ganglion, a difference in the point of origin of the main axon might have little effect. If input signals are carried, however, and if firing of the main axon relies on the summation of input from several branches, the separation of branching points could be very important. Obviously, to understand functional consequences of branch point differences, one will have to understand in detail the neuronal circuitry in which each cell is involved.

It should also be noted that these comparisons relate to a single point of time in the lifespan of these lobsters, which were four to five years old. We do not know how fixed the structure of neurons remains. It is possible that initially the structure of a cell is prescribed exactly but that during growth and maturation, changes may occur.

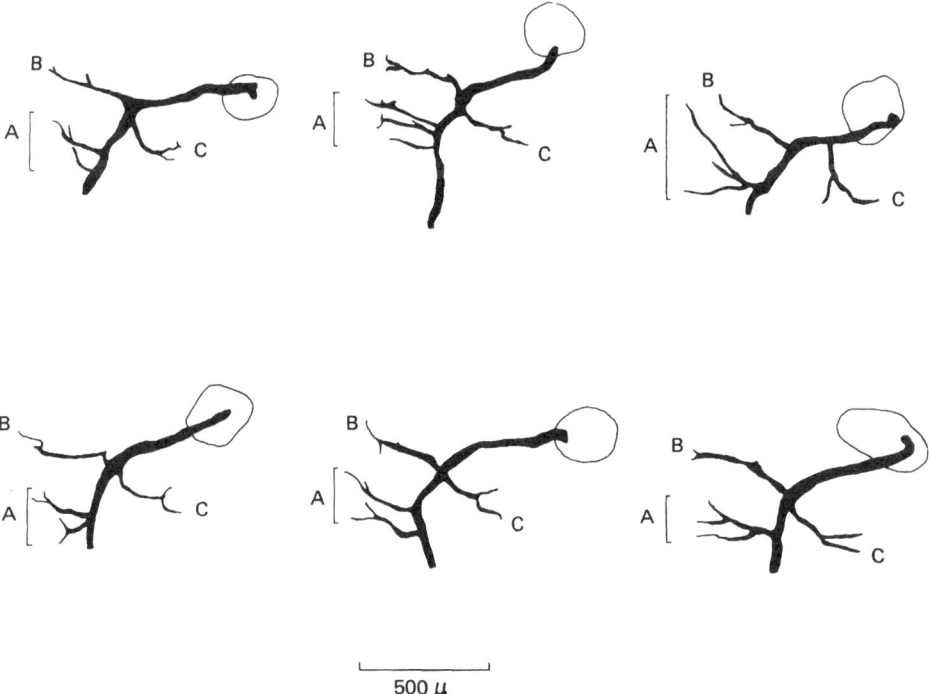

500 μ

Fig. 6. Simplified plan projections of the cell I_2 from abdominal ganglia of six different lobsters. A, B, and C represent branching regions to be compared in these cells.

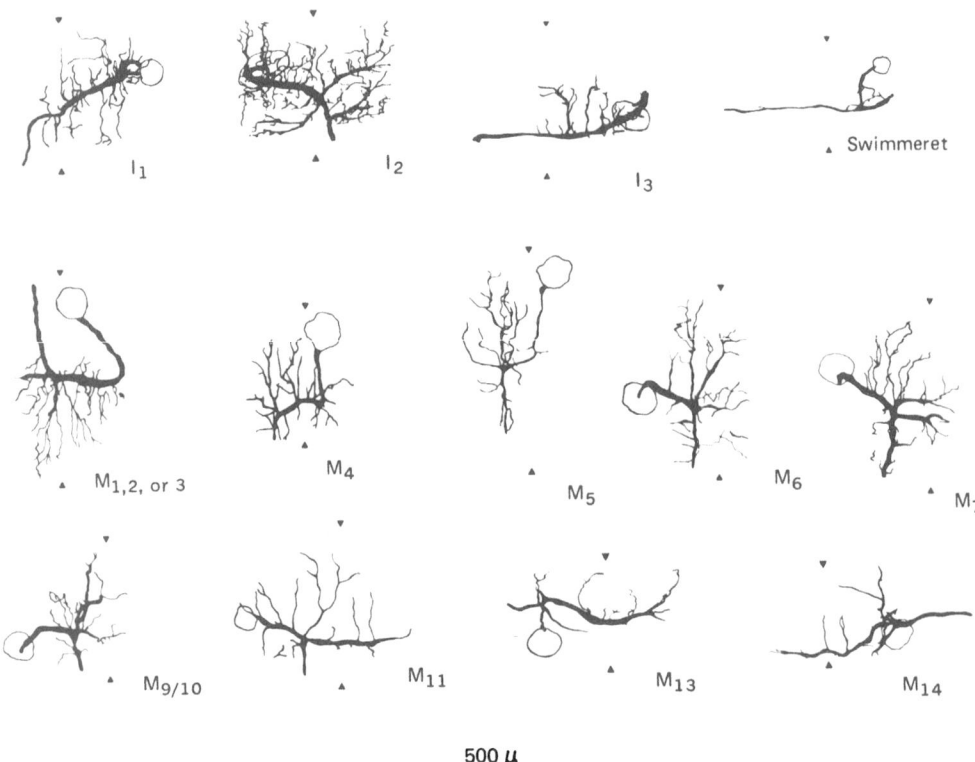

I_1 I_2 I_3 Swimmeret

$M_{1,2, or 3}$ M_4 M_5 M_6 M_7

$M_{9/10}$ M_{11} M_{13} M_{14}

500 μ

Fig. 7. Horizontal plane projections (plans) of identified cells. The arrowheads mark the mid-line.

Branching Patterns of Different Cell Types

Using the map and methods described by Otsuka et al. (1967) and our extension of the identification methods to axons emerging from the third root (see Table 3 above), we set out to inject and compare the identified cells in lobster abdominal ganglia. The results are illustrated in Fig. 7. Some of the cell types shown were injected only once, but most were injected at least twice, and the general reproducibility of shape seen for I_2 held in the other cases as well. The most striking feature of the set of cell plans is that the branching pattern of each type of cell is unique. Each cell, therefore, can be recognized by its geometric shape. This result emphasizes the concept of the uniqueness of identified cells in invertebrate ganglia. Not only are peripheral connections to muscles distinctive, but the anatomy of the intraganglionic branching pattern is also distinctive. This situation should be contrasted with the nervous systems of higher animals where a striking feature of the anatomy is that often there are many cells within an area having similar branching patterns. In these cases, it is necessary to use functional means to distinguish between morphologically homogeneous populations of cells.

Several features of the branching patterns of lobster abdominal ganglion cells deserve comment. The first involves the sidedness of the branches. Some cells seem to be equally concerned with both sides of the ganglion and send branches very widely through the neuropil of both sides (I_2, $M_{1/2/3}$, M_4, M_{11}, swimmeret cell, M_{13}, M_{14}); others are more restricted to one side of the neuropil, sending few processes across the midline (I_1, I_3, M_5, M_6, M_7, $M_{9/10}$). In those neurons sending branches into both sides, the symmetry of the branches about the midline is striking. Note particularly I_2 and $M_{1/2/3}$. There is no apparent correlation between the bilaterality of the branching pattern and whether the main process of the cell crosses the midline.

The existence of these two classes of neurons suggests interesting behavioral and physiological speculations. The symmetrical relationship between neuropil branches of some cells on both sides of a ganglion, coupled with the bilateral symmetry of cell pairs, may be a way of introducing a maximum safety factor in coordinating the muscular activity of both sides of the abdomen. Intuitively, one feels that the activity, at least of the fast abdominal musculature, should be strictly synchronized to give the most powerful swimming stroke. On the other hand, if symmetry is related to synchrony, the finding of a class of neurons not arranged in this way is surprising. The existence of electrical coupling between pairs of cells is obviously relevant here, and electrical coupling has been demonstrated between pairs of the nonsymmetrical cells (I_1) as well as between symmetrical cells (I_2) (Otsuka et al., 1967). Since all cells send branches to, or across, the midline of the ganglion, electrical contacts could be made with contralateral partners at the midline and activity could be coordinated in this way. In one case this was explored by Dr. Paula Orkand. Serial thin sections were cut for electron microscopic examination of the entire contact region between the main axons of two I_2 cells as they cross the midline (see Fig. 4). No anatomical feature which could be associated with electrical synapses was seen; in fact, throughout the region the cell processes were separated by a layer of connective tissue. In this case, the location of the electrical synapse between symmetrical cells remains unidentified. This result emphasizes the need for electron microscopic verification of regions indicated by light microscopy as potential sites of contact between cells.

Since the axis of movement of the normal abdomen is defined by the segmental pivot points, which allow no appreciable lateral motion, there is no obvious need for unilateral activity. However there may be times in the life of the lobster when fast asymmetric movements are important. For example, during molting, when the abdomen is soft, a special set of neurons may be left uncoupled to aid the animal in escaping from its shed exoskeleton (we are indebted to Dr. C. R. Slater for this suggestion).

Another interesting possibility that we saw after examination of the branching patterns of the identified neurons, is that one might make tentative functional subdivisions of the neuropil based upon the destinations of cell branches. For example, there are regions of neuropil where branches of neurons to tonic muscles are found, but never branches of neurons supplying phasic muscles. We have made a preliminary attempt to extend this analysis by making a movie film of serial 10 μm sections through a ganglion. This was done by aligning and registering photographs of sections (as above) and using the register marks to orient photographs for cinematography. The movie served as a visual aid for identification of axon tracts and other gross features of ganglionic organization. After viewing the film in forward and reverse directions several times, it was possible to pick out distinctive neuropil regions. These regions were often subgroups of fibers which traversed the entire length of the ganglion and emerged in the connectives on either side. Once we had set up criteria for recognizing these regions (axon size, position in ganglion, neighboring groups, etc.), we returned to an examination of the serial sections prepared from dye-injected cells.

One group of axons found in the connectives runs through the entire ganglion and lies immediately above the medial giant fibers. These axons are visited often by branches of $M_{1/2/3}$ and M_4, M_5, M_6, M_7, $M_{9/10}$, the motor nerves of the fast flexor muscles, and I_3 the fast extensor inhibitory neuron; they are avoided by branches of I_2, the inhibitory cell to the fast flexors, and by M_{13} and M_{14}, motoneurons to fast extensors. This suggests that this bundle of presumed interneurons may carry fast flexion signals (Fig. 8). As might be expected, the nerve cells innervating slow flexor muscles do not send branches into this region. Such a neuropil region may contain "command fibers" as described by Wiersma and Ikeda (1964) and Kennedy et al. (1966). Their physiological experiments demonstrate that, in stimulating single axons dissected from connectives, one can simultaneously excite excitatory neurons to one type

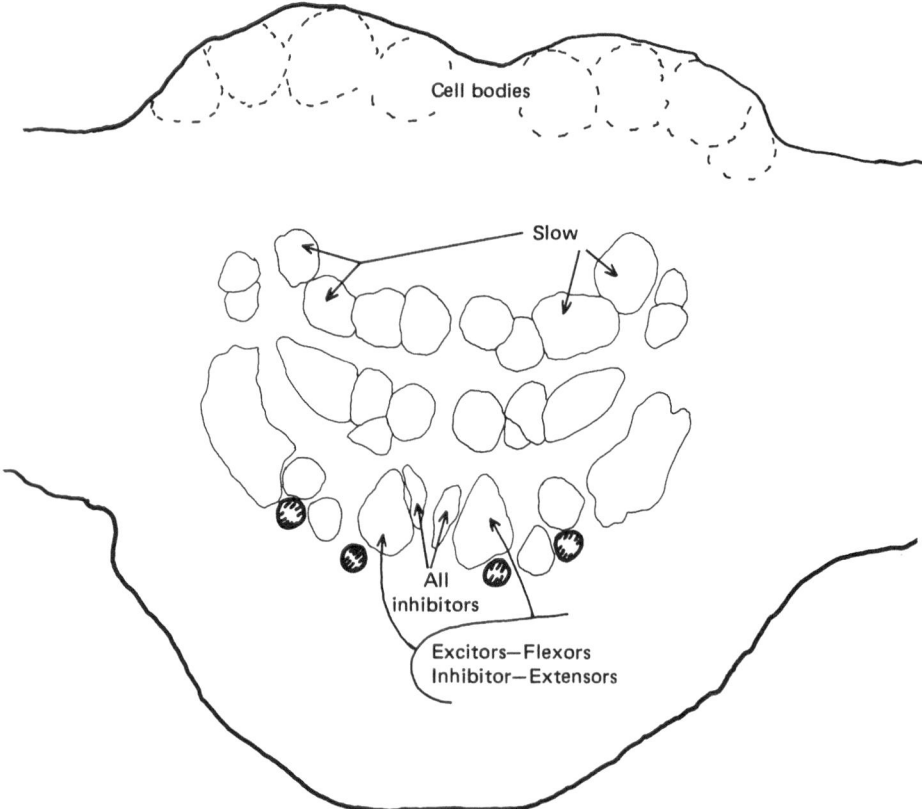

Fig. 8. A map of neuropil regions visited by processes of physiologically identified cells. The circular profiles in the middle of the figure represent the approximate positions of groups of axons passing through the ganglion. These have been mapped by examination of movie films of serial photographs of ganglionic sections (see text). Three regions are indicated: one which receives branches of slow system neurons; one which receives branches of excitatory neurons to fast flexors and the inhibitory neuron to the fast extensors; and one to which all inhibitory cells send branches.

of muscle (e.g., flexors) and the inhibitory neuron to the antagonistic muscle set (extensors) and inhibit the inhibitory and excitatory neurons to reciprocal sets of muscles.

We have been able to make preliminary assignments of function to several other regions of the neuropil as well (Fig. 8). For example, neurons of the tonic system of muscles send branches to regions high in the neuropil while fast muscle neurons do not; one region receives branches of all inhibitory fibers, and swimmeret muscle neurons project to a particular position.

The combination of the techniques of dye injection with a knowledge of the physiological function of injected cells thus may provide a powerful tool for unraveling the complexity of the neuropil. While this type of analysis is in a very primitive state at present, it seems clear that it may serve as the basis for a rational neurophysiological examination of the neuropil.

Conclusion

Our original question as to the reproducibility of the structure of the same cell in different individuals has been answered very clearly. At a gross level, the structure of a cell is very distinctive, as distinctive as a fingerprint. The same major branches of a cell are consistently found sending smaller branches to the same destination in the neuropil. However, the exact details of the branching patterns are *not* identical. The interpretation of these differences is not possible at present, and would probably require a complete understanding of the development and function of the ganglion. We suspect that the lobster is not the ideal animal for such an undertaking.

Acknowledgments. Supported by PHS grants NS 02253 and NS 07848 and research career development award HD-5899. We thank Dr. E. J. Furshpan for providing us with complete details of unpublished experimental techniques and for his help throughout the course of these studies. E. Maier, F. Foster, K. Fischer, R. B. Bosler, M. LaFratta, and J. LaFratta, and J. Gagliardi provided technical assistance. D. Cox, as usual, provided expert secretarial help.

References

Alexandrowicz, J. S.: Muscle receptor organs in the abdomen of *Homarus vulgaris* and *Palinurus vulgaris*. Q. J. microsc. Sci. *92*, 163–199 (1951).

Allen, E. J.: Studies on the nervous system of Crustacea. I. Some nerve elements of the embryonic lobster. Q. J. microsc. Sci. *36*, 461–482 (1894).

Erlandson, R. A.: A new maraglas, D.E.R. 732, embedment for electron microscopy. J. Cell Biol. *22*, 704–709 (1964).

Hess, R., and A. G. Pearse: Labelling of proteins with cellulose reactive dyes. Nature, Lond. *183*, 260–261 (1959).

Kennedy, D., W. H. Evoy, and J. T. Hanawalt: Release of coordinated behavior in crayfish by single central neurons. Science *154*, 917–919 (1966).

——, and K. Takeda: Reflex control of abdominal flexor muscles in the crayfish. II. The tonic system. J. exp. Biol. *43*, 229–246 (1965).

Kerkut, G. A., and R. J. Walker: Marking individual nerve cells through electrophoresis of ferrocyanide from a microelectrode. Stain Technol. *37*, 217–219 (1962).

Kravitz, E. A., A. O. W. Stretton, J. Alvarez, and E. J. Furshpan: Determination of neuronal geometry using an intracellular dye injection technique. Fedn Proc. Fedn Am. Socs exp. Biol. 27, 749 (1968).

Otsuka, M., E. A. Kravitz, and D. D. Potter: The γ-aminobutyric acid (GABA) content of cell bodies of excitatory and inhibitory neurons of the lobster. Fedn Am. Socs exp. Biol. 24, 399 (1965).

——— ——— ———: Physiological and chemical architecture of a lobster ganglion with particular reference to gamma-aminobutyrate and glutamate. J. Neurophysiol. 30, 725–752 (1967).

Pitman, R. M., C. D. Tweedle, and M. J. Cohen: Branching of central neurons: intracellular cobalt injection for light and electron microscopy. Science 176, 412–414 (1972).

Purves, D., and U. J. McMahan: The distribution of synapses on a physiologically identified motor neuron in the central nervous system of the leech: an electron microscopic study after the injection of the fluorescent dye Procion yellow. J. Cell Biol. 55, 205–220 (1972).

Retzius, G.: Zur Kenntniss des Nervensystems der Crustaceen. Biol. Untersuch., N.F. 1, 1–50 (1890).

Stretton, A. O. W., and E. A. Kravitz: Neuronal geometry: determination with a technique of intracellular dye injection. Science 162, 132–134 (1968).

Thomas, R. C., and V. J. Wilson: Precise localization of Renshaw cells with a new marking technique. Nature 206, 211–213 (1965).

Wiersma, C. A. G., and K. Ikeda: Interneurons commanding swimmeret movements in the crayfish Procambarus clarki (Girard). Comp. Biochem. Physiol. 12, 509–525 (1964).

Discussion

Dr. Robert Shlaer (Northwestern University): Does the injected dye only fill cells by diffusion or by an active process? If not by an active process, perhaps Procion can be bonded to something such as a small protein to make it fill cells actively.

I would like to comment that in looking at the fine dendritic detail in the cerebral cortex, much of it seems to be the result of processes going around other structures—blood vessels, cell bodies, and so on.

It might be that the individual differences in fine morphology are simply an accommodation to differences in the positions of larger structures.

Dr. Kravitz (Harvard): We really didn't do very much chemistry to find out how Procion dyes are moving, because of a technical problem.

These dyes will bond covalently to proteins; therefore, if after injection you grind the tissue up and find dye bound to protein, you don't know whether it bound in the extract after homogenization, or whether it was bound to proteins within the cell. We think dye probably binds to protein rather slowly, and that is why we see it spreading down the axon. For the most part we imagine it moves by diffusion, although we haven't really rigidly tested that.

We did do one or two experiments of putting in compounds that are supposed to disrupt microtubules but we didn't see anything very convincing. The idea was to see if there was any indication of an active movement down fine branches.

I agree with your comment that some of the geometry of neurons could have a trivial origin.

Dr. Cohen (Yale): The term "diffusion" is used repeatedly, but what we mean is that the stain is migrating by an unknown process. It might be useful to use compounds which affect intracellular transport.

CHAPTER 3

The Chemistry of Reactive Dyes and its Relevance to Intracellular Staining Techniques

C. V. Stead

The Procion dyes used so successfully for intracellular staining were originally developed for the textile industry, and little information on their chemistry has been readily available to neurobiologists. The initial part of this chapter deals with the rationale behind the use of Procion dyes in the textile industry and describes the basic chemistry of the dyes. This provides a framework for extension of their use to intracellular staining. Specific questions relating to intracellular staining are dealt with in the second part of the chapter.

Procion Dyes in the Textile Industry

The purchaser of a colored textile material demands that the textile will withstand various treatments during its useful life without losing its color. One of the aims of dyestuffs chemistry is to design dyestuffs which will be retained on the fiber when the textile is subjected to these various treatments. An important treatment is washing, and to achieve fastness to washing there must be an interaction between the dye and the fiber to cause the dye to be retained on the fiber and not desorbed during the washing process. In this interaction the physical and chemical nature of the fiber is of prime importance. Up to 1956, four types of interaction were exploited to achieve fastness to washing. These were:

(1) *Salt linkage.* Formation of an ionic bond between a basic center in the fiber and an acidic group on the dye or vice-versa. Applicable to fibers containing basic or acidic groups, e.g., wool, nylon, polyacrylonitrile.

(2) *Solid solution.* Formation of a solid solution of a water-insoluble dye in the fiber. Applicable to fibers which are plastic in nature, e.g., nylon, polyester, cellulose acetate.

(3) *Hydrogen bonding.* Formation of hydrogen bonds between suitable sites on the fiber and dye. Applicable to fibers containing groups capable of forming hydrogen bonds, notably cellulose.

(4) *Insolubilization.* Formation of aggregated of insoluble dye within the fiber pores. Applicable to open-structured fibers, notably cellulose.

The first two of these had presented a satisfactory solution to the problems of dye retention on all fibers apart from cellulose. They were not applicable to cellulose since this fiber contains neither acidic nor basic centers and is not plastic in nature.

It is amorphous and very hydrophilic, the only protruding, potentially reactive groups being alcoholic hydroxyl groups. Therefore, interactions (3) and (4) were employed in the dyeing of cellulose, but both of these had deficiencies. Mechanism (3) afforded the dyer easy processing but gave only limited fastness to washing, whereas mechanism (4) gave satisfactory fastness at the expense of much more difficult processing. A solution to this problem had long been perceived in the formation of a covalent bond between the dye and the fiber, and the early literature contains numerous methods for the preparation of colored cellulose derivatives which contain such a bond. These were, however, not practical dyeing procedures since they invariably relied on inert solvents and vigorous conditions to cause bond formation. An acceptable dyeing process would have to be carried out in aqueous medium and the prospect of reacting upon the alcoholic hydroxyl groups of cellulose in such a hostile medium presented a most discouraging situation. As a consequence, the problem remained until Stephen and Rattee (British Patents 797, 946; 798, 121) showed that dichloro-s-triazinyl dyes would react with cellulose under mild conditions in water. This work led to the marketing, in 1956, of the first members of what is now the Procion MX range of reactive dyes.

In the preparation of a typical Procion MX dye (I) either a simple dye containing an amino group is

condensed with cyanuric chloride to give a dichloro-s-triazinyl dye or the condensation is carried out on a dyestuff intermediate, the subsequent processing of which yields the desired dye. Looking closer at the condensation stage, the reason for the high reactivity of cyanuric chloride is the contribution of canonical forms such as (II) to the structure causing a deficiency of

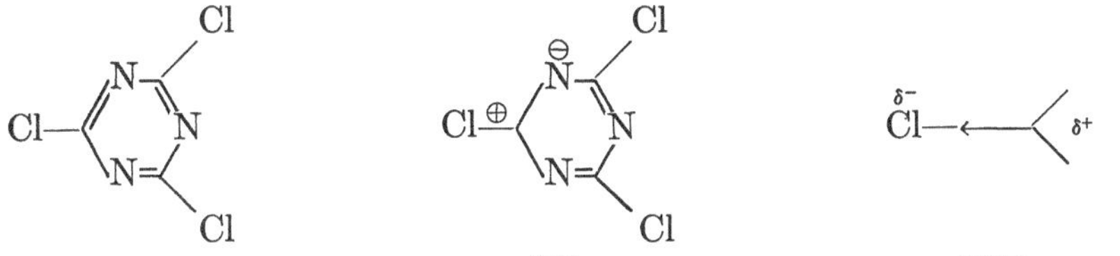

(II)   (III)

electrons on the ring carbon atoms. The greater electronegativity of chlorine compared with carbon also polarizes the carbon chlorine bond (III), thus further contributing to the build up of positive charge on the ring carbon atoms. This makes the molecule susceptible to nucleophilic attack as shown:

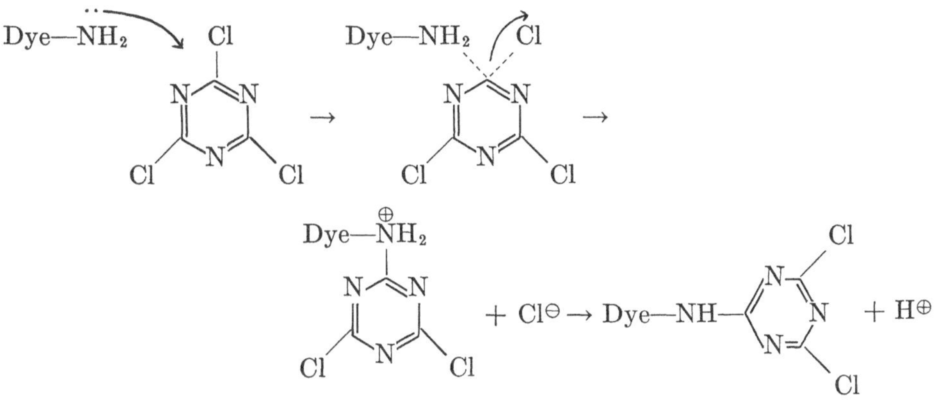

This reaction occurs readily when a neutral solution of the amino-containing dyestuff in water is added to a suspension of cyanuric chloride in water at 0–5°C. Dilute alkali is run in to maintain the pH at 5–6 and finally raise pH to 7, the reaction going smoothly to completion in 15–30 min.

The Procion MX dye structure is composed of two distinct parts, the chromogen and the reactive system, these being joined through the —NH— grouping. The chromogen contributes the color. Nothing is required of it above the need for it to initially contain a suitable group, usually amino, to which the reactive system can be attached. This freedom of choice removes the need to select large dye structures capable of extensive aggregation which had previously been a characteristic of cotton dyes. This enables very simple dyes to be chosen purely on the basis of shade and properties such as fastness to light. One result of this is that far brighter dyeings on cotton can be achieved with reactive dyes than with other types of dyestuff. In practice the azo, anthraquinone, and phthalocyanine chromogens predominate in the ranges of reactive dyes: all contain a number of sulfonic acid groups to confer on the dyestuff the water solubility which is required in a cotton dye. Typical skeleton structures employed to obtain various shades are shown in Table 1. The reactive system in the Procion MX dyes is the dichloro-s-triazinyl group and in this unit the same influences which cause the high reactivity of cyanuric chloride persist although they are muted by feedback of electrons into the ring from the adjacent amino group thus:

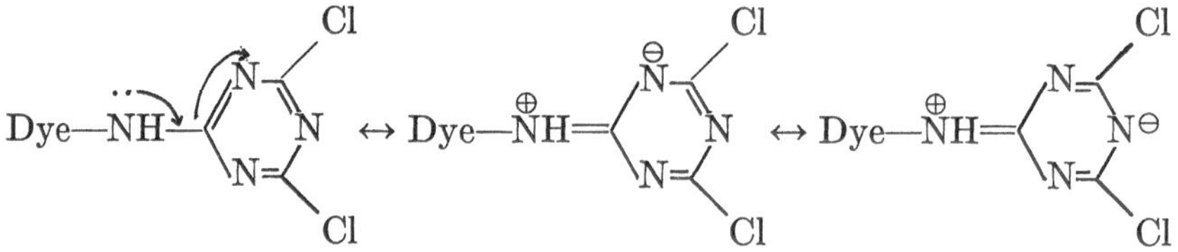

This diminishes the tendency to build up positive charge on the ring carbon atoms and decreases the level of reactivity compared with cyanuric chloride. Under acid condi-

Table 1. Typical Reactive Dye Structures

Skeleton structure	Shade
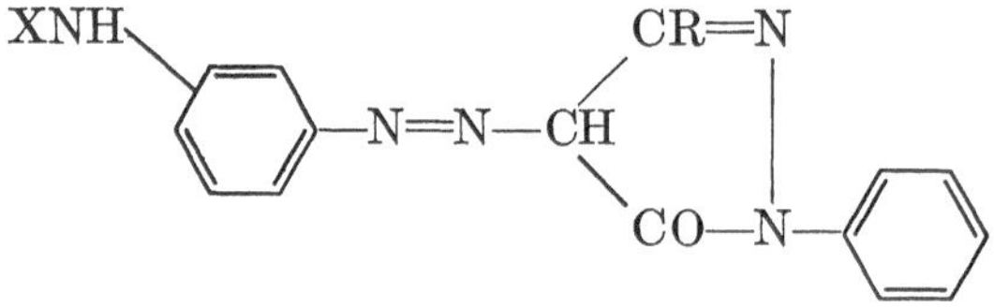	Yellow
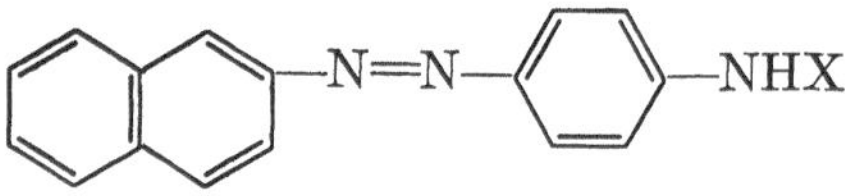	"
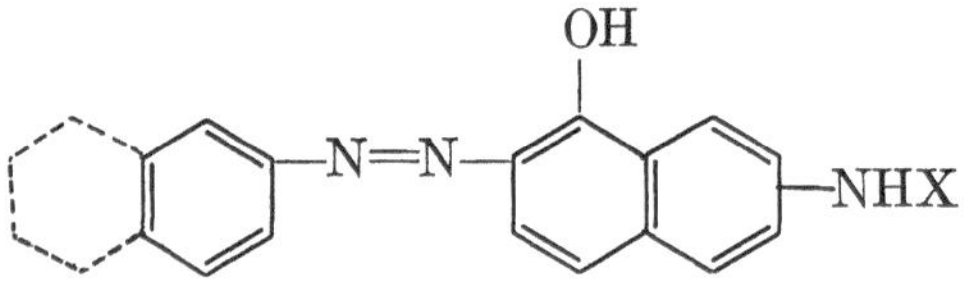	Orange, Red
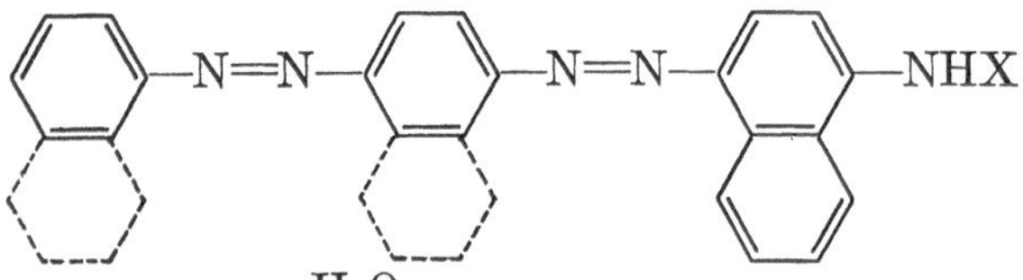	Brown
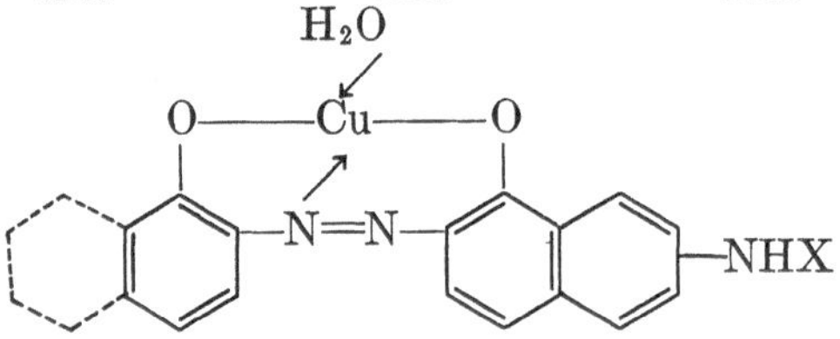	Rubine, Purple, Blue
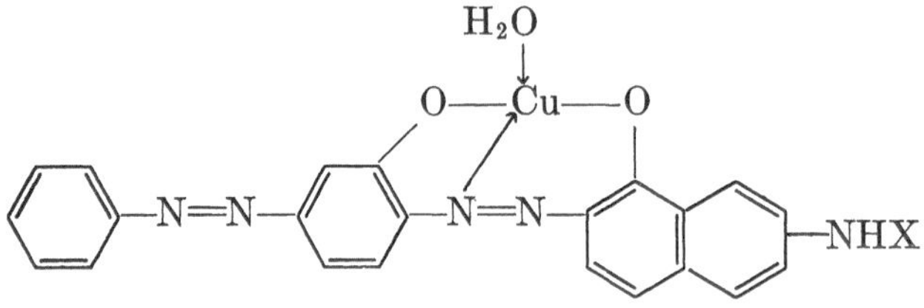	Navy Blue

Table 1. Typical Reactive Dye Structures

| Skeleton | Shade |

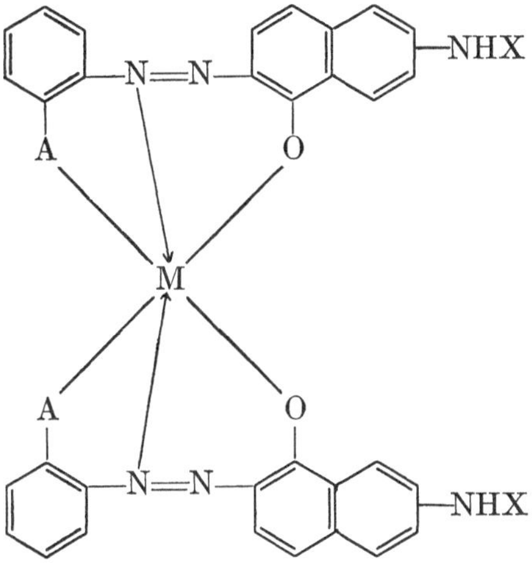

Skeleton

A = O or COO
M = Cr or Co

Brown, Gray, Black

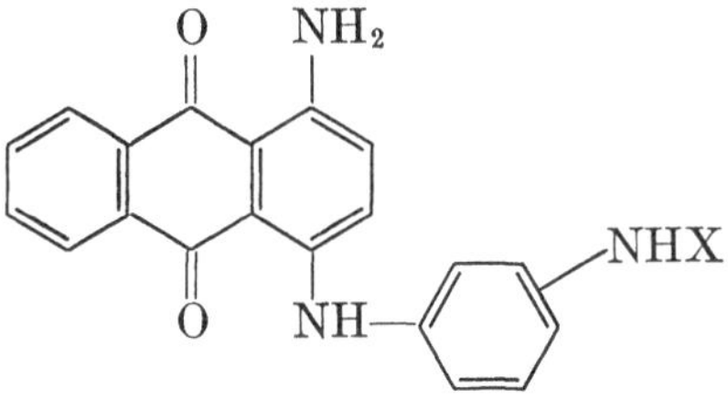

Bright Blue

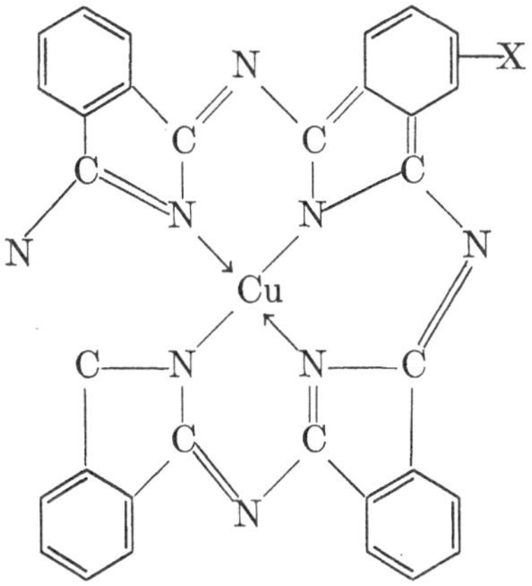

Turquoise

X = Reactive group
Sulfonic acid groups and other irrelevant substituents are omitted.

tions protonation can occur, negating this effect and causing the lability of the chlorine atoms to increase markedly. The result is autocatalytic hydrolysis leading to an unreactive hydroxy-*s*-triazinyl species. This effect is prevented by incorporating a buffer into the isolated dye powder to ensure stability during its shelf life. The dye powder as marketed therefore consists of dye, buffer (usually mixed phosphates), innocuous diluent (often salt) and minor amounts of surface active agents added to make the powder non-dusty.

Although less reactive than cyanuric chloride, the dichloro-*s*-triazinyl dyes are still readily susceptible to nucleophilic attack at temperatures in the 40–50°C range, depending upon the individual dye chosen. It is upon the facility for this mode of reaction that the methods devised for dyeing cotton and viscose with these dyes depend. The main method of application is the cold batchwise dyeing method. In this process the dye is dissolved in water and, since nucleophiles are not present to any sensible extent at neutral pH values, the solution is quite stable. The cloth is immersed in the solution and salt is then added, causing the dye to be driven from the solution by a common ion effect. The planar shape of the dye molecule and its capacity for forming hydrogen bonds with the cellulose encourages adsorption by the fiber and thus the dye is exhausted from the liquor to the cloth without chemical change taking place. When these purely physical processes have been satisfactorily accomplished, alkali, usually sodium carbonate, is added to raise the pH to about 10.5. At this pH the necessary nucleophiles are generated and rapid attack on the dye occurs thus:

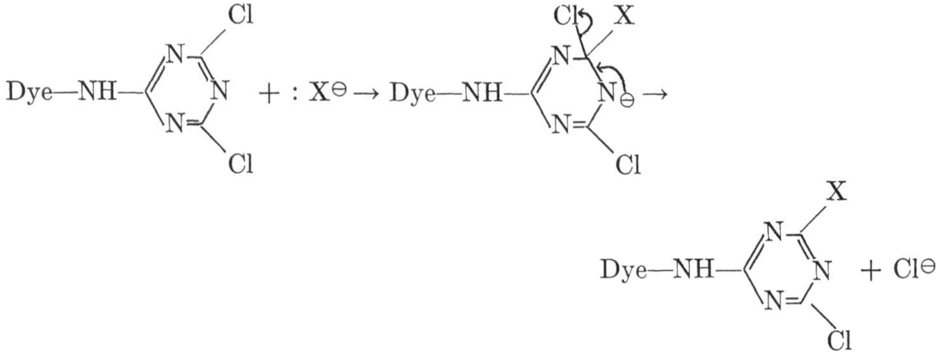

In this sequence the attacking species : $X^{\oplus}$ can be either Cellulose—$O^{\oplus}$ leading to fixation of the dye on the fiber or $OH^{\ominus}$ leading to hydrolysis, producing loose dye which is washed out of the cloth at the end of the dyeing process.

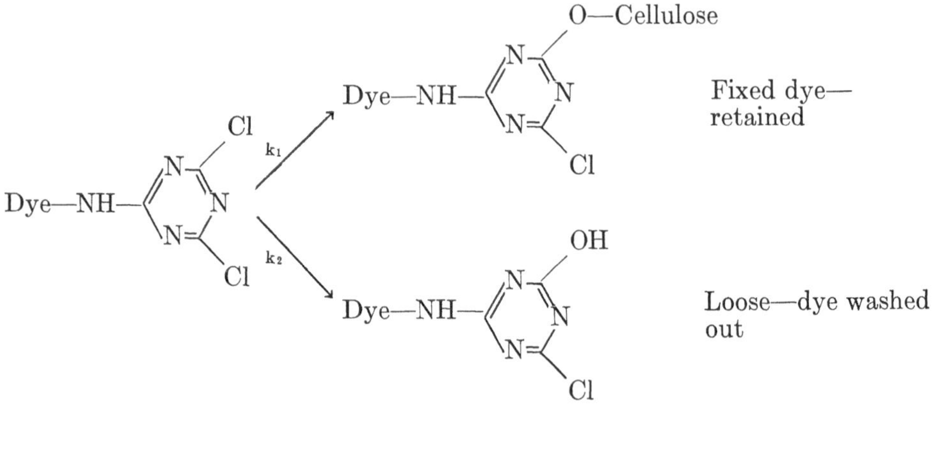

The rate constants for these two reactions are similar and thus there is no inherent pref-erence for fixation. All other things being equal, the large excess of water present—between 5 and 30 ml are used for each gram of cloth—should lead to hydrolysis being the overwhelming reaction. Two effects swing the process in favor of fixation. These are the fact that the dye has been adsorbed by the fiber, causing the concentration of dye in the fiber phase to be up to 500 times greater than the concentration of dye in solution, and, secondly, the ready ionization of the C6 primary hydroxy group in cellulose which causes the concentration of Cellulose—$O^{\ominus}$ ions to be about 25 times greater than the concentration of hydroxyl ions inside the fiber. These two effects completely out-weigh the vast excess of water present and ensure that fixation predominates over hydrolysis.

In conventional batchwise dyeing the exhaustion stage takes 15–30 min and the fixation stage 45 min. The versatility of the Procion MX dyes does, however, allow wide variations in conditions and by suitable modifications of procedure either of these stages can be speeded up or slowed down. One variation is the pad-batch technique in which a concentrated dye solution and alkali are simultaneously fed into a shallow trough through which the cloth is running. Excess liquor is squeezed out and the cloth, damp with alkaline dye solution, is then stored for 2–6 hr at room temperature. The physical ab-sorption of the dye by the cloth is rapid due to the very small volume of dye solution employed, thus telescoping the exhaustion stage and allowing virtually the whole of the dyeing time for the fixation to occur.

In the dyed cloth, the dyestuff molecule can be shown to be attached by a covalent bond to the cellulose fiber by a number of methods. The dye cannot be removed by treatment with boiling polar solvents such as pyridine; if the dye is an azo compound, the azo linkage can be reduced leaving an intermediate amine joined to the cloth which can then be chemically converted back to the same or a different dye *in situ*. The bond formed between the dye and the fiber has the characteristics of a rather stable ester and is quite impervious to the mild alkaline hydrolysis conditions met with in domestic washing treatments. Indeed, some of the bonds in the fiber are more easily broken than the dye-fiber bond, allowing conclusive proof of dye-fiber reaction by enzymic degrada-tion of the cellulose and isolation of dye specimens with fiber fragments still attached.

The dichloro-*s*-triazinyl dye has two equally reactive chlorine atoms, but it is note-worthy that this does not give it a second chance of achieving fixation. If hydrolysis then occurs in the alkaline medium employed, the resulting chloro-hydroxy-*s*-triazine ionizes (IV), imparting a full negative charge to the system; feedback of this into the ring elimi-

(IV)

nates the positive charge on the ring carbon atoms and completely stops nucleophilic attack. In the fixed dye molecule (V) the remaining chlorine atom still possesses a degree of reactivity although this is diminished by the presence of the electron-repelling —$O^{\ominus}$–Cellulose grouping which functions as shown. Under prolonged dyeing times, this chlorine atom can be hydrolyzed (VI) or replaced by a further Cellulose—O$^{\ominus}$ ion cross-linking the fiber as in (VII).

Structurally analogous to the fixed form (V) are the chloro-alkoxy-*s*-triazinyl dyes (VIII) which are most easily prepared by reaction of an alkoxy-dichloro-*s*-triazine, particularly 2-methoxy-4,6-dichloro-*s*-triazine with an amino containing dye at 20–30°C. Dyes of this type figure to a minor extent in the Procion H and Cibacron ranges of reactive dyes. They are less important than the more readily prepared chloroamino-*s*-triazinyl dyes (IX) which form the backbone of these ranges. These are invariably prepared by condensation of a dichloro-*s*-triazinyl dye with ammonia or an amine at a temperature of 20–50°C

and at such a pH that an appreciable amount of free base is present. Thus with arylamines a pH of 5–6 is appropriate whereas with ammonia or a strongly basic alkylamine a pH of about 9 is required. Due to the introduction of the electron-repelling alkoxy or amino group into the system, these dyes display a lower level of reactivity than the dichloro-*s*-triazinyl dyes. In practical terms the temperature necessary to cause reaction with cellulose rises from the 20–40°C range to about 80°C, and generally slightly more alkaline conditions, about pH 11, are required. Dyes of the Procion H and Cibacron ranges can therefore be applied by dyeing techniques similar to those described above with the exception that the operating temperature is about 80°C.

The application methods described so far have been batchwise dyeing techniques in which the whole piece of cloth emerges from the process uniformly colored. The lower level of reactivity exhibited by the monochloro-*s*-triazinyl dye does, however, make them ideally suited to a different type of application process, namely a printing process. In this, the dye is dissolved in water to produce a concentrated solution to which the alkali is added followed by a thickening agent, producing a stiff paste which is chemically quite stable at room temperature. This paste is printed onto the cloth in a pattern by means of an embossed roller, and the cloth then heated to a temperature in the 100–150°C range either by steaming or by dry heat. At these temperatures rapid reaction of the dye with the cloth occurs, 5–15 sec being a sufficient length of time to complete the reaction, and the dye is fixed in the desired pattern. The process is completed by washing the cloth to remove that portion of the dye which has hydrolyzed along with the excess alkali and the thickening agent. A significant advance has been to increase the percentage of applied dyestuff which undergoes fixation by the use of dyes with two reactive systems. This is the particular feature of the Procion SP dyes, some of which are of the general pattern (X). In this range

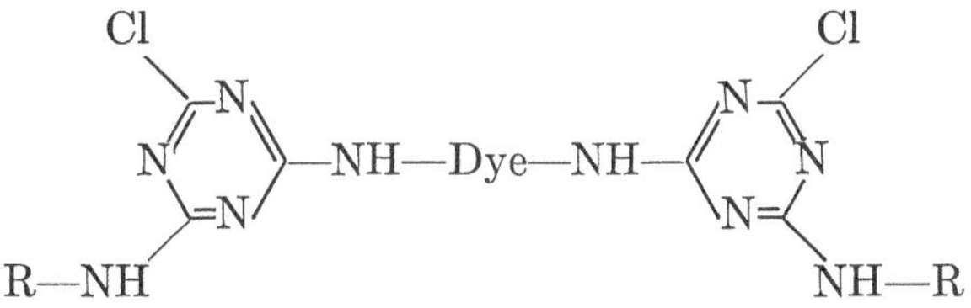

(X)

careful attention has been paid to building in structural features to disrupt the planarity of the molecule in order to minimize hydrogen bonding between traces of unfixed dye and the fiber and thus facilitate its removal in the final washing stage.

The success of the chloro-s-triazinyl dyes led to an intensive search for other reactive systems which could be used in the dyeing of cellulose. Many such systems are now available. The most closely related to the chlorotriazines are those based on other chloro-heterocyclic systems. Typical are the Reactone (Geigy) and Drimarene (Sandoz) dyes which contain a trichloropyrimidinyl-amino system (XI). Variants wherein the reactive system is attached to the chromogen via a carbonylamino grouping are the Levafix E (Bayer; XII), Elisiane (C.F.M.C.; XIII) and Solidazol (Cassella; XIV) dyes, formed by condensation of an amino-containing dye with the appropriate chloroheterocyclic acid chloride.

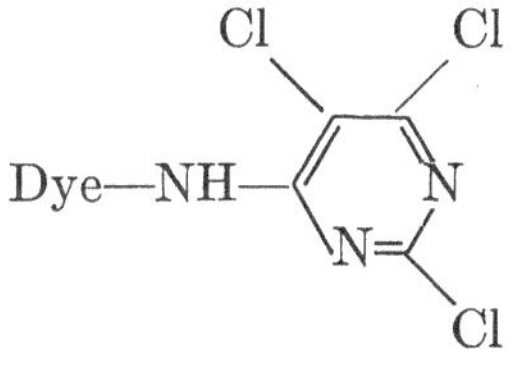

(XI)

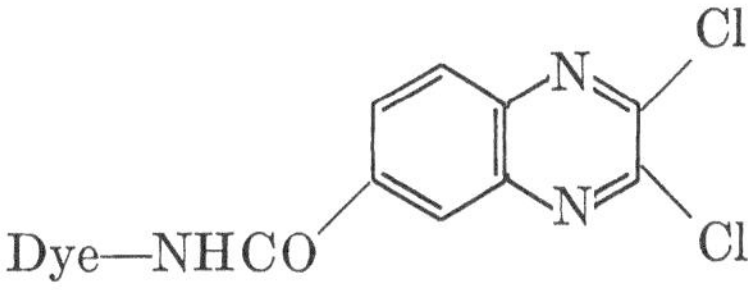

(XII)

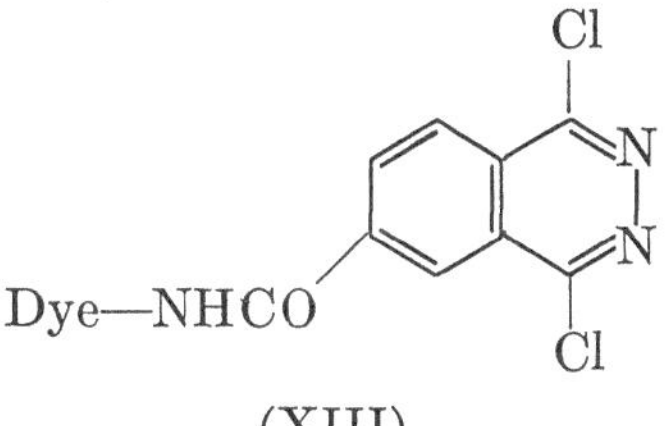

(XIII)

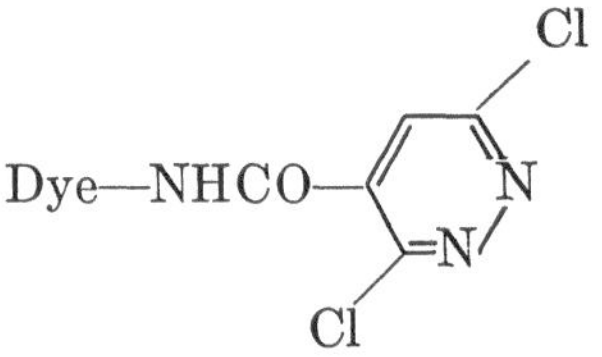

Dye—NHCO—(XIV structure with Cl groups)   Dye—N(XV structure with Cl, O groups)

(XIV) 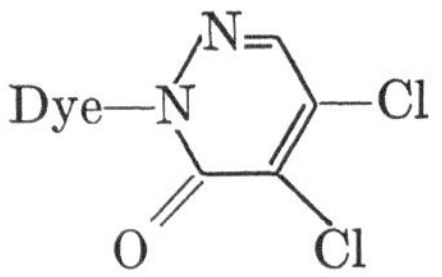   (XV)

In the Primazin P (B.A.S.F.) dyes a 4,5-dichloropyridazone system (XV) is incorporated. These reactive systems do not give the high level of reactivity obtained with the dichloro-s-triazines; they are more comparable with the monochloro-s-triazines. This is due to the presence of three ideally placed nitrogen atoms in the s-triazine ring inducing, on the ring carbon atoms, a far greater positive charge than that which is induced by the two hetero nitrogen atoms contained in these alternatives. Charge density distributions calculated for the various nitrogen heterocycles clearly illustrate this point.

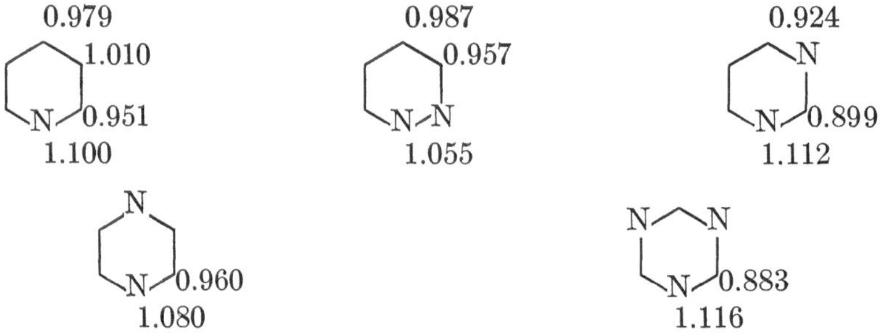

0.979   1.010   0.951   1.100

0.987   0.957   1.055

0.924   0.899   1.112

0.960   1.080

0.883   1.116

In the case of the pyrimidine based systems, reactivity can be boosted to the level of the dichloro-s-triazines by means of an electron-attracting group attached to the 5-position. This occurs in the case of the 5-cyanodichloropyrimidinyl dyes (XVI) and, to a lesser extent, with the Reactofil (Geigy; XVII) dyes. An alternative method of raising reactivity arises by varying the leaving group. This can be most conveniently achieved by forming a quarternary ammonium salt from a monochloro-s-triazine and a suitable

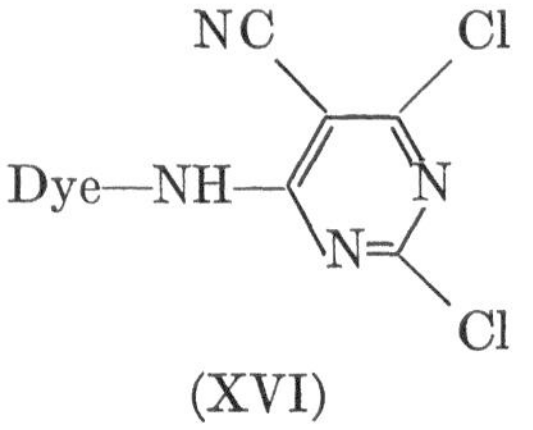

Dye—NH—(XVI structure with NC, Cl, Cl groups)   Dye—NHCO—(XVII structure with Cl, Cl groups)

(XVI) 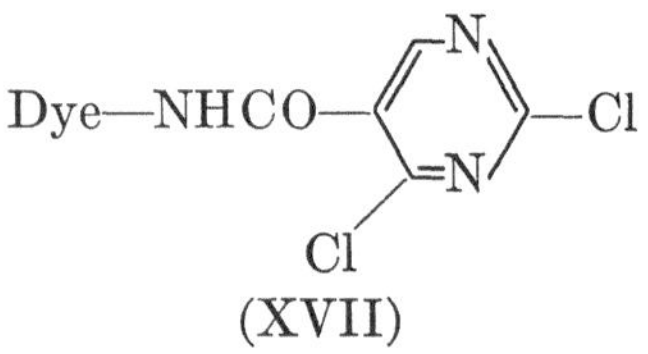   (XVII)

base. In these compounds the full positive charge on what is to be the leaving group makes nucleophilic replacement extremely facile and reaction with cellulose under alkaline conditions will occur at temperatures in the 20–50°C range.

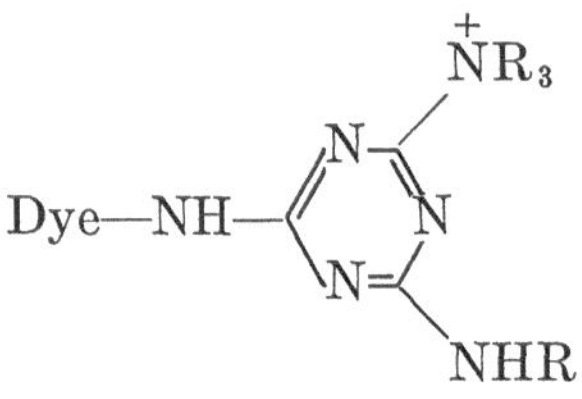

Dye—NH—(XVIII structure with $\overset{+}{N}R_3$, NHR groups)   Dye—NH—(XIX structure with Cl, CH_3, SO_2CH_3 groups)

(XVIII) 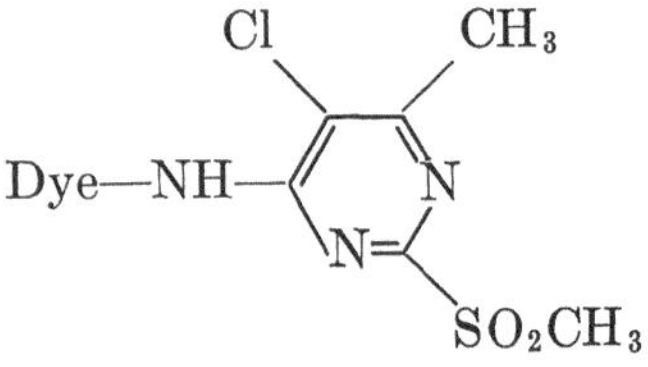   (XIX)

$$\text{Dye—NH—} \underset{(XX)}{\overset{\text{Cl} \quad \text{F}}{\bigcirc}}$$

The tertiary amines which can be used must present a very sterically accessible nitrogen atom. Notable are pyridine and the strongly basic trimethylamine and 1,4-diazabicyclo [2,2,2] octane which form the quaternary salts so readily that they can be used as dyebath additives to promote cold dyeing properties in monochloro-*s*-triazinyl dyes. Other leaving groups which have been used recently to enhance reactivity are methylsulfonyl in the Levafix P (Bayer; XIX) dyes and fluorine in dyes of the type (XX).

A second general class of reactive dyes functions through a nucleophilic addition process, thus differing from the halogenoheterocyclic dyes in which fixation is achieved by a substitution mechanism. The reactive system is usually a short unsaturated aliphatic side chain attached to the chromogen via an electron-attracting group, or a precursor which generates this system during the application process. Typical are the Remazol (Hoechst) dyes which contain a B-sulfatoethylsulfone group. When treated with alkali this group loses sodium hydrogen sulfate generating a vinyl sulfone (XXI). In this system the electron-attracting sulfone grouping makes the terminal carbon atom electron deficient and causes attack by ionized cellulose, or hydroxyl ions, to occur at this site. The small alkyl reactive system leads to only weak physical adsorption by the fiber and, consequently, these dyes are almost exclusively used in printing applications, where their weak physical attraction for the fiber makes washing off of unfixed dye easy. The Remazols are more reactive than the monochloro-*s*-triazines but less so than the dichloro-*s*-triazines.

$$\text{Dye—}\overset{\text{O}}{\underset{\overset{\oplus}{\underset{\ominus}{\text{O}}}}{\overset{\|}{\text{S}}}}\text{—CH}_2\text{—CH}_2\text{—OSO}_3\text{Na} \xrightarrow{-\text{NaHSO}_4} \text{Dye—}\overset{\text{O}}{\underset{\overset{\oplus}{\underset{\ominus}{\text{O}}}}{\overset{\|}{\text{S}}}}\text{CH}\!\!=\!\!\text{CH}_2$$

$$(\text{XXI})$$

$$+ \ \bar{\text{O}}\text{—Cellulose} \rightarrow \text{Dye—}\overset{\overset{\ominus}{\text{O}}}{\underset{\overset{\oplus}{\underset{\ominus}{\text{O}}}}{\text{S}}}\!\!=\!\!\text{CH—CH}_2\text{—O—Cellulose}$$

$$\updownarrow$$

$$\text{Dye—}\overset{\text{O}}{\underset{\overset{\oplus}{\underset{\ominus}{\text{O}}}}{\overset{\|}{\text{S}}}}\text{—}\overset{\ominus}{\text{CH}}\text{—CH}_2\text{—O—Cellulose} \xrightarrow{+\text{H}^{\oplus}}$$

$$\text{Dye—}\overset{\text{O}}{\underset{\overset{\oplus}{\underset{\ominus}{\text{O}}}}{\overset{\|}{\text{S}}}}\text{—CH}_2\text{—CH}_2\text{—O—Cellulose}$$

This high level of reactivity for a printing dye leads to diminished stability of the alkaline print paste and hence they are often applied by a two-stage printing technique. In such an application a neutral print paste containing the dye is printed onto the cloth which is then dried and subsequently contacted with alkali, e.g., by running through a hot alkaline bath, to cause dye-fiber reaction to occur.

There are numerous examples of this type of reactive system, all of which function in a similar manner. Thus in the Primazin (B.A.S.F.; XXII) dyes the B-sulfatopropionyl-amino group reacts via an acrylamido entity and in the Levafix (Bayer; XXIII) and B-chloroethylsulfamoyl (XXIV) dyes the generated reactive intermediate is an ethylene imine derivative rather than an activated alkylene species.

$$\text{Dye—NHCOCH}_2\text{CH}_2\text{OSO}_3\text{Na} \rightarrow \text{Dye—NHCOCH=CH}_2 \rightarrow$$
$$\text{(XXII)}$$
$$\text{Dye—NHCOCH}_2\text{CH}_2\text{O—Cellulose}$$

$$\text{Dye—SO}_2\text{NHCH}_2\text{CH}_2\text{OSO}_3\text{Na}$$
$$\text{(XXIII)} \qquad \downarrow$$
$$\overset{\text{CH}_2}{\diagup}$$
$$\text{Dye—SO}_2\text{N} \qquad \rightarrow \text{Dye—SO}_2\text{NHCH}_2\text{CH}_2\text{O—Cellulose}$$
$$\diagdown$$
$$\overset{\text{CH}_2}{\uparrow}$$
$$\text{Dye—SO}_2\text{NHCH}_2\text{CH}_2\text{Cl}$$
$$\text{(XIV)}$$

This latter group is the one most readily attached to the copper phthalocyanine nucleus and is to be found in a number of phthalocyanine-based Procion turquoise dyes. The most widely useful of these aliphatic reactive systems remains however, the B-sulfatoethyl-sulfone system, but this type of system does have a fundamental drawback in the need to first synthesize an intermediate containing the reactive group rather than simply to attach the reactive system to one of the wide variety of chromogens available that contain an amino group, as in the case of the chloroheterocyclic dyes. One way of circumventing this is to prepare instead of a sulfone a B-sulfatoethylsulfonylamino system by reaction of a dyestuff containing an amino group with carbyl sulfate (XXV), itself obtained by the action of sulfur trioxide on ethylene. Certain of the more recently introduced Remazol dyes contain this grouping.

$$\text{Dye—NH}_2 + \begin{matrix} & \text{O} & \\ \diagup & & \diagdown \\ \text{CH}_2 & & \text{SO}_2 \\ | & & | \\ \text{CH}_2 & & \text{O} \\ \diagdown & & \diagup \\ & \text{SO}_2 & \end{matrix} \rightarrow \text{Dye—NHSO}_2\text{CH}_2\text{CH}_2\text{OSO}_3\text{H}$$
$$\text{(XXXV)}$$

$$\begin{matrix} \text{CF}_2\text{—CF}_2 \\ | \qquad | \\ \text{CH}_2\text{—CHCOCl} \\ \text{(XXVI)} \end{matrix} \qquad\qquad \begin{matrix} \text{C}_6\text{H}_5\text{SO}_2\text{CH}_2\text{CH}_2\text{COCl} \\ \\ \text{(XXVII)} \end{matrix}$$

In a similar manner, the compounds (XXVI) and (XXVII) have attracted attention for the one-step introduction of a reactive group, the latter being used in the preparation of the Solidazol (Cassella) dyes.

Reactive dyes have made their main impact on the dyeing of cotton and viscose rayon but the principle of causing a dye to covalently bond on the fiber can, of course, be extended to other fibers which contain a suitable reaction site. Wool and nylon, both of which contain amino groups, are obvious candidates, and ranges of reactive dyes designed for application to both of these fibers have been marketed. The problems encountered have differed from those encountered in the dyeing of cellulose. With cellulose the very hydrophilic nature of the substrate makes penetration of the dye into the fiber easy while the need to react upon an alcoholic hydroxy group makes controlled pH changes a vital factor in achieving fixation. Conversely, the wool fiber is covered by a very hydrophobic sheath, the epicuticle, which makes penetration both difficult to achieve and irregular. The dye is usually driven into the fiber by dyeing at the boil and must then remain mobile within the fiber for some time to give an even coloration. Reaction with the fiber amino and, possibly, thiol groups which will occur very readily at neutral pH values at the temperatures employed must, therefore, be prevented from being overly rapid and thus prematurely immobilizing the dye. To meet this criterion, a reactive system having a very low level of reactivity must be chosen. In the Procilan range of premetallized dyes for wool, the feebly reactive acrylamido group is utilized. In the dyeing of nylon the fastness to washing obtained with water-soluble anionic dyes is excellent, and here a reactive system would be a totally unnecessary expense. The disperse dyes, which penetrate and dye the fiber more evenly than the acid dyes have, however, some deficiency in wash-fastness properties. The employed dyeing temperature of 100°C is, again, sufficiently high to cause only moderately reactive groups to combine with the fiber amino groups; such groups are used in the Procinyl range of disperse dyes for nylon. These technical considerations place the level of reactivity aimed for in wool and nylon dyes far below that required for the cold dyeing of cellulosic fibers.

Procion Dyes as Intracellular Stains

When considering dyes for intracellular staining techniques it will be of paramount importance to operate under the mildest possible conditions. This automatically selects the Procion MX dyes as the preferred class of reactive dyes, and in their application a rough parallel exists between the staining technique and the cold batchwise dyeing process. Both involve transport of the dye via an aqueous medium to the reaction site, a nucleophilic substitution reaction at the site and finally the washing away of unfixed dye. The ready solubility of the Procion MX dyes in cold water enables a 5–10% (w/v) (or even higher) aqueous solution of dye to be prepared for injection into the system. With extinction coefficients of 30,000 being common among dyes, this will represent a considerable visual effect. In a staining technique there will be an important difference from batchwise dyeing in that there will be no easy way of impelling the dye to the reaction site analogous to the salting-on stage of the conventional dyeing process. Rather, as in the pad-batch technique, the mobility of the dye within the system must be relied on to carry it spontaneously to the reaction site. This will be facilitated by employing dyes of small molecular size and the best results would therefore be anticipated with the simple yellow, orange, and red monoazo dyes, where aggregation effects in solution are at a minimum. The possibility of limited hydrogen bonding between these dyes and the substrate will also be helpful in localizing the dye on the substrate and enhancing its chances of achieving fixation.

Of the possible groups on the substrate which could be considered as a reaction site, only the amino group appears likely to enter into reaction. To cause reaction with a hydroxyl group, a pH in the 10-11 region would be required to ionize the hydroxyl group, and this condition would seem unacceptable. However, a free amino group on the substrate will be a potent nucleophile and will readily react with the dye at room temperature without the need to raise the pH. Even here, some knowledge of and control over the internal pH of the system would be advantageous since at the normal pH value strongly basic amino groups could be shielded from reaction by salt formation. There also exists the very real possibility that the hydrochloric acid generated in the condensation will upset the internal pH of the system, again perhaps deactivating the amino groups through salt formation. Incorporation of a suitable buffer (e.g., mixed phosphate or sodium diethylmetanilate) or even a mild alkali might be helpful in preventing these effects.

The members of the Procion MX range of dyes are, of course, chosen for their aesthetic value in coloring cellulosic textiles. Selection of a dye for intracellular staining will not be guided by this requirement but will rest solely on the ease of detection within the dyed specimen. With a dyestuff the obvious choice is to rely on visual detection by light microscopy, but with suitable alternative structures at least two other possibilities arise. The desire for color having been removed, there is no reason why any water-soluble organic compound carrying a dichloro-*s*-triazinyl group and possessing a suitable, readily identified feature should not be used. Thus a dichloro-*s*-triazinyl derivative of a fluorescent compound could be prepared and used; indeed, the compound (XXVIII) has been developed by Chayen et al. (1971) for reaction

$$OC_2H_5$$

(XXVIII)

with sterols in essentially nonaqueous medium, allowing easy chromatographic identification of these natural products. In this context the fluorescent behavior of Procion yellow MX4R (XXIX) is remarkable since this compound does not show

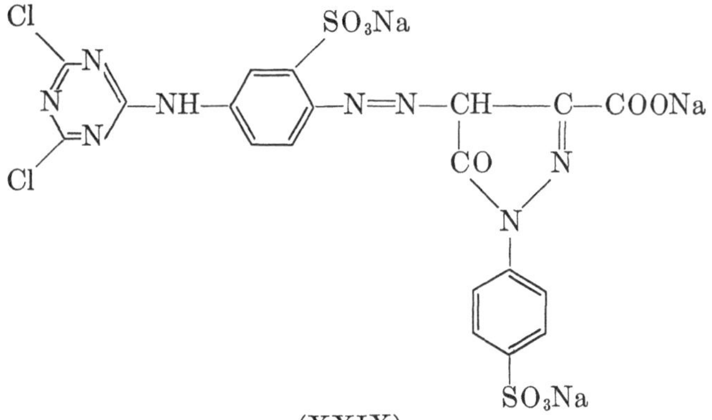

(XXIX)

fluorescence when used in textile dyeing. Besides utilizing fluorescence, the prospect of incorporating a metal atom into the reactive entity for subsequent detection by electron microscopy deserves attention.

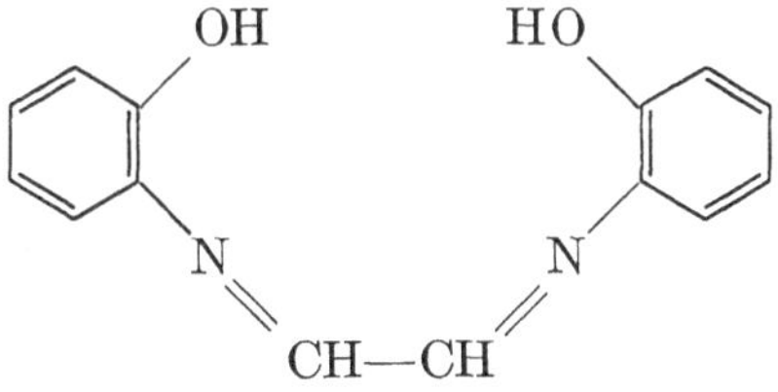

(XXX)

In available reactive dyestuffs, copper, chromium, and cobalt are commonly used to form stable metal complexes. Staining with dyes of this type would locate such an atom on the substrate surface. These nuclei would probably be too small for location by electron microscopy but again there is no reason why a suitable complexing system carrying the required solubilizing groups and a dichloro-s-triazinyl group should not be synthesized for the attachment of a much heavier metal atom to the substrate surface. Thus the system (XXX) described by E. Bayer (1964) for forming complexes with uranium could, suitably modified, possibly serve as a basis for such a reagent. Further details on reactive dyes can be found in Beech (1970).

References

Bayer, E.: Structure and specificity of organic chelating agents. Angew. Chem. International Edition 3, 325–332 (1964).

Beech, W. F.: Fibre-reactive dyes. London: Logos, 1970.

Chayen, R., S. Gould, A. Harell, and C. V. Stead: 1-Ethoxy-4-(dichloro-s-triazinyl) naphthalene: a fluorescent reagent for corticosteroids. Analyt. Biochem 39, 533–535 (1971).

Vickerstaff, T.: Reactive dyes for textiles, J. Soc. Dyers and Colourists 73, 237–247 (1957).

Discussion

Dr. Hugh Rowell (Berkeley): Dr. Stead, last night in conversation you said that it would be possible to tailor the solubility of Procion dyes. Could you say more about that?

Dr. Stead (I.C.I., Manchester): Yes, willingly. We do tailor the solubility for our own ends. We want the dye to be sufficiently soluble to make up solutions of any desired strength. We actually aim at about 6% solubility. A higher order of solubility would diminish the efficiency at the absorption stage for textile dyeing resulting in less dye fixing and more being lost to the dyeing process.

The dye structure that I showed early on has two sulphonic acid groups. It would be very easy to raise that, which would markedly enhance the solubility. This might be very helpful in your applications and would be well worth trying.

Dr. Rowell: Is this a difficult job? Can a chemist do this readily?

Dr. Stead: It would be quite easy. One would start from a different intermediate (aniline-2,5-disulfonic acid instead of aniline-2-sulphonic acid) and then follow the same stages and conveniently isolate the final highly soluble dye by pouring the aqueous solution into about ten times its own volume of acetone. The heavily charged dye entity would precipitate beautifully. It could be filtered, dried and kept for as much as three or four weeks, I would say, without any problem.

A buffer would help if you don't object to it. It would enable you to keep the sample for a considerable period of time.

Dr. Brian Mulloney (University of California, San Diego): We notice in using Procion dyes, that if you make up a solution and keep it over a period of time, say a week or two, it seems to polymerize and form a gel. We therefore prepare a fresh solution for each biological application. Does the origin of the problem lie in bad technique on our part, or is this a general property of the dyes of which everybody should be aware?

Dr. Stead: I think it is a general property rather than bad technique.

Hydrolysis would happen even in the solid phase (which contains 5 or 10% moisture usually) without the buffer. When the dye is dissolved in water and left for a week or so, slow hydrolysis will gradually knock out the buffer. Hydrolysis will then occur, rapidly rendering the dye useless. Certainly in the laboratory we would invariably prepare a fresh dye solution for experimentation.

You could keep a solution overnight under refrigeration for use the next day or even the day after, but I wouldn't keep it a week.

Dr. R. Llinás (Iowa): Dr. Stead, what would you recommend as the simplest technique to increase the spreading of Procion dye once it has been injected intracellularly? Let us say, some treatment which would activate Procion binding to protein only after it has diffused freely inside a neuron for some 24 hours.

Dr. Stead: I think that possibly the best way to try would be to use a monochloro triazinyl (Procion H) dye. This is of lower reactivity. It could be injected, allowed to diffuse, and then an injection of either trimethylamine solution or, better still, a solution of 1,4-diazabicyclo [2,2,2] octane. This would boost the level of reactivity of the dye and cause it to react readily.

The difficulty, as I see it, will be putting two solutions one after the other in the same cell.

Dr. Llinás: Is it possible to introduce your activating compound extracellularly after injection, and let it diffuse across the neuronal membrane to activate dye binding intracellularly?

DR. STEAD: Yes, you could inject the dyestuff and subsequently immerse the cell in the activating compound as part of the fixation procedure.

DR. KRAVITZ (Harvard): Just a comment on that. Of the various dyes that Tony Stretton and I have found to be useful, Procion yellow was an M dye while most of the others were H dyes, so that the H dyes might be used in just the way that Dr. Stead suggests. We found that generally dye leakage did not occur during fixation. It occurred during dehydration. Thus one might be able to inject H dye, let it diffuse, fix it and finally add trimethylamine to the alcohols used for dehydration.

DR. DUBIN (Colorado): What about lowering pH during the process so that you could change the reactivity of the dye?

DR. STEAD: This would certainly be a feasible approach. Lowering the pH by injection of acid would de-activate the amino groups through salt formation. After diffusion reactivity could be restored by raising the pH by alkali injection. If this could be done the result would be a useful process.

DR. DUBIN to Dr. Kravitz: You did your fixation at low pH. You said it worked the best.
Did you ever try to change to a high pH, at the end of the experiment to increase reactivity?

DR. KRAVITZ: What we thought was happening was that after 24 hours or so, whatever binding was going on had taken place, but we are not certain of that.
We think that what Dr. Stead described, dye adsorbing onto the protein, slowly filling the cell, and then very slowly reacting with amino groups, is going on all the time.
Dr. Stead's suggestion presumably would be to lower the pH during the migration period but still keep the preparation alive. One thus would lower dye reactivity and possibly get it to migrate much further.

DR. BENNETT (Einstein): I am sure there are many people here who have had the experience of injecting Procion and watching the cell die.
Is there anything that is likely to be in the dye that you haven't mentioned?
What is the buffer, for example? Are there likely to be such things as heavy metals or divalent ions?

DR. STEAD: No. The buffer is usually a mixture of phosphates. I don't know what effect that will have on the cell.

ANONYMOUS: What about the dedusting agent which I believe is dodecyl benzene?

DR. STEAD: This could be dealt with extremely easily in the commercial product by stirring with acetone in which the dedusting agent will dissolve. The dye will be quite insoluble. It could therefore be filtered off and washed with acetone, a few times more. Then the dedusting agent would be completely removed.
I do believe, however, that Wilson Diagnostics material doesn't have dedusting agents in it.

DR. KRAVITZ: I think among the batches of dye that we had, the ones with dedusting agent were turbid.
The dedusting agent could easily be removed by centrifugation.
Most of the dye used for injection is free of the dedusting agent.

DR. STEAD: Do the cells still die even with this?

DR. KRAVITZ: Well, of course, it is an alkylating agent and is binding to amino groups. I think that ultimately it has to kill cells. It is just a matter of how fast. If you could slow the reactivity down, that would help.

DR. DAN HARTLINE (University of California, San Diego): I was wondering just how steep the temperature dependence of bonding is because I guess most of those of us who work on invertebrates put the material in the refrigerator while the dye is migrating.

I wonder if you could reduce the bonding during migration even further by lowering the temperature more.

DR. STEAD: I would say that the temperature dependence of the bonding is pretty steep. I am afraid I can't give you a quantitative answer. At refrigerator temperatures the bonding would be slowed, but, assuming diffusion is important to dye migration, wouldn't this also slow down the diffusion rate? If the diffusion were not slowed, then refrigeration would be a good idea. Perhaps you could give it some time at room temperature during the late stages. Would the specimen be taken straight out from the refrigerator and fixed?

DR. HARTLINE: Some people would use a cold fixative.

DR. STEAD: I would prefer to see the material left for a while at room temperature just prior to fixation to make sure reaction was complete. The ideal temperature will be about room temperature.

DR. KRAVITZ: That might be more of a problem to people who are working in mammals, of course. Has anyone who is injecting cells in mammals tried lowering the temperature during diffusion?

DR. LINDSTRÖM (Göteborg): After injecting cells we usually perfuse our animals with cold saline and leave the spinal cord in the refrigerator overnight before we fix the tissue. I have no exact figures, but this procedure certainly improves the staining of the axons (see Chapter 14).

DR. LEO LIPETZ (Ohio State): Does the dye as packed have any detergent or wetting agent included with it?

DR. STEAD: No, just the dedusting agent. You don't need a wetting agent with such a heavily sulfonated dye because it is going to dissolve in water.

DR. ELOISE KUNTZ (Michigan State): Is there any information as to how much photo-dynamic destruction you may get from using the dye?

If there is photodynamic action, I would expect that, if people would illuminate the tissues with wavelengths of light that are not absorbed by the dye, death of the cell could be minimized.

DR. STEAD: Yes, that is probably true. In addition dyes do fade under the influence of light, and this effect varies from dye to dye. Procion yellow MX4R is of moderate light fastness and will, like any dye, fade eventually. In your examination very high energy radiation is used and you are simulating an extremely fast fading process. I believe that Dr. Van Orden's paper will discuss this question.

I would just make one further point concerning decomposition. Destruction of an azo compound takes place often by a reductive mechanism in the presence of light, and any substrate which will yield reducing species on illumination will be bad for the dye. Some such process may be operating here, particularly in the presence of ultraviolet light.

DR. KRAVITZ: It really is fun to make Procion dyes, and it is quite easy to do. Tony Stretton and I synthesized a Procion dye with fluorescein amine bound to the cyanuric chloride ring. As indicated in our paper, this wasn't particularly useful.

But the general synthesis of the dye is easy. The steps are essentially quantitative and almost anyone with any experience using a balance and some flasks can go ahead and start making some of these dyes.

DR. STEAD: I have put a few details about the times, temperatures and pH in my paper which will help in this direction.

DR. T. OGDEN (Utah): I wonder if anybody has had experience with Procion going bad on the shelf, and whether there are some tests that you might apply to find out whether it is going to fluoresce when it is injected into a cell.

DR. STEAD: The normal mode of decomposition entails loss of the reactive chlorine atoms through hydrolysis. With a buffered dye kept under laboratory conditions, no appreciable decomposition should occur over a long period of time.

As a safeguard, when a sample is received the chloride ion content of the powder could be estimated; a further determination of chloride ion on a sample after boiling for 2 hr with N sodium hydroxide could be done and from the difference the labile chlorine content established. Subsequently, say every year, these figures could be checked.

Should hydrolysis occur, the dye will no longer react with the substrate and will be quite useless.

CHAPTER 4

Principles of Fluorescence Microscopy and Photomicrography with Applications to Procion Yellow

L. S. Van Orden III

The recently introduced Procion yellow intracellular staining technique relies on fluorescent microscopy for visualization of stain-filled neurons. This chapter will outline the fundamentals of this type of microscopy and present quantitative data on the fluorescence of Procion yellow.

Fluorescence depends upon the absorption of light, usually ultraviolet (300–400 nm) or blue-violet (400–500 nm). The fluorescence excitation spectrum is generally similar to the absorption spectrum of a compound. Fluorescent light is emitted at lower energy than that of the exciting light, hence it is of longer wavelength (usually 50–200 nm greater than the excitation wavelength). The duration of fluorescence emission is of the order of nanoseconds or less, while any more persistent emission is arbitrarily termed phosphorescence. Fluorescence intensity is greatly enhanced by lowering the ambient temperature, a phenomenon which should be studied more fully in connection with fluorescence microscopy. The fundamental principles of fluorescence and their application to biological studies are reviewed by Udenfriend (1969).

A typical fluorescence microscope consists of a high intensity light source equipped with filters to absorb wavelengths above the excitation maximum, a darkfield substage condenser, and a barrier filter to absorb any exciting light which may pass through the specimen. Since the fluorescent object becomes the light source, refractile errors are diminished and the resolution limits of fluorescence microscopy (approximately 0.1–0.2 μm) are somewhat better than in transmission microscopy. However, because light is emitted from the specimen in all directions, the fluorescent "signal" is quite weak in comparison with the "noise" of the exciting light. Elimination of this "noise" constitutes the major accomplishment of successful microscopy.

Fluorescence Microscopes

Light Sources

Most fluorochromes used in biomedical research have excitation maxima in the 350 to 500 nm range. Compounds with excitation maxima near one of the mercury emission wavelengths (365, 405, 436, or 546 nm) are optimally excited by high pressure mercury light sources. Even fluorochromes with higher excitation maxima (e.g., Procion yellow = 470 nm, fluorescein isothiocyanate = 470 to 480 nm) have sufficiently broad excitation spectra that the excitation by the 436 nm mercury line is usually adequate. When mercury sources are unsuitable, high pressure xenon light sources are available which have more energy in the range between 436 and 546 nm (the mercury lines). Mercury light sources are less expensive than xenon systems, although lamp life may be shorter and mercury intensity is much less constant.

Since mercury light has a discontinuous spectrum, it is easy to remove unwanted wavelengths by the use of filters. Filters such as the Schott BG-3 or BG-12 are suitable for excitation from 400 to about 450 to 470 nm; the Schott UG-1 covers the 350 to 400 nm range but has a slight transmission in the red portion of the spectrum. Interference filters are extremely useful to isolate one mercury line for virtually monochromatic excitation, but they have a rather low transmission—usually 30 to 40%—and must be inserted in the light path at a point such that the rays are perpendicular to the filter.

In addition to the filters just described, all light sources should include one or more infrared-absorbing filters (e.g., Schott BG-38), as the heat produced by the high intensity source is considerable (see Table 1).

Specimen Illumination

Illumination via a substage darkfield condenser is employed almost exclusively. Low power dry objectives are used with dry condensers, which have a reasonably large field of illumination. Oil immersion objectives require immersion condensers, which have a smaller area of illumination but a higher light intensity. When objectives of 10x and lower are employed there is little value in a darkfield condenser. Care should be taken to exclude air bubbles when immersion condensers are used since they will degrade the image quality.

The uniformity of illumination of the field by substage condensers is poor, particularly at low magnifications. A relatively new incident light illumination system developed by scientists in Russia (Brumberg, 1959) has been described by Ploem (1967) and refined to the point that it is commercially available through E. Leitz, Inc., Carl Zeiss, Inc. and C. Reichert, A. G. The heart of this illumination system is a dichroic (which should really be called "dichromatic") interference mirror which preferentially reflects light of the desired excitation wavelength down through the objective onto the specimen; emitted fluorescent light is selectively transmitted by the dichroic mirror into the tube lens system of the microscope. Fig. 1 illustrates such a system, which has a variety of advantages: (1) uniform specimen illumination, with light beam focussed and centered by the objective, which also functions as a condenser; (2) capability of reduction of the illumination to very small areas, thus reducing photodecomposition; (3) reduction of background due to light scatter; (4) alternate or simultaneous fluorescence and bright field (e.g., phase contrast) illumination; (5) better geometry for quantitative fluorescence microscopy; and (6) applicability to very thick or opaque specimens. This system is not particularly efficient at objective magnifications under about 20× due to the low numerical aperture of such lenses (see Table 1).

Table 1. Suggested Filter Combinations for use with Dichroic Mirror Vertical Illumination (data summarized from Walter, 1970)

Excitation wavelength	Excitation filters[a] (always include 4 mm. BG-38)	Dichroic mirror[b]	Integral barrier filter[b]	Optional slide barrier filter[c]	Fluorescence emission
Ultraviolet (340–400 nm.)[d]	2 mm UG-1	TK-400	K-400	K-430, 460, 470, or higher	Violet to Blue (425–470 nm.)
Violet to Blue (400–440 nm.)	3–6 mm BG-3 (or TAL 405, or KP 25)[e]	TK-455	K-460	K-460, 470, or higher	Blue (450–490 nm.)
Blue (440–490 nm.)	3–6 mm BG-12 (or TAL-480)[e]	TK-495	K-495	K-460, 470, 490, or higher	Green to Yellow (500–585 nm.)
Green (490–550 nm.)	2 nm BG-36 (+TAL-546)[e]	TK-580	K-580	K-570, 590, or 610	Yellow, Orange or Red (575–700 nm.)

[a] Schott A. G. filter numbers are indicated.

[b] In the vertical illuminator after Ploem, E. Leitz Inc.; Leitz filter numbers are indicated, the numerical designation representing the emission wavelength cutoff (i.e. 50% transmission at that wavelength and above.)

[c] E. Leitz Inc. numbers are indicated, representing cutoff wavelength as above.

[d] Light of wavelengths less than 340 to 360 nm is not passed by glass optics to any useful extent.

[e] Interference filters; Leitz filter numbers are indicated, the numerical designation representing the wavelength transmitted.

Objective Lenses

Microscope objectives are corrected for spherical and chromatic aberration. Objectives most suitable for high resolution transmission microscopy are not necessarily optimal for fluorescence applications. Apochromatic objectives are highly corrected for both aberrations and are best for color photomicrography; they yield the best definition when transmitted light is blue or green. However, apochromatic lenses contain many elements and layers of cement, and this results in considerable lens fluorescence when these objectives are exposed to high intensity light sources, as in fluorescence microscopy.

Fluorite objectives are generally recommended for fluorescence microscopy, as they have appreciably less autofluorescence than apochromatic lenses, and are nearly as well corrected for chromatic and spherical aberration. Since fluorescence microscopy implies an attempt at attaining monochromatic illumination, correction for spherical aberration is usually more important than for chromatic aberration.

Focussing and Specimen Optics

Since the depth of focus of immersion darkfield condensers is quite small, it is important to have the correct microscope slide thickness (usually about 1.2 mm). Cover

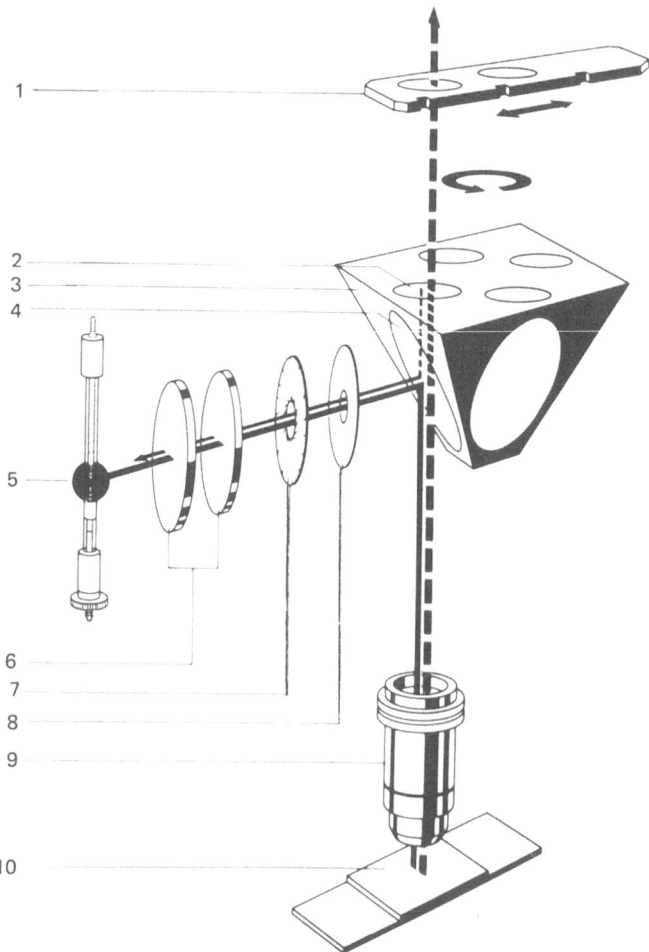

Fig. 1. Incident light illuminating system for fluorescence microscopy, after Ploem (1967). (1) optional slide barrier filter, which may be omitted under some conditions; (2) barrier filter in the turret integral with a dichroic (dichromatic) mirror; (3) rotating turret to permit selection of correct dichroic mirror and integrated barrier filter; (4) dichroic mirror; (5) light source, high-pressure mercury or xenon; (6) heat-absorbing and excitation filters; (7) field diaphragm; (8) stray light stop; (9) objective lens which serves both as a condenser to focus the exciting light onto the specimen and as an objective to collect the fluorescent light; and (10) specimen on microscope slide. Solid line indicates excitation light, most of which is reflected onto the specimen by the dichroic mirror. Dashed line indicates emitted fluorescent light, most of which is passed by the dichroic mirror and barrier filters to the microscope tube lens system. Slightly modified, courtesy of Dr. Manfred Nahmmacher, E. Leitz Inc., Rockleigh, N.J.

slip thickness is also critical, as microscope objectives are corrected for specific cover slips, usually No. 1 or $1\frac{1}{2}$; objectives which are adjustable for cover slip thickness variations are also available.

the binocular and the monocular (camera attachment) eyepieces, so very precise focus

Most trinocular microscopes are not in focus in the same plane when viewed through

should be obtained through the monocular when taking photographs. Specimen stage drift and out-of-focus conditions due to vibration sometimes present difficulties with the relatively long exposures required in fluorescence photomicrography.

Barrier Filters

Barrier filters are used to eliminate exciting wavelengths which have not been absorbed by the specimen. The majority of such filters have extremely low transmission below the nominal cutoff (usually the 50% transmission wavelength) and quite high transmission above that wavelength. Leitz and Zeiss filters are designated so that the cutoff wavelength is implicit in the filter number.

Interference filters are unsatisfactory as barrier filters because of the relatively low transmission at the nominal wavelength, the transmission of other orders of light as well as the nominal wavelength, and the problem of insertion of the filter at a point in the light path at which the rays are perpendicular to the filter. Interference filters are of some use in certain microspectrofluorometric applications (the "poor man's monochromator") but were really designed for bright field transmission microscopy and microphotometry (see Table 1).

Microfluorometry and Microspectrofluorometry

Quantitative fluorescence microscopy enables the objective comparison of fluorescence intensities of several specimens. This technique is usually unnecessary for evaluation of intracellular staining with fluorescent dyes. One might suppose, however, that the microfluorometric examination of a cell would provide information on the amount of Procion yellow injected. The histologic requirements for quantitative fluorescence measurements are described by Ruch (1970).

Microspectrofluorometry is applicable to the differentiation of true fluorescence from light scatter or for distinguishing wanted fluorochromes from autofluorescence. Unfortunately the human visual system is liable to misinterpretations of color under some circumstances. Photographic emulsions can also render color inaccurately. When autofluorescent structures in tissue appear of a color similar to that of an injected fluorochrome the final differentiation may require microspectrofluorometry. Van Orden (1970) has described a reasonably simple instrument for microspectrofluorometry.

Fluorescence Photomicrography

Selection of Filters

Ideal appearance of a preparation on visual inspection does not guarantee an informative photomicrograph. The differences in spectral sensitivity of the retina and the film contribute to this, as do exposure time and psychophysical phenomena related to emphasis of the most interesting color.

Both excitation and barrier filters must be selected with some thought for the spectral sensitivity of the film, which is usually different from that of the eye, and frequently greater in the blue-green wavelengths employed for excitation. Thus an interference barrier filter may be more satisfactory for photomicrography than for viewing.

One should be alert to possible fluorescence of the barrier filter itself, although this is usually encountered only with ultraviolet radiation (which would be markedly reduced by the glass microscope optics); it is a possible cause of high background and "fogged" prints.

Selection of Film

Most fluorescence photomicrography in the United States is carried out with Kodak film, and valuable information on the use of this material for fluorescence work may be found in Kodak Technical Publication M-27 (1968). A format larger than 35 mm usually necessitates excessive exposure times and should be avoided.

It is worth repeating that photographic emulsions are relatively sensitive to ultraviolet and blue light, thus the background due to exciting light may be much greater in a photomicrograph than would be assumed upon visual microscopy. For this reason daylight-type color films are preferable to films balanced for tungsten illumination, since the daylight films are formulated to be used with light sources which have dominant blue components. Since fluorescence emission is usually weak, a relatively high speed film is preferred. For transparencies, Kodak High Speed Ektachrome (ASA 160) is popular. The speed of this film may be increased to as much as ASA 640 by special processing (see Kodak Technical Publication M-27). In the author's laboratory, high speed GAF Anscochrome (ASA 200 or ASA 500) gives better spectral fidelity than Ektachrome. Anscochrome ASA 500 film may be increased to ASA 1000 by using a special processing kit. In all cases, the grain of the film increases with the speed, and this reduces image quality. Color reversal films may be used but have a lower speed than transparency films.

While color reproductions are the most informative, publication cost is often prohibitive. Black-and-white slides and prints are frequently adequate and have the additional advantages of simple processing and manipulation of contrast to reduce background and emphasize fluorescent structures. Most fluorescence photomicrography can be carried out with Kodak Tri-X panchromatic film (ASA 400), which has acceptably-low graininess when developed with Kodak Microdol-X (dilution 1:3). For the highest resolution, if slow speed can be tolerated, Kodak Panatomic-X (ASA 40) can be used. A new fine grain film with good characteristics is H & W Control VTE Panchromatic film (ASA 80) (manufactured by Agfa-Gevaert and distributed in the U.S.A. by H & W Control, St. Johnsbury, Vt.).

Determination of Exposure

The usual microscope photometers are not sufficiently sensitive to detect the weak light emitted from fluorescent specimens. Furthermore, since contrast between the specimen fluorescence and tissue background is important, it is inadvisable to use integrating photometers for determining the exposure time. Trial exposures over a wide range of times are usually necessary. Exposure times for Procion yellow specimens might be in the range of 30 sec to 3 or 4 min with color films of ASA 160.

Reciprocity Failure

A sequence of trial exposures may reveal a nonlinearity with differences in object fluorescence; this is due to failure of reciprocity (reciprocity is the inverse relationship between light intensity and exposure time which permits the halving of exposure time if intensity of illumination is doubled). With long exposures the reciprocal relationship does not hold, since the sensitivity of the photographic emulsion decreases. In color fluorescence photomicrography this represents a special problem due to different reciprocity characteristics of the various color-sensitive emulsion layers and the consequent differences in color rendition as a function of exposure time. Each manufacturer provides specific information on exposure compensation to alleviate reciprocity failure.

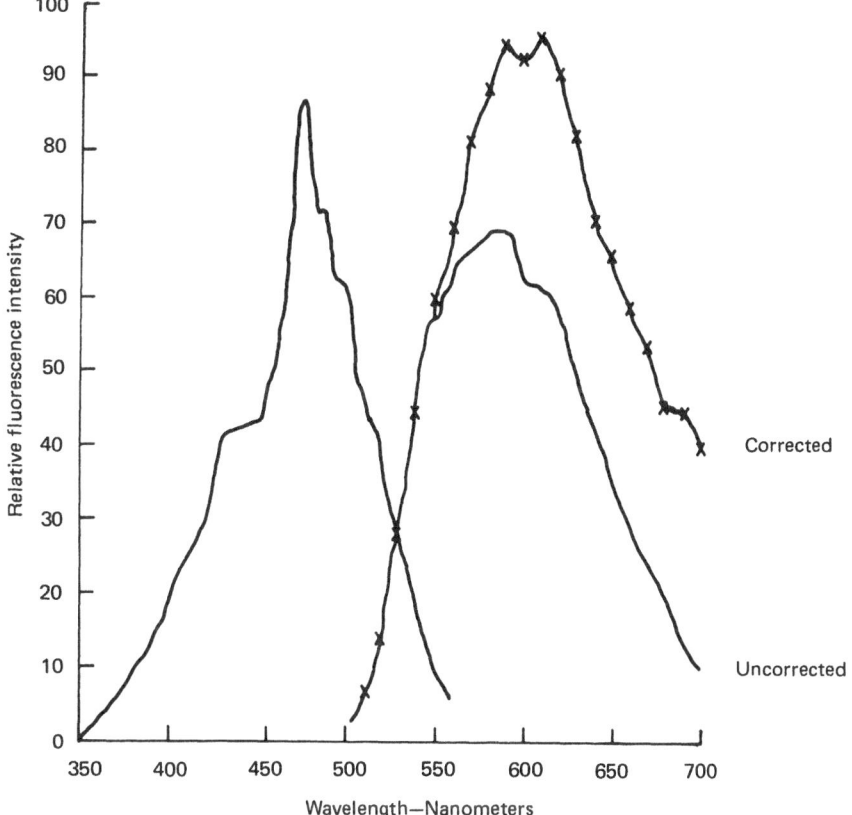

Fig. 2. Fluorescence excitation and emission of Procion yellow in *Necturus* optic tectum neurons. The excitation spectrum was not corrected but the microspectrofluorometer employed registers within 5 nm of reported excitation maxima of known compounds in this spectral region. The uncorrected and corrected emission spectra are shown, to illustrate the rather large amount of fluorescence of longer wavelengths which is not appreciated in uncorrected spectra.

Applications to Procion Yellow

Excitation and Emission Spectra

Excitation and emission spectra of *Necturus* optic tectum neurons, injected with Procion yellow and fixed by perfusion with 10% formalin (material provided by R. Baker and C. Nicholson) were measured in a microspectrofluorometer (Van Orden, 1970). The excitation maximum was 467 to 475 nm (Fig. 2). which is sufficiently close to the 436 nm mercury line to guarantee satisfactory excitation. The emission

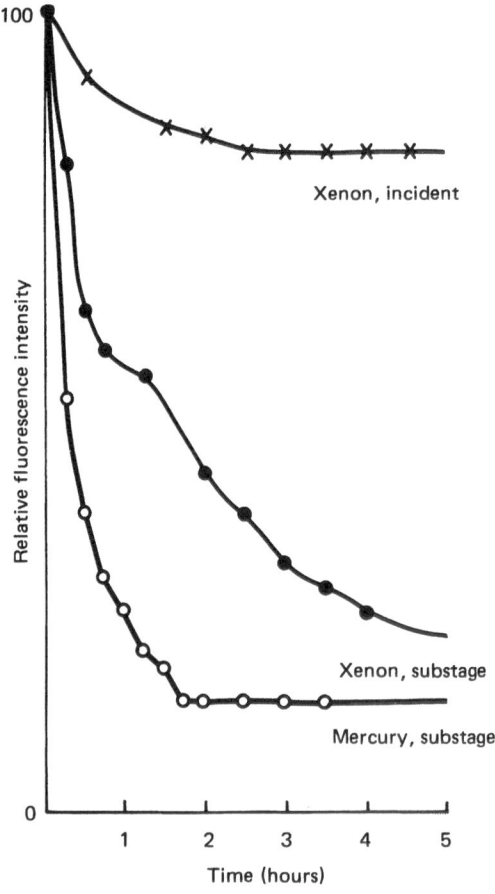

Fig. 3. Photodecomposition (fluorescence fading) of Procion yellow with exposure to exciting light. *Necturus* optic tectum neurons containing Procion yellow were irradiated with high pressure mercury (200 watts) or xenon (150 watts) light in a microspectrofluorometer with a conventional substage oil immersion darkfield condenser, or with xenon light through an incident illumination system (Ploem/Leitz); the light was in all cases passed through BG-38 and BG-12 filters, and fluorescence emission measured at 580 nm.

maximum was approximately 580 nm (600 nm corrected for instrument response) (Fig. 2).

Photodecomposition

Fluorophores vary enormously in times of photodecomposition, when irradiated at their excitation maxima, from a few seconds to many hours. Specimens of Procion yellow-injected neurons (see section on *Excitation and Emission Spectra*, above) were exposed to light from high pressure mercury or xenon light sources for times up to 5 hr, with excitation filters optimal for fluorescence microscopic viewing (4 to 6 mm Schott BG-12). Fluorescence of fluorochromed neurons was measured in a microspectrofluorometer at 15 to 30 min intervals, at 580 nm (uncorrected emission maximum). The rates of photodecomposition are illustrated in Fig. 3. Fluorescence fading was greatest with substage mercury excitation, less pronounced with substage xenon excitation, and minimal with xenon excitation employing incident light. This is probably a result of the additional interference filter which is part of the latter illumination system. Interference filters might be added to substage illumination systems if prolonged observation is desired.

References

Brumberg, E. M.: Fluorescence microscopy of biological objects using light from above. Biophysics *4*, 97–104 (1959).

Ploem, J. S.: The use of a vertical illuminator with interchangeable dichroic mirrors for fluorescence microscopy with incident light. Z. wiss. Mikrosk. *68*, 129–142 (1967).

Ruch, F.: Principles and some applications of cytofluorometry. In: Introduction to Quantitative Cytochemistry. Ed. G. L. Weid and G. F. Bahr. Vol. II, pp. 431–450. New York: Academic Press, 1970.

Udenfriend, S.: Principles of fluorescence. In: Fluorescence Assay in Biology and Medicine. Vol. II, pp. 1–41. New York: Academic Press, 1969.

Van Orden, L. S. III: Quantitative histochemistry of biogenic amines: a simple microspectrofluorometer. Biochem. Pharma. *19*, 1105–1117 (1970).

Walter, F.: Fluoreszenzmikroskopie in Biologie und Medizin. Leitz-Mitt. Wiss. u. Techn. *V*, 33–40 (1970).

CHAPTER 5

Procion Yellow as a Marker for Electron Microscopic Examination of Functionally Identified Nerve Cells

D. Purves and U. J. McMahan

A detailed knowledge of the number, arrangement, and types of synaptic connections between nerve cells is one of the prerequisites for understanding the integrative function of individual neurons. A great deal of information about the general arrangement of connections between populations of cells in the central nervous system of vertebrates has already been obtained by using metallic impregnation methods for light and electron microscopy. For further analysis of synaptic contacts between nerve cells, it would be useful to develop methods for marking specific cells for microscopic examination after they have been studied by electrophysiological techniques. In recent years microelectrodes have been widely used to inject various marking substances into cells of known function in order to determine their geometry by light microscopy (Stretton and Kravitz, 1968; Lux et al., 1970; Pitman et al., 1972a). It is difficult, however, to resolve the synaptic connections of specifically marked cells at the light microscopic level both because of their small size and their occurrence, in many cases, in spatially complex neuropils. One approach to this problems is to combine electrophoretic marking techniques with electron microscopy, where the details of cellular connections can be clearly resolved. For a fine structural marker one would like to have a substance that not only makes the nerve cell and all of its processes easily identifiable in the electron microscope but also does not distort the cell or obscure its organelles, especially synaptic structures such as vesicles and membrane thickenings. In addition, the marker should be visible in the light microscope so that specific portions of a nerve cell can be selected for electron microscopic examination and relationships to other structures recognized.

Pitman et al. (see Chapter 6) in their studies of neurons in the cockroach have shown that the injection of cobalt chloride holds great promise for use in electron microscopic studies. Our approach has been to study specific cells with both physiological and fine structural techniques by adapting the electrophoretic injection of the fluorescent dye Procion yellow to the requirements of fixation and staining for electron microscopy. We applied this method to nerve cells in the segmental ganglion of the medicinal leech, in particular to the "large longitudinal motoneuron." This cell was chosen because its function and many of its synaptic connections with cells in the same and adjacent ganglia are known (Nicholls and Baylor, 1968; Stuart, 1970; Nicholls and Purves, 1970). By combining dye injection with light and electron microscopy we have been able to establish where synapses are situated upon the cell, estimate their minimum number, and identify at least two anatomically distinct synaptic types. The details of our methodology and its effect on the fine structure of leech neurons will be the substance of this chapter. A comprehensive presentation of our findings may be found elsewhere (Purves and McMahan, 1972).

Technique of Injection and Fixation

· Isolated segmental ganglia were pinned in a shallow bath of physiological saline and viewed with darkfield illumination. The large longitudinal motoneuron could be recognized by its size and position in the ganglion, and its identity confirmed by recording its characteristic action potential with a dye-filled microelectrode.

The electrodes contained 6% (w/v) Procion yellow M4RAN (Stretton and Kravitz, 1968; Purves and McMahan, 1972) and were selected for resistances between 200–400 MΩ. Those with lower resistances often damaged the cells so that they became markedly distorted, while those with higher resistances did not deliver dye fast enough. The dye was injected by passing 500 msec pulses of hyperpolarizing current at 1 Hz for 20–60 min; the current ranged from 1×10^{-8} to 5×10^{-8} A. Under these conditions the cell became appreciably yellow within 2–3 min and was deeply yellow at the end of the injection period. The dye was allowed an additional diffusion time of 30 min to 5 hr; in general, we used times in the range of 1–2 hr since the processes with which we were concerned are within a few hundred microns of the cell body.

Ganglia were then fixed for 30 min with 0.8% glutaraldehyde and 1.0% paraformaldehyde in phosphate buffer at pH 7.4. They were postfixed in phosphate buffered 1.0% osmium tetroxide for 90 sec to 4 min with agitation, and stained for 2 hr at 4°C with 0.5% uranyl acetate buffered with tris maleate. After dehydration with a graded series of aqueous ethanol solutions, preparations were embedded in Epon 812 or Maraglas and cured for 45–90 min in a vacuum oven at 90°C.

Location of Injected Neurons for Observation in the Electron Microscope

Because the segmental ganglion of the leech is small, injected cells are easily recognized with standard fluorescence optics in intact embedded preparations when fixed with aldehyde alone (Purves and McMahan, 1972). However, additional treatment with osmium tetroxide, which is necessary both for optimal tissue preservation and staining for electron microscopy, blackens the whole ganglion so that the cells cannot be seen. By varying the time of osmium tetroxide treatment we found that Procion fluorescence can be detected in sections with brief osmium exposures. In general, 3 min in osmium tetroxide results in both adequate fluorescence in the light microscope and sufficient contrast in the electron microscope. Shorter times (90 sec) were chosen when maximum fluorescence was desired (e.g., in observing fluorescence in thin sections; see below).

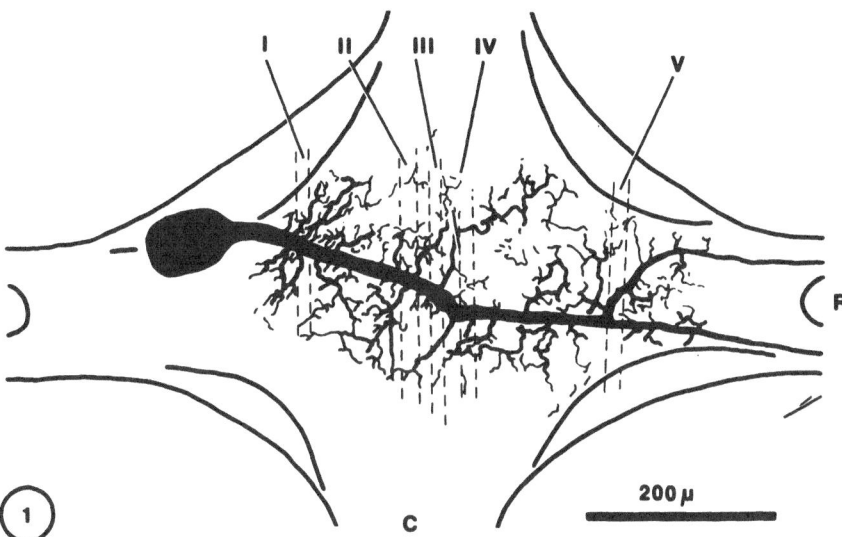

Fig. 1. Light microscopic reconstruction of an injected large longitudinal motoneuron. The cell body, as is true of other neurons in the leech ganglion, is located superficially; a single main process enters the neuropil, where synaptic connections are made. Throughout its course in the neuropil the main process gives off numerous branches; it then bifurcates to run out the contralateral roots (R) to innervate the longitudinal musculature of the body wall segment. Dotted lines indicate the plane of section in subsequent figures; regions between dotted lines (Roman numerals) show the approximate locations of the sections shown in Fig. 5. C = posterior connective.

Three methods were used for locating fluorescent profiles seen with the light microscope for study with the electron microscope. Since each was particularly useful for certain purposes, they are described in some detail.

The most generally useful approach was to cut transverse thick sections (3 μm) through the periphery of the ganglion until the fluorescent profile of the cell body was encountered. The plane of section in all experiments is indicated in Fig. 1, in which the position and branching pattern of one of the large longitudinal motoneurons within the ganglion is illustrated. One could stop at any point and cut thin sections for examination in the electron microscope. Thin sections were routinely mounted on 75 mesh parlodion coated grids and counterstained with lead citrate or uranyl acetate and lead citrate. By comparing the position of a fluorescent profile to nearby landmarks, as well as noting its shape, the same profile was readily found in adjacent thin sections viewed with the electron microscope; the identity could usually be confirmed by characteristic changes in the dye-injected cell described below. Examples are shown in Figs. 4 and 5 where the cell body of an injected motoneuron and its main process have been located at selected points within the ganglion.

A second technique was to examine the same thin section in both fluorescence and electron microscopes. Gold sections (about 1000 Å) were transferred by wire loop from the knife–boat to slides that had previously been immersed in 1.0% glycerol in absolute alcohol and had been allowed to dry. When sections dried on these slides, the film of glycerol prevented them from sticking firmly so that sections could be subsequently floated off by placing a drop of water on them. After photographing the sections with fluorescence optics, they were transferred to grids, stained, and examined in the electron microscope. This method was useful in determining the nature of structures too small to appear in both thick and adjacent thin sections. For example, in our early studies of light microscopic sections of injected motoneurons we saw very fine fluorescent profiles, one or a few microns in length and often less than a micron in diameter, protruding from the main process of the cell as it traversed the neuropil. Initially, we were unsure whether these fine profiles corresponded to some real feature of the cell or whether they were an artifact due to leakage of dye. By examining the same section in both light and electron microscopes, as shown in Fig. 2, it was clear that these were in fact small processes filled with dye. Such processes turned out to be of special interest because we found that most of the synaptic contacts on the motor cell occur on them (see also Figs. 5 and 6).

A third approach was useful in studying selected details of the injected cell within the neuropil, such as a medium-sized branch of the main process at some distance from it (see Fig. 1). In this case we cut a series of 3 μm sections from appropriate regions of the neuropil, placed them on glycerol-treated slides and examined them by fluorescence microscopy. We then photographed those sections containing the type of branch we wanted to study and re-embedded them for thin sectioning. This was done by transferring the thick section to a drop of water in an aluminum weighing dish and evaporating the drop by gentle heating. The section was then mounted on a small cube of cured Maraglas by placing a drop of epoxy cement (Ross Chemical and Manufacturing Co., Detroit, Michigan, U.S.A.) on the section and then the cube of Maraglas on top of this. After curing the epoxy cement for 4 hr at 60°C, the aluminum was peeled away, leaving the 3 μm section glued to the surface of a block which could be trimmed in the usual way. Usually 20–30 thin sections were recovered from the

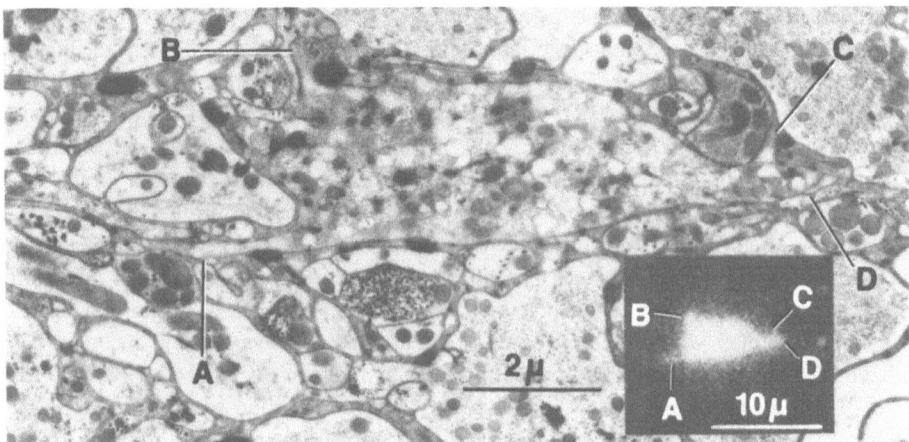

Fig. 2. Electron micrograph of a cross section of an injected motor cell main process with several spines (A, B, C, D). Inset is the same section photographed in the fluorescence microscope before transfer to grid for examination with the electron microscope.

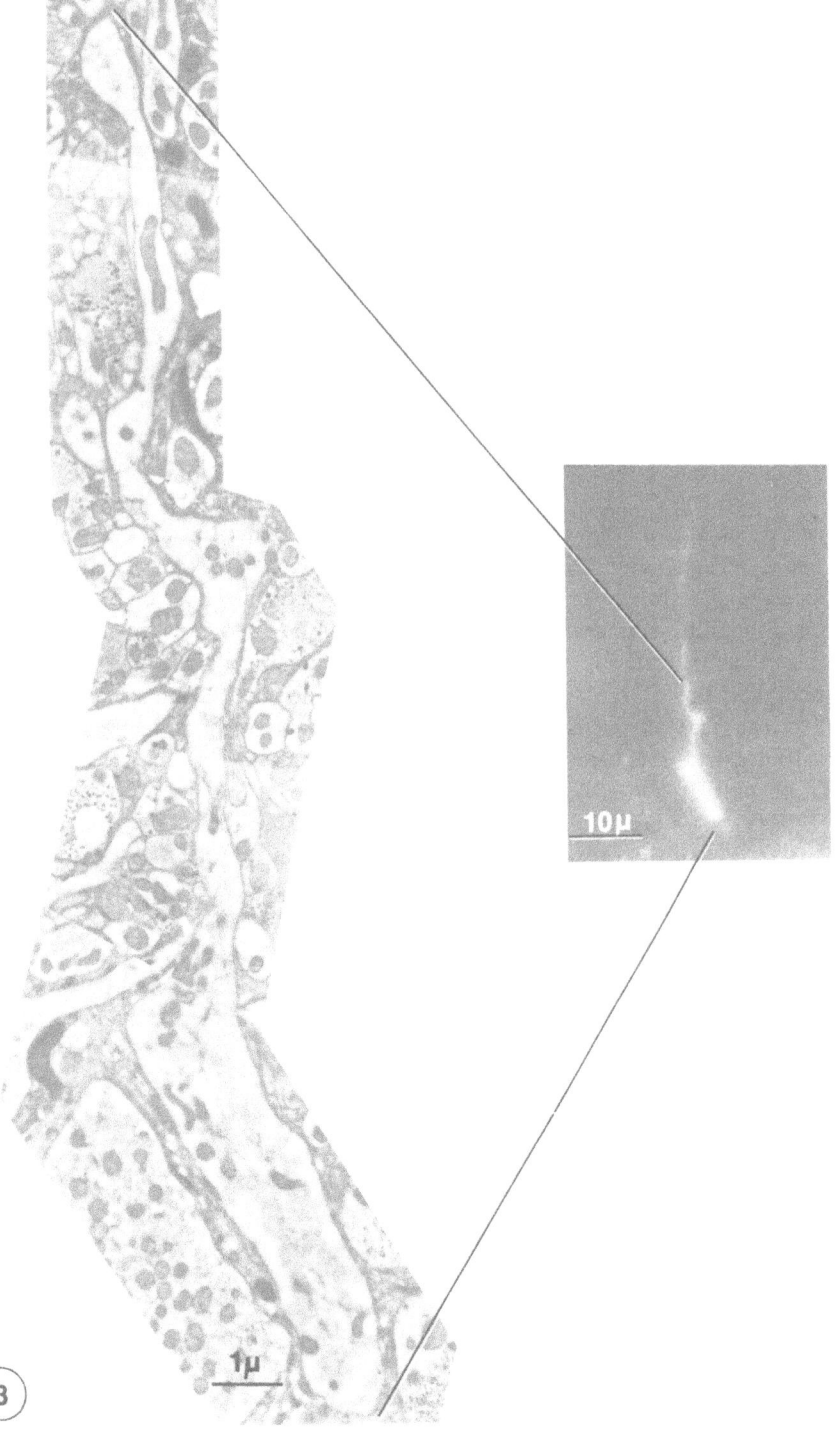

Fig. 3. Light micrograph: a branch extending from about 25 μm to about 70 μm from the main process seen in 3 μm section with fluorescence optics. Electron micrograph: following re-embedment and thin sectioning of the original thick section, the same branch was found in the electron microscope and a montage constructed by combining micrographs of several thin sections; lines indicate the extent of the branch included in the montage.

original thick section. An example of this technique is shown in Fig. 3; the inset is a fluorescence micrograph of the original thick section, while the electron micrograph is a montage constructed from several thin sections obtained after resectioning. The lines between the two photographs show the extent of the branch appearing in the thick section which has been reconstructed in the montage.

Changes in Fine Structure Owing to Dye Injection

In general, we found that while the injection of Procion yellow does cause alterations in some organelles, most features of the cell are preserved. These changes were greatest in the cell body, presumably owing to the trauma of impalement and subsequent electrophoresis, and because the concentration of the dye is highest here. Typical changes in the fine structure of a dye-filled cell body are shown in Fig. 4, where an injected motoneuron (Fig. 4A, B, and D) is compared with a cell having the same cytoplasmic characteristics and in the same position on the opposite side of the ganglion (Fig. 4C and E). The uninjected cell is presumably the homologous, contralateral motoneuron, although in the absence of a marker its identity can only be tentative. The injected cell is distinguished by characteristic small vacuoles in its cytoplasm, apparently caused by distortions of endoplasmic reticulum and mitochondria (Fig. 4D). In addition, the ground substance of the cytoplasm (and of the nucleus) is more homogeneous and darker than that of the control cell. Other structures such as the nuclear envelope, dark bodies, and granular vesicles remain largely unchanged by the injection procedure (Fig. 4D and E).

These changes become less intense as the cell's main process is followed away from the cell body, but they are appreciable even at distances of 100–200 μm and are helpful in confirming the identity of the injected profile. This is seen in Fig. 5, which shows the main process at increasingly distant points within the neuropil (see Fig. 1 for locations). These changes, however, cannot be reliably detected in medium-sized or smaller branches; the medium-sized branch in Fig. 3, for example, although well filled with dye, appears only marginally different from other uninjected branches of similar size.

Synapses Occurring on the Injected Motoneuron

Synapses in the ganglia of the leech resemble those found in vertebrates; the presynaptic element contains numerous vesicles while the postsynaptic component is characterized by a dense band of material on the cytoplasmic side of the plasma membrane (Coggeshall and Fawcett, 1964; Purves and McMahan, 1972). Numerous presynaptic profiles were apposed to the motor cell as it traversed the ganglion. There was no indication that the injection procedure significantly altered the structure of the synapse. Examples of two morphologic types of presynaptic structures (based on their size and the arrangement of synaptic vesicles) making contact with this cell are shown in Fig. 6.

The large longitudinal motoneuron was not seen to make presynaptic contacts with processes of other cells (nor has evidence for such synapses been found in the physiologi-

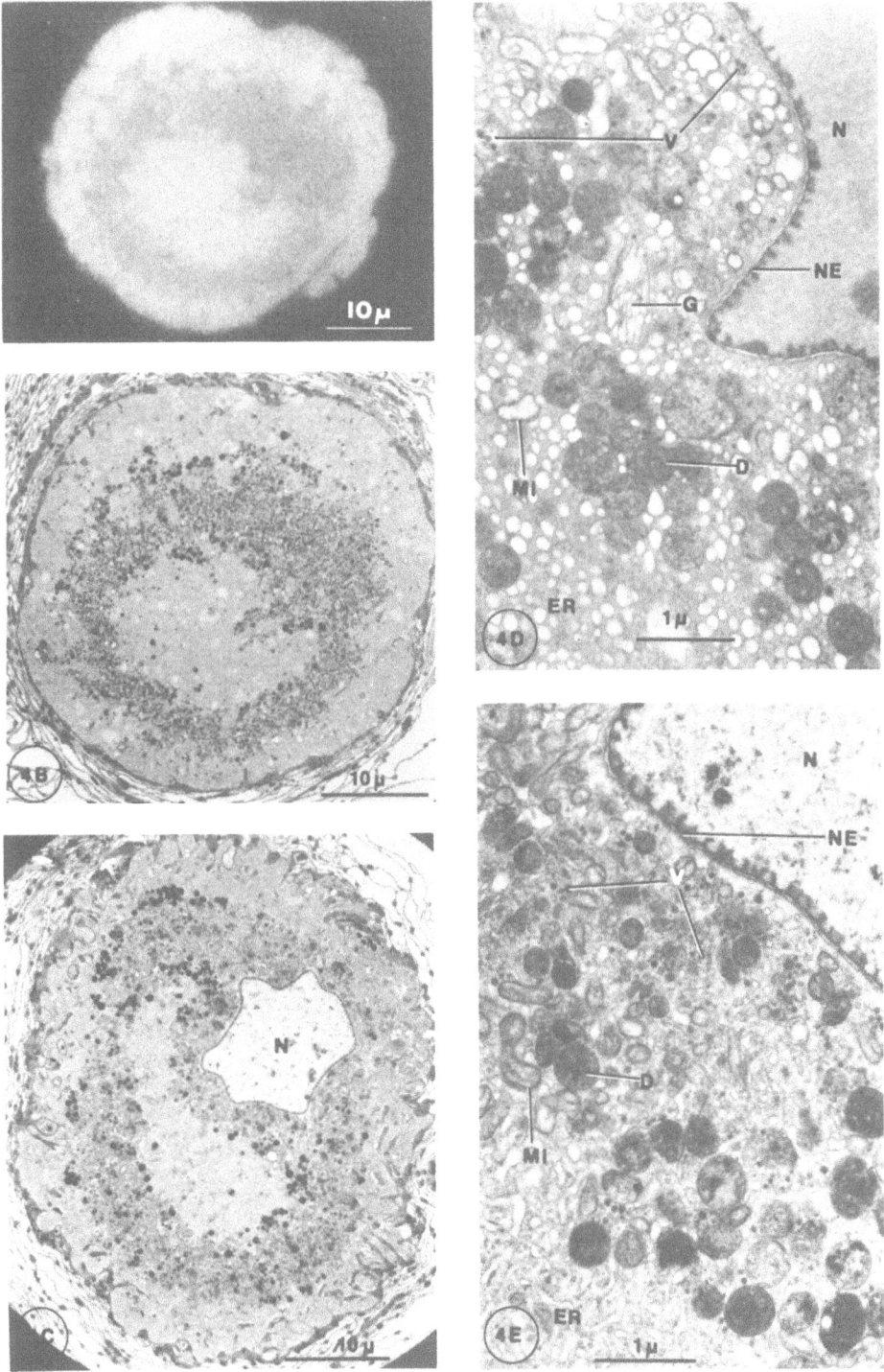

Fig. 4. A. Fluorescence micrograph of injected motor cell (3 μm section). B. Low-power electron micrograph of same cell. C. Uninjected cell from same preparation. N = nucleus. D. At higher power, injected cell appears vacuolated because of swollen endoplasmic reticulum (ER) and disrupted mitochondria (MI). Other structures (dark bodies = D, vesicles = V, nuclear envelope = NE, Golgi apparatus = G) are readily recognized. E. Higher power view of uninjected cell for comparison with (D).

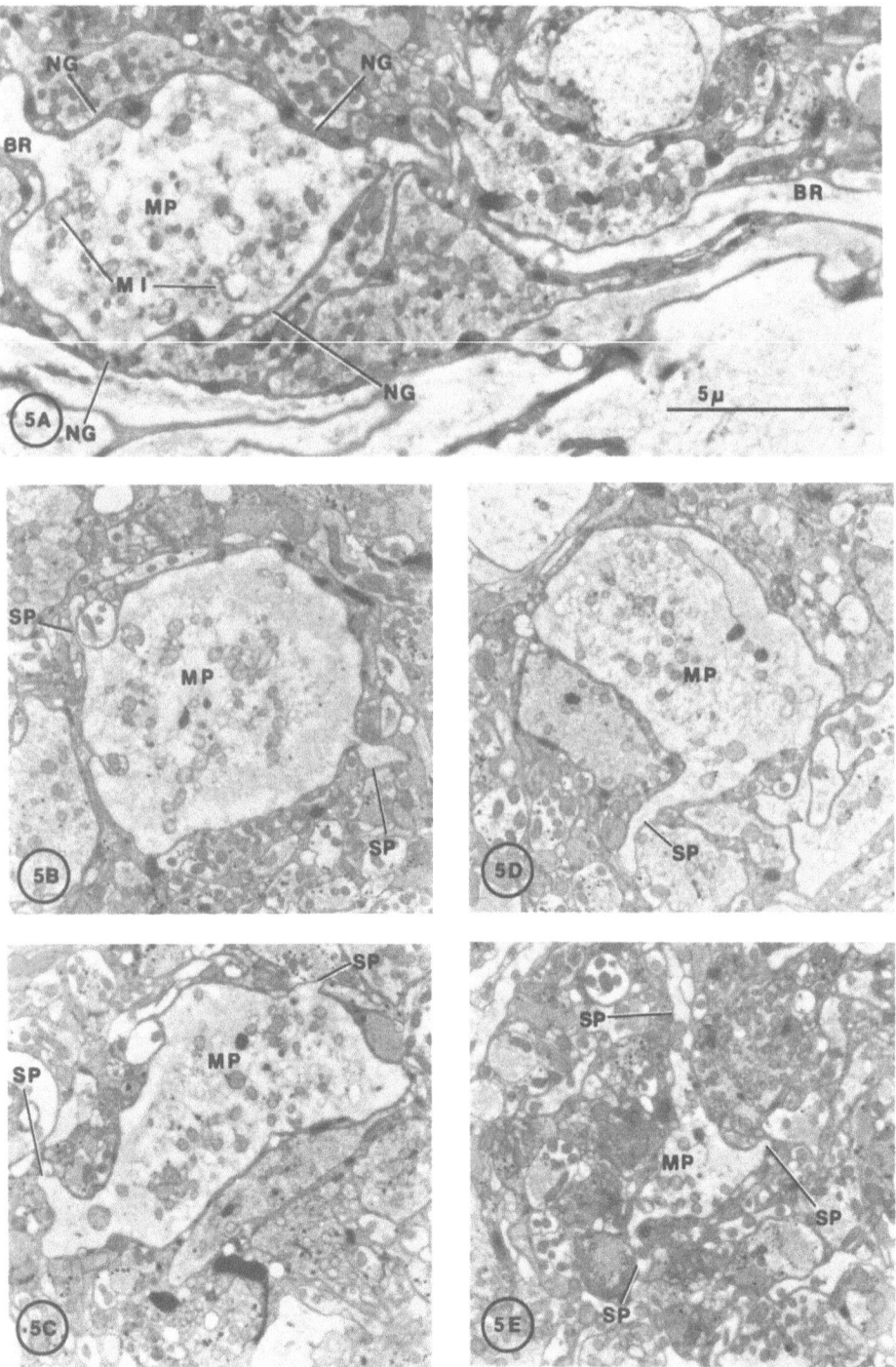

Fig. 5A–E. Injected motor process at selected points in the neuropil. A. Main process of injected motor cell (MP) in the initial portion of its course in the neuropil (region

cal studies done so far on this cell). To examine the question of whether dye injection directly alters presynaptic structure, in another series of experiments we injected a sensory cell known to make chemically transmitting contacts on other cells in the same ganglion. Here we could find fluorescent profiles filled with large numbers of vesicles which again did not appear different from uninjected profiles containing similar vesicles. Thus both pre- and postsynaptic structures are adequately preserved after dye injection.

Limitations of Procion Yellow as a Marker for Electron Microscopy

Since nerve cells when viewed in the electron microscope are not marked throughout all of their processes by the dye injection, it was necessary in the present study to routinely correlate fine structure with fluorescent profiles. Some of the limitations of this technique should be emphasized. One drawback that might be particularly important for the study of some other nervous systems is that the method probably requires the use of small pieces of tissue, since a successful result depends on the ability of osmium tetroxide to fix and stain the specimen adequately before quenching the fluorescence. With leech ganglia (about 0.3 by 0.8 mm) fixation was routinely adequate once we had determined the appropriate exposure times. With shorter exposures to osmium tetroxide (90 sec), however, details in the central part of the neuropil tended to be poorly defined. A second limitation is the relative difficulty of locating small fluorescent profiles in the electron microscope. While the main process within the neuropil was easily recognized in the electron microscope on the basis of the morphological changes described, branches could be identified as belonging to the injected cell only if they were followed in continuity, if they were selected from thick sections that had been mounted and sectioned for the electron microscope, or studied in the same thin section. Sectioning thick sections was a useful method for studying isolated branches, but it was time-consuming, as was examination of the same thin section in both light and electron microscopes. Correlating very small isolated profiles by either method was difficult because of distortions caused by resectioning and the heating of the plastic by the electron beam.

In spite of these limitations, we believe this technique will be generally useful in examining the specific arrangement of synaptic contacts on identified cells in the leech neuropil and in other preparations where intracellular recording is feasible.

Acknowledgements. This work was supported by USPHS Grants Nos. 5-T1 MH 07084 and NS-02253.

I in Fig. 1). Two large branches (BR) can be followed out of field. Main process and branches are surrounded by an investment of neuroglial processes (NG). Note the disruption of mitochondria (MI) and clumping of cytoplasmic material: these features made it possible to recognize larger injected profiles in the neuropil with only occasional reference to the fluoresence microscope. B–E are cross sections from the same preparation. Ventral is top, anterior to the right. B. Main process as it nears midline (region II, Fig. 1) showing similar cytoplasmic changes. C. Main process at midline (region III, Fig. 1). D. Main process in contralateral neuropil (region IV, Fig. 1). E. Posterior branch of main process beyond bifurcation (region V, Fig. 1). Cytoplasmic changes still present, but less obvious.

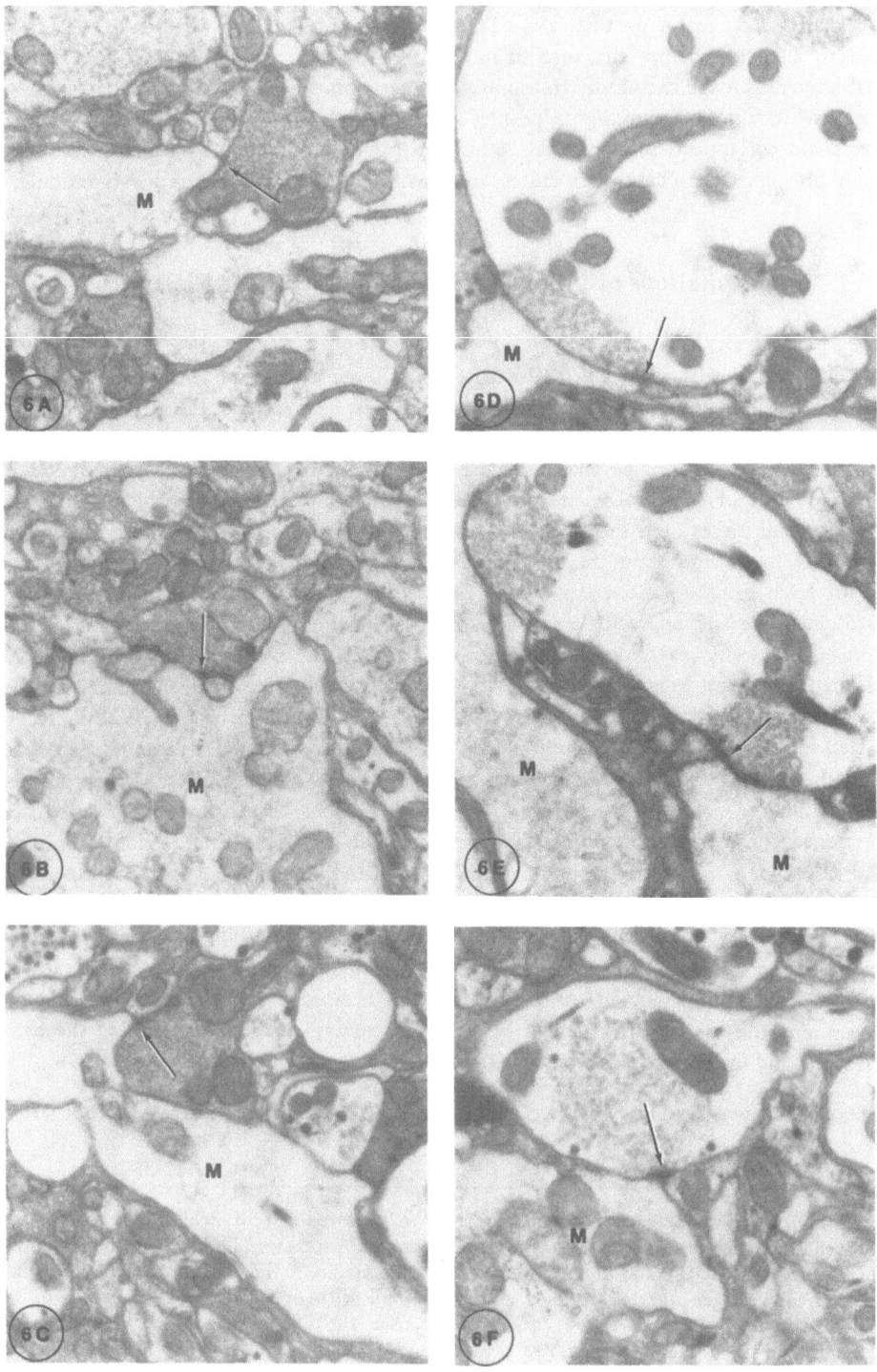

Fig. 6. Presynaptic profiles in contact with processes of injected motor cell. A–C. Examples of a smaller type of presynaptic profile densely packed with agranular vesicles on

References

Coggeshall, R. E., and D. W. Fawcett: The fine structure of the central nervous system of the leech *Hirudo medicinalis*. J. Neurophysiol. *27*, 229–289 (1964).

Lux, H. D., P. Schubert, and G. W. Kreutzberg: Direct matching of morphological and electrophysiological data in cat spinal motoneurons. In: Excitatory Synaptic Mechanisms. Ed. P. Anderson and J. K. S. Jansen. pp. 189–198. Oslo: Universitetsforlaget, 1970.

Nicholls, J. G., and D. A. Baylor: Specific modalities and receptive fields of sensory neurons in CNS of the leech. J. Neurophysiol. *31*, 740–756 (1968).

———, and D. Purves: Monosynaptic chemical and electrical connexions between sensory and motor cells in the central nervous system of the leech. J. Physiol., Lond. *209*, 647–667 (1970).

Pitman, R. M., C. D. Tweedle, and M. J. Cohen: Branching of central neurons: intracellular cobalt injection for light and electron microscopy. Science *176*, 412–414 (1972).

Purves, D., and U. J. McMahan: The distribution of synapses on a physiologically identified motor neuron in the central nervous system of the leech: an electron microscopic study after the injection of the fluorescent dye Procion yellow. J. Cell Biol. *55*, 205–220 (1972).

Stretton, A. O. W., and E. A. Kravitz: Neuronal geometry: determination with a technique of intracellular dye injection. Science *162*, 132–134 (1968).

Stuart, A. E.: Physiological and morphological properties of motoneurones in the central nervous system of the leech. J. Physiol., Lond. *209*, 627–646 (1970).

Discussion

DR. COHEN (Yale): I am interested in the criteria that you use to define a synapse in the neuropil of the leech. In many cases there appeared to be vesicles without any accompanying subsynaptic densities.

What are the structural criteria for a synapse in a central neuropil? In the arthropod, we are finding vesicles in many places that don't look to be synaptic when using the classical vertebrate criteria.

DR. MCMAHAN (Harvard): Certainly it is not clear at present what role the various structural specializations at synapses play in synaptic transmission. The fact that we find membrane thickenings in the leech while they are absent in arthropods is not very surprising. Indeed, there is broad variation in the appearance of these specializations within a single vertebrate nervous system. We did observe some cases in the leech where thickenings appeared to be absent but I suspect this often may have been due to the plane of section because in studies where we used serial sections, thickenings absent in one section, where there were clustered synaptic vesicles, would appear in another.

DR. L. LARRAMENDI (Chicago): This is clearly a very important method for correlating information at the light and electron microscopic levels. Could you indicate the limitations of this procedure?

---

motor cell process (M). In A, synapse (arrow) is on short branch about 2 μm from main process; in B, synapse is on spine of main process, while in C synapses are on a larger branch approximately 17 μm from the main process. D–F. Examples of larger type of presynaptic profile in which vesicles are generally clustered at the synaptic specialization. In D the synapse is on a short branch about 4 μm from the main process, while in E and F the synapses are on short, broad spines of the main process.

DR. MCMAHAN: One of the limitations is that it is necessary to use small pieces of tissue in order to be able to fix briefly with osmium tetroxide and not quench the fluorescence.

Another is that it is very difficult to find in the electron microscope structures less than 1 μm in diameter identified in the fluorescence microscope. This stems from both the complexity of the tissue and distortions in the plastic caused by heat from the beam. Those are the two principal drawbacks. Perhaps Dr. Purves can give some others.

DR. PURVES (University College): I would like to emphasize that many of the branches or spines we studied are much less than a micron in diameter. To locate them, one depends on the fact that one dimension of the profile has a length greater than 1 μm. Thus it is possible to examine very fine structures but an isolated process that has no dimension greater than 1 or 2 μm is impossible to locate with certainty.

DR. M. OBERDORFER (Wisconsin): I think in this problem of correlation between semi-thin and thin sections, one might use the new high voltage electronmicroscopy technique.

If the profiles are truly denser in these Procion yellow-filled thin sections, you might be able to see an increased density in a 1–1.5 μm section and still use the same thin and semi-thin under the high voltage electron microscope as you did with the light microscope.

DR. MCMAHAN: As one proceeds further and further away from the cell body along the main process of the cell the increased cytoplasmic density and swollen mitochondria seen in the cell body become less and less obvious. Three hundred μm away the process looks very much like any other process, and so these changes are not reliable markers at all.

DR. LLINÁS (Iowa): I would like to ask whether you found small neuronal processes filled with the vacuolar cytoplasm that were not filled with Procion yellow.

DR. MCMAHAN: No. The cytoplasmic distortions are unique features of the injected cells.

DR. OGDEN (Utah): I wonder if anyone has used any of the heavy metal Procion dyes. Are they any more electron dense than regular Procion dyes?

DR. KRAVITZ (Harvard): We have injected them, but we have never done electron-microscopy. They were less satisfactory in terms of migration.

CHAPTER 6

The Form of Nerve Cells:
Determination by Cobalt Impregnation

Robert M. Pitman, Charles D. Tweedle,
and Melvin J. Cohen

Form and function are perhaps more closely related in neurons than in any other cell type. A major approach to understanding the cellular events underlying behavior concentrates on relatively simple invertebrate systems that give rise to specific behavioral acts. The general procedure followed is (1) to identify the individual neurons that generate the behavior, and (2) to analyze the particular pattern of connections between neurons to determine the properties of the circuit that might directly govern the form of the behavioral act. Initial studies dealt primarily with identification of nerve cell bodies because of their relative accessibility. This led to the production of neuron soma maps for several invertebrates (Hughes and Tauc, 1962; Otsuka et al., 1967; Nicholls and Baylor, 1968). One such cell body map for the cockroach is shown in Fig. 1 (Cohen and Jacklet, 1967). The distribution of about 50 bilateral pairs of motoneurons is defined in relation to the nerve trunks through which their axons leave the ganglion.

The development of the Procion yellow fluorescent dye injection technique by Stretton and Kravitz (1968) extended the identification of individual nerve cell bodies to the major neurites of the cells deep within the neuropil. It also made possible the specific anatomical identification of a cell which had been subjected to detailed intracellular electrophysiological analysis. The success of the Procion yellow technique led us to attempt its use in examining the factors that control the formation of connections between neurons (Cohen, 1970). However, with insect material, we found that this procedure presented several difficulties. It was not consistently possible to obtain good whole mount preparations where the entire dendritic tree could be visualized within the ganglion. It became necessary to attempt reconstruction of the dendritic tree from serial sections and this proved both time consuming and inexact. We felt that the finer processes, where much of the synaptic activity was probably taking place, were not adequately resolved. Lastly, the fluorescent dye is not electron dense, and we, like many others were interested in observing the synaptic connections on the stained cells using the electron microscope.

In order to satisfy better some of the above requirements, we began investigating other reagents that could be uesd for microelectrode studies and then injected into a neuron to stain the finest branches of the dendritic tree. The stain must also be

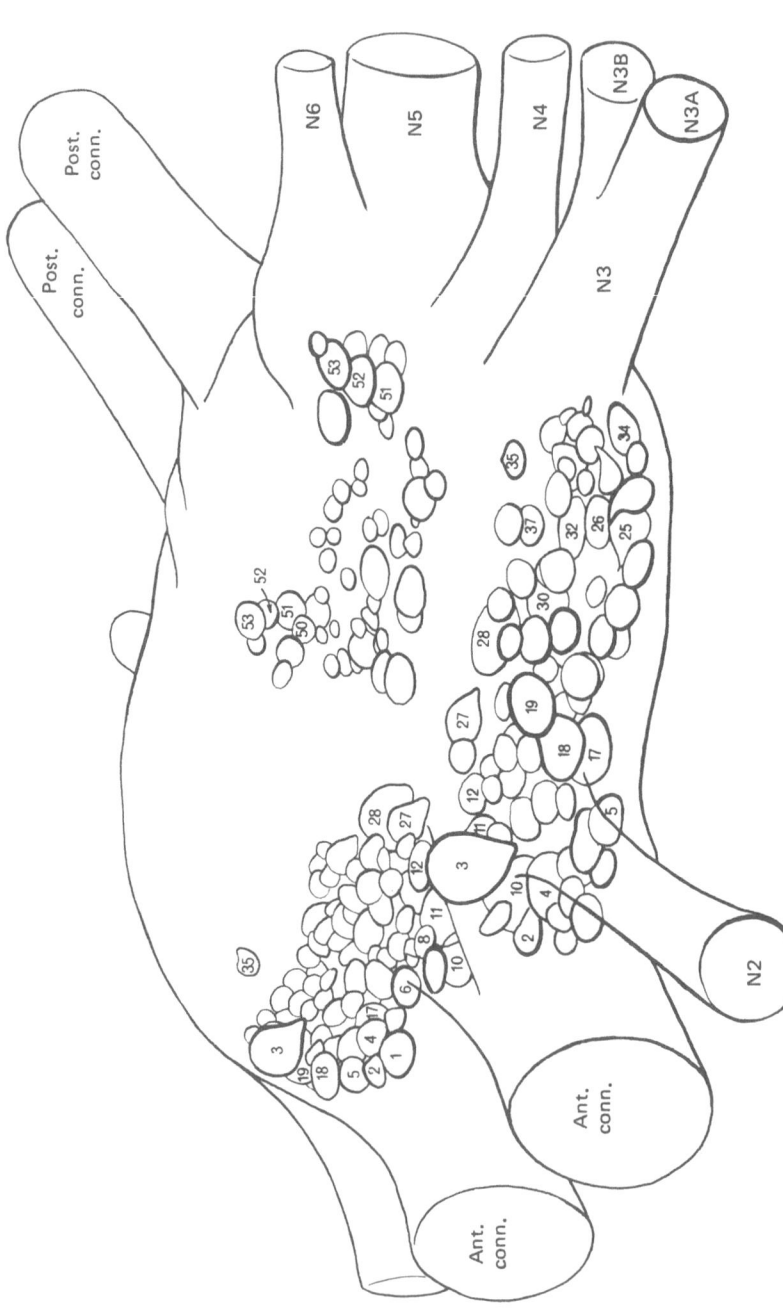

Fig. 1. A perspective drawing of the metathoracic ganglion of the cockroach *Periplaneta americana* showing the location of some of the identified motor nerve cell bodies. Members of a bilaterally matched pair of cells have the same number on each side of the ganglion. Note in particular cell 3 whose axon exits via the ipsilateral nerve 4 and cell 28 whose axon leaves through ipsilateral nerve 5. *Ant. conn.*, anterior connectives; *Post. conn.*, posterior connectives; *N2* to *N6*, peripheral nerve trunks. The ganglion is approximately 1 mm long. (Modified from Cohen and Jacklet, 1967).

electron dense. This led to the recent development in our laboratory of the cobalt sulfide staining technique by Pitman et al. (1972a). The cobalt procedure permits intracellular electrical recording from a neuron using glass capillary microelectrodes. The cobalt can then be passed into the neuron from the electrode, and it appears to stain the finest branches of the dendritic tree. With proper treatment, an insect ganglion can be cleared to allow observation of the three-dimensional geometry of the dendritic branches in whole mount preparations. The stain is also visible in the electron microscope, thus permitting analysis of synaptic structures on the stained cell. This paper is concerned with the procedure and results obtained with the cobalt staining technique as it has developed during the past year in our laboratory.

Intracellular Cobalt Injection

All results presented in this paper are from the ventral nerve cord of the cockroach *Periplaneta americana*. The location of specific metathoracic motoneurons discussed here, in particular cells 3 and 28, can be seen in the cell body map shown in Fig. 1.

Our method makes use of the reaction of cobaltous chloride with ammonium sulfide to form a black precipitate of cobalt sulfide within the neuron (Pitman et al., 1972a). In the original report, we mentioned that it should also be possible to use other soluble metallic chlorides which form sulfide precipitates. Since our initial paper, we have found iontophoretic injection to be superior to pressure injection for insect neurons and have used the iontophoretic procedure exclusively for the material presented in this study. Concentrations of cobalt chloride ranging from 1 to 50 mM are best for electrical studies to be followed by staining. Using glass capillary electrodes filled with this concentration of cobalt chloride, we have recorded overshooting action potentials from the cell bodies of dorsal neurons in the thoracic and abdominal ganglia (Pitman et al., 1972a). We have also been able to record overshooting action potentials with these electrodes in other insect central neurons whose somata have been rendered more electrically responsive either by axotomy or by treatment with colchicine (Pitman et al., 1972b).

After electrical recording, cobalt is ejected from the electrode into the cell by passing positive current pulses of 5×10^{-8} A, 0.5 sec in duration, at 1 Hz for a 1 hr period. The preparation is then immersed for 30 min in 10 ml of insect saline containing 0.05 to 0.1 ml of 100% ammonium sulfide solution. The tissue is fixed for 2 hr in the following solution modified from Karnovsky (1965): 4.5% paraformaldehyde, 1.1% glutaraldehyde, and 4.5% of 0.15M phosphate buffer. Dehydration is accomplished by using alcohols buffered with 0.15M phosphate to prevent loss of the stain, which is soluble in acid. The tissue is then cleared in reagent grade creosote and the whole mount viewed immersed in creosote in a shallow depression slide as seen in Figs. 2 through 6. If sections are desired for light or electron microscopy, the tissue is taken from the creosote into two changes of 100% acetone for 1 hr each. It is then moved into a mixture of 25% Spurr embedding medium in acetone for 1 hr and run up in increasing concentrations of the plastic according to the procedure described by Spurr (1969).

In preparations to be viewed in whole mount or sectioned for light microscopy,

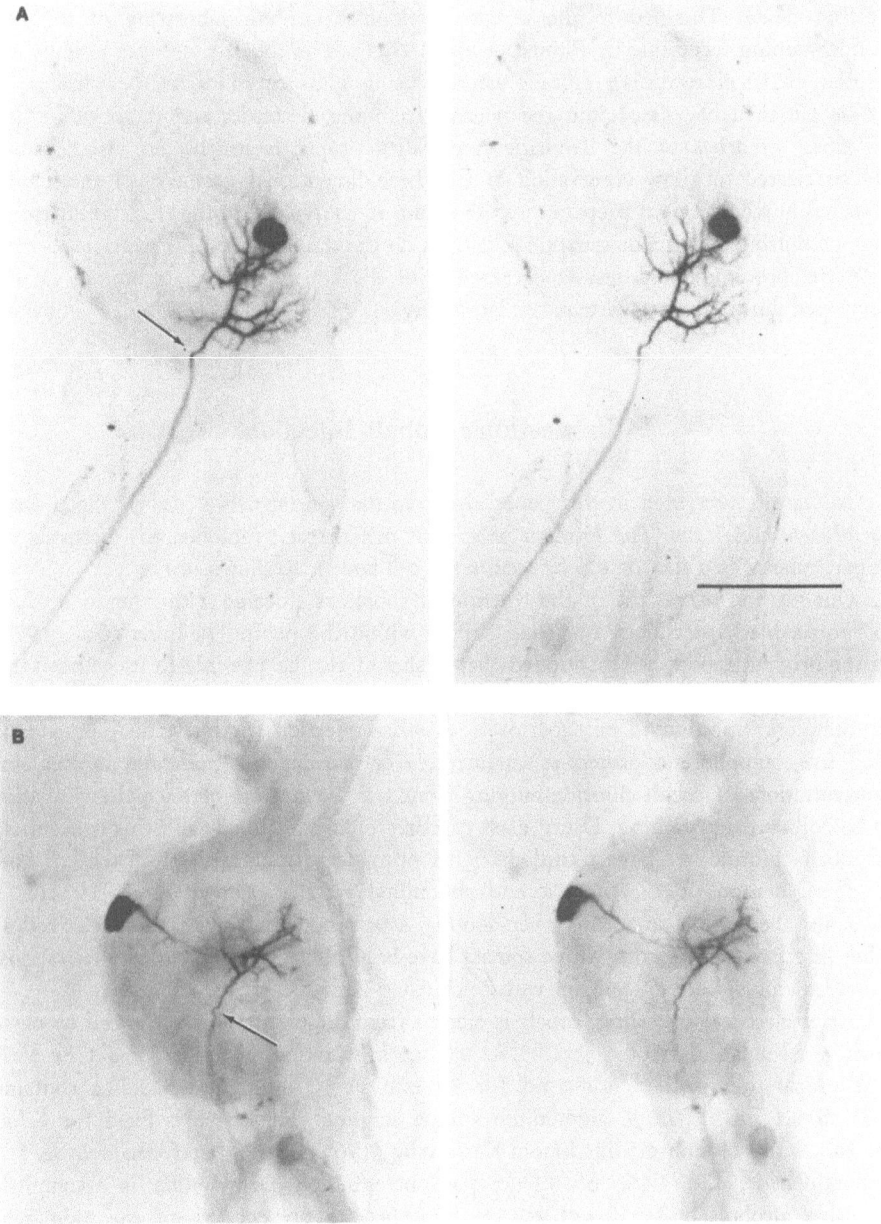

Fig. 2A and B. Stereophotographs of the metathoracic ganglion showing cell 28 stained by iontophoretic injection of cobalt into the soma. A. Dorsal view of the cleared whole mount preparation. The ventrally placed cell body is seen through the 1 mm depth of the ganglion with the initial process running vertically toward the dorsal surface. The characteristic medial and lateral branches of the dendritic tree can be seen throughout the depth of the ganglion. The arrow indicates a typical crook just as the axon leaves the ganglion via nerve 5. The posterior edge of the ganglion is to the left. B. The same preparation seen in A has been cut in the median sagittal plane and the view is "on the half shell" from the cut medial surface looking laterally toward the periphery of

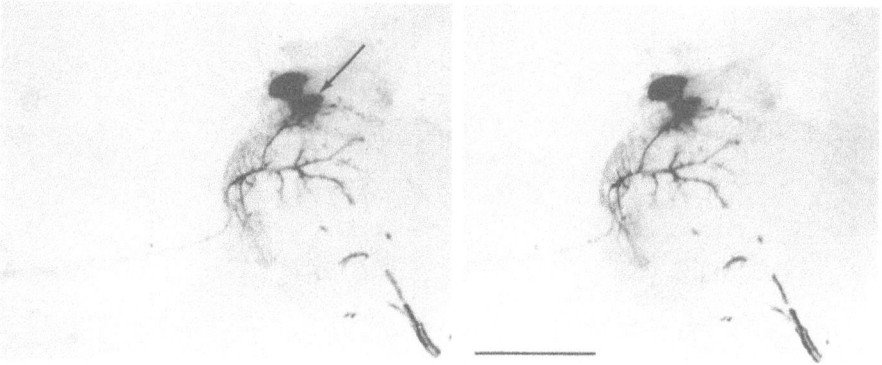

Fig. 3. Stereophotographs of the metathoracic ganglion showing the left cell 3 stained by intracellular iontophoretic injection of cobalt into the soma. Dorsal view with the anterior connectives seen at the upper right corner. The initial process drops vertically from the cell body for a short distance and then swings laterally and posteriorly towards the dorsal surface to give off the major medial and lateral components of the dendritic tree. The axon continues from the main lateral dendritic branch to leave the ganglion via nerve 4. The large mass close to the cell body and indicated by the arrow is a highly branched dendritic "brush" that is characteristic of this particular motoneuron. The linear structures in the lower right are tracheoles. Calibration: 0.5 mm.

no post osmication was used. We have found, however, that for electron microscopy some post osmication is necessary to preserve vesicular structure in this insect tissue. Fifteen minutes of immersion in a solution of 1 part 2% osmium tetroxide and 1 part 0.15M phosphate buffer is adequate. The coloring of the tissue is still light enough to allow the stained processes to be seen in the block face, and the fixation is sufficient to allow some of the presumed synaptic vesicles to be seen in the electron microscope. The electron micrographs in Figs. 7 through 9 were prepared from tissue treated in this manner.

Iontophoresis Through Cut Nerve Trunks

The technique used by Iles and Mulloney (1971) to move Procion yellow into a neuron by iontophoresis through the cut nerve trunk also works extremely well with the cobalt (see Chapters 7, 10, and 15). We have used this procedure to stain motor nerve cell bodies in central ganglia by driving the cobalt up through the cut ends of axons in peripheral nerve trunks. We have also used this method to stain

the ganglion. The cell body is on the ventral surface and the large right anterior connective is seen at the top of the picture. Note the two fine short branches on the initial process within 100 μm of the soma. The same crook in the axon mentioned above is also visible at the arrow as the axon leaves the ganglion. Calibration is 0.5 mm and is the same for both figures. (Fig. A modified from Pitman et al., 1972a).

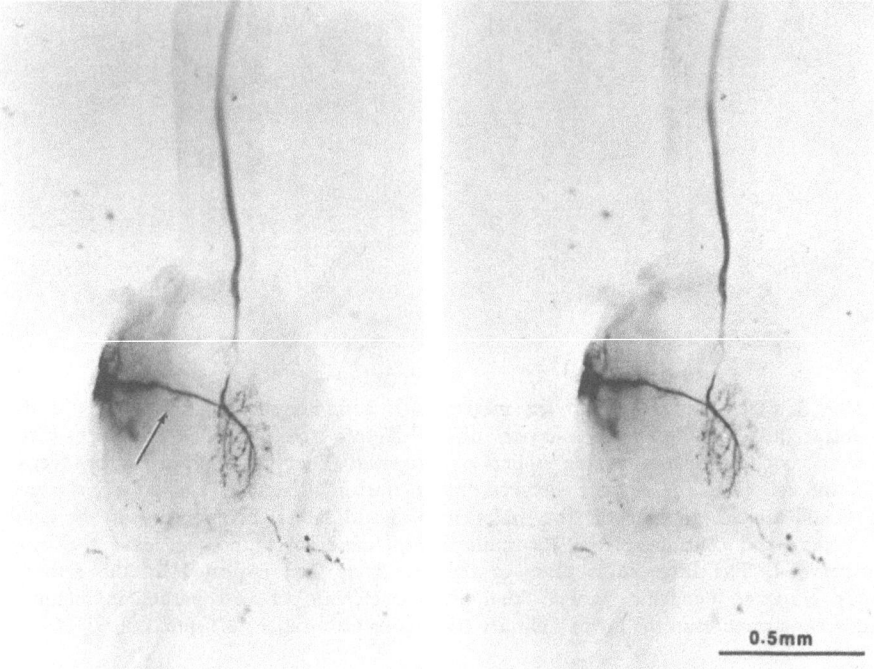

Fig. 4. Stereophotographs of the 6th abdominal ganglion showing a giant fiber stained by iontophoretic injection of cobalt into the cell body. Dorsal view of the whole cleared ganglion showing the giant fiber running anteriorly toward the top of the picture in the right neural connective between the 6th and 5th abdominal ganglia. The cell body lies at the left edge of the ganglion and sends its initial process transversely to cross the midline. There a branch runs anteriorly to become a giant fiber that travels the length of the nerve cord. The posterior branch from the initial process gives rise to the main dendritic tree of this cell. The arrow indicates a region along the initial process that seems covered with very fine branches. Calibration: 0.5 mm.

giant fibers and other interneurons within the CNS by moving cobalt into the system through the cut ends of central connectives as seen in Fig. 5. This is a particularly beautiful method of visualizing the course of the giant fibers and their collaterals because the cobalt will traverse almost the full length of the cord in these large diameter units, The particular parameters used for iontophoresis through the ventral cord are described in the legend of Fig. 5. When passing the cobalt into peripheral nerve trunks, the time during which the current is applied may vary from 1 to 15 hr depending upon the preparation. This technique works well for localizing central nerve cell bodies and frequently also stains the dendritic tree within the neuropil.

Three-Dimensional Visualization

It is our impression that the clarity, resolution, and beauty of the whole mount cobalt preparation, cleared in creosote and viewed with a stereo-dissecting microscope,

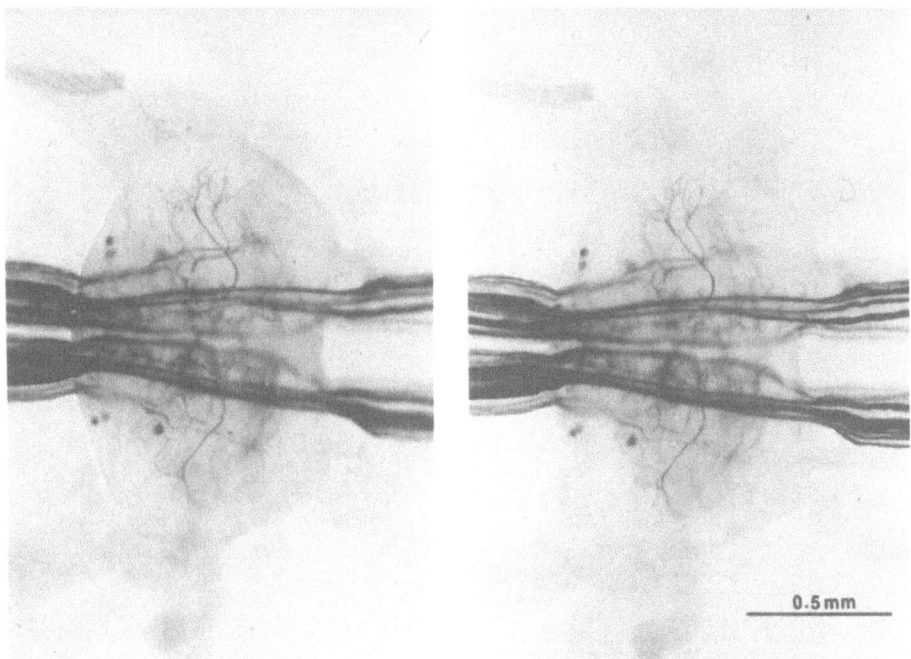

Fig. 5. Stereophotographs of a cleared metathoracic ganglion seen from the dorsal surface. Cobalt was introduced into the neurons by iontophoresis through the ventral cord which was cut posteriorly between the 3rd and 4th abdominal ganglia and anteriorly at the circumesophageal connectives. The cut abdominal connectives were placed in a pool of 100mM cobalt chloride; the cut ends of the esophageal connectives were placed in a separate pool of saline. The cobalt chamber was made the anode and the saline chamber the cathode in a circuit through with 1×10^{-6} A of current was passed for 15 hr. The cobalt traversed the length of the nerve cord between the 3rd abdominal ganglion and the 1st thoracic ganglion, staining primarily the giant fibers and their collaterals. The small posterior cell bodies are most likely interneurons. The anterior connectives are to the right and the posterior connectives to the left. Calibration: 0.5 mm. (Courtesy of V. Pakalnis).

approaches that of the classic Golgi preparations of Trujillo-Cenoz and Melamed (1970). The individual motoneurons shown in Figs. 2 and 3 have dendritic ramifications that range through most of a quadrant of the ganglion. In order to see the entire dendritic tree in its full three-dimensional array within the ganglion, we must look at processes a few microns in diameter over a distance approaching the 1 mm depth of the ganglion. This represents a severe photographic problem involving high resolution over a great depth of field. The stereo photographs shown in this paper are a compromise between maximum depth of field and adequate resolution to represent the dendritic ramification of a single neuron within the ganglion. They were taken in the following manner. The front objective lens of a Zeiss Stereo IV dissecting microscope was removed and replaced with a Zeiss 63 mm focal length Luminar lens designed for macrophotography. The binocular body was replaced with a monocular tube supporting a Zeiss 35 mm Attachment Camera. The preparation was mounted in creosote on a depression slide

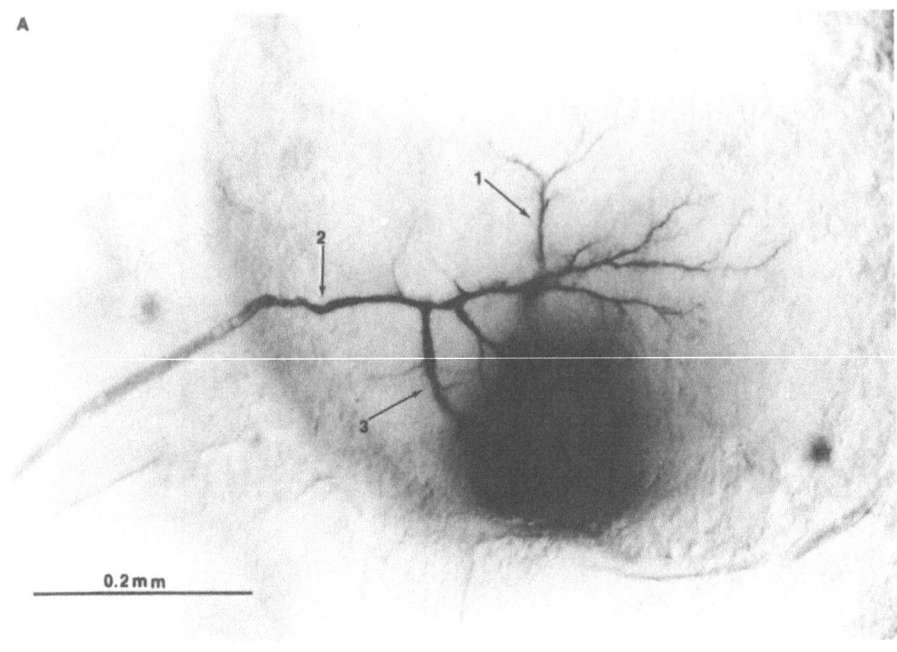

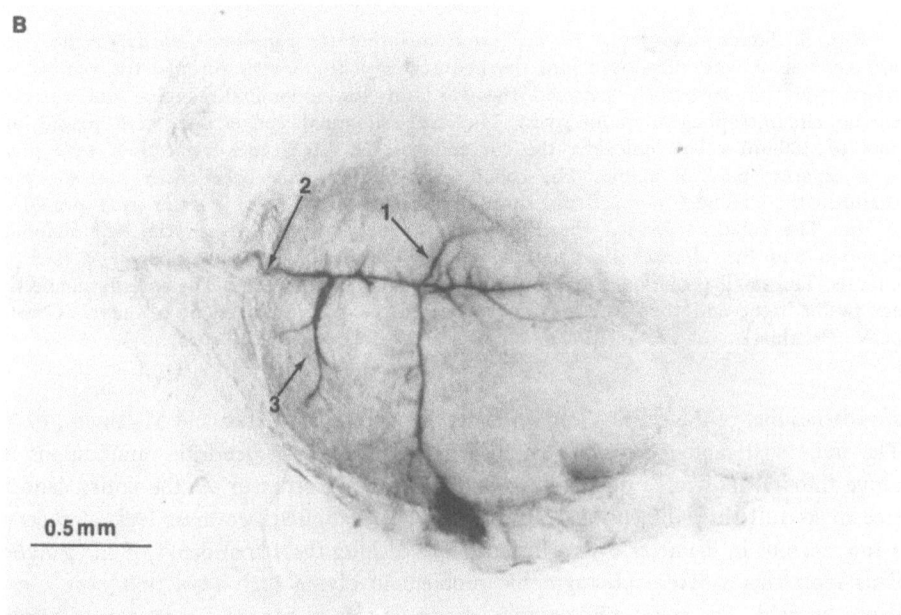

Fig. 6A, B and C. Photographs taken through the compound microscope of the right cell 28 in the metathoracic ganglion stained by intracellular injection of cobalt chloride. Arrows with the same number indicate an identical part of the neuron in different photographs. All pictures are taken from the same preparation. A. Dorsal view of the intact cleared ganglion immersed in creosote. Posterior is to the left. The major branches of the dendritic tree are clearly seen. Arrow 2 indicates the crook where the axon leaves the ganglion through right nerve 5. The cell body and the initial process are out of

and placed on a tilt table. The material to be photographed was placed as close to the axis of rotation of the table as possible. One member of the stereo pair was taken with the table tilted from the horizontal 5° to the left; and the other member taken with the table tilted an identical amount to the right. The photographs were then mounted while viewing them with stereo glasses so that common points in the pair were approximately 70 mm apart. The mounted stereo pair can then viewed with standard stereo glasses.

With some preparations where the neurities and cell body are confined to one side of the ganglion, it is useful to take two sets of stereo pairs; one set in the dorso-ventral axis and the other from the cut medial surface of the ganglion. Fig. 2B is taken from a preparation cut in the median saggital plane and viewed from the cut medial surface looking outward toward the periphery of the ganglion. With this combination of two stereo views taken at a 90° difference in viewing angle, we feel that all the stained branches of the dendritic tree can be seen.

Observations with the Compound Microscope

For greater resolution of the detailed structure of the main dendritic branches, the brightfield compound microscope is used both with whole mount and sectioned material. Fig. 6A shows a dorsal view of the intact ganglion in which cell 28 was stained. The picture is focused on the main dendritic tree near the dorsal surface, and even

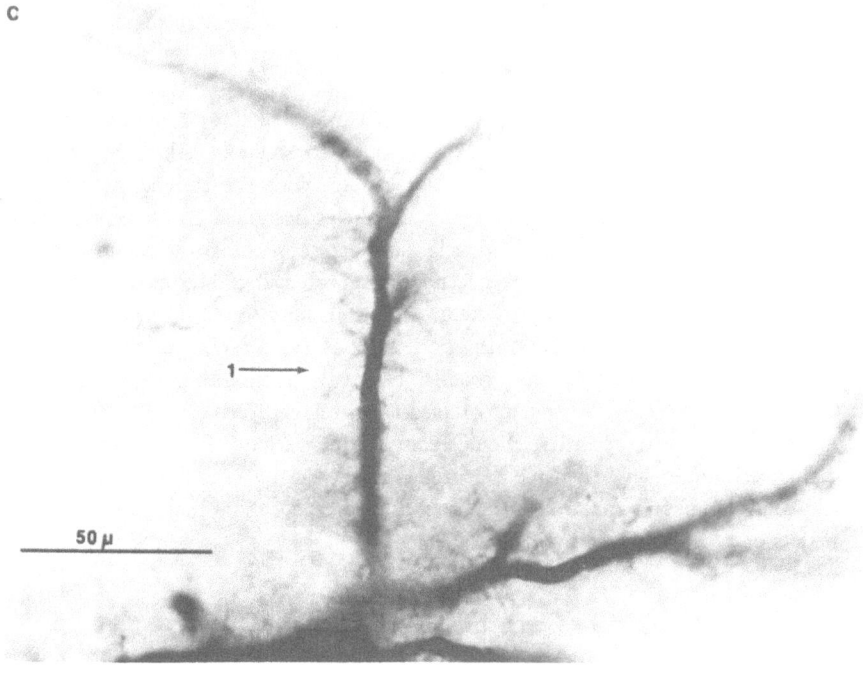

focus at this magnification. B. The preparation was embedded in plastic and 200 μm section cut in the dorso-ventral plane. The cell body is seen at the ventral surface and the axon leaves the ganglion posteriorly through nerve 5. C. Higher magnification view of the dendritic branch indicated by arrow 1 (dendrite opposite the arrow).

at this magnification, fine side branches are seen on the main dendritic trunks. The preparation was then embedded in plastic and a 200 μm section cut by hand as seen in Fig. 6B At still higher magnification in Fig. 6C, extremely fine short branches less than 2 μm in diameter are seen along a main dendritic shaft.

Such a 200 μm section may be re-embedded in plastic and 1 μm sections can then be cut of a precisely identified dendritic branch for observation with light microscopy. Thin sections for electron microscopy can be cut from the same block face and the ultrastructure of a defined segment of dendrite and its surroundings examined.

Electron Microscopy

The electron micrographs in Figs. 7 through 9 are taken from a preparation in which cell 28 was stained with cobalt. All sections were taken in the region of the small branches that come off the initial segment within 200 μm of the soma. These branches are best seen in Fig. 2B. The tissue was immersed in osmic acid for 15 min and was light enough in color to distinguish the darkly stained neurites against the light brown background of the ganglion. The actual process being sectioned for electron microscopy could thus be directly observed in the block face while sectioning.

The cobalt precipitate takes a variety of appearances within the stained cell. It may be uniformly dense within a process as seen in Fig. 7. In some situations it is dispersed and tends to accumulate along the inner surface of the plasma membrane as in Fig. 8. The very small processes of less than 0.5 μm in diameter will sometimes be solid and black, similar to the appearance indicated by Blackstadt (1970) for Golgi stained material viewed in the electron microscope.

The defining characteristics of possible presynaptic structures that may contact the dendrites of cells stained with cobalt are still open to some question. Several profiles in Figs. 7 through 9 labeled (p) make close contact with the stained processes. They have vesicles that appear similar to the synaptic vesicles described for vertebrate material. Clear synaptic densities in the stained cell, such as those labeled (s) in Fig. 9, are not frequently seen. They have not been generally reported in the thoracic and abdominal portions of the arthropod nervous system. In our opinion, the ultrastructural nature of synapses in the regions of the arthropod nervous system outside the brain is still not well defined. Our finding that colchicine greatly increases the electrical activity that can be recorded from the somata of insect central neurons (Pitman et al., 1972b) now makes it possible to record routinely from a wide variety of insect central neurons. Using this technique with identified neurons, it should be possible to examine more precisely central synaptic function in arthropods and to correlate these functional studies with details of synaptic ultrastructure.

General Commentary

Density of Fine Branches

One of the striking features revealed by the cobalt stain is the prevalence of fine short branches on the major neurites within the neuropil. In the light microscope, as seen in Fig. 6, these branches appear to be 0.5 to 1 μm in diameter and 2–5 μm

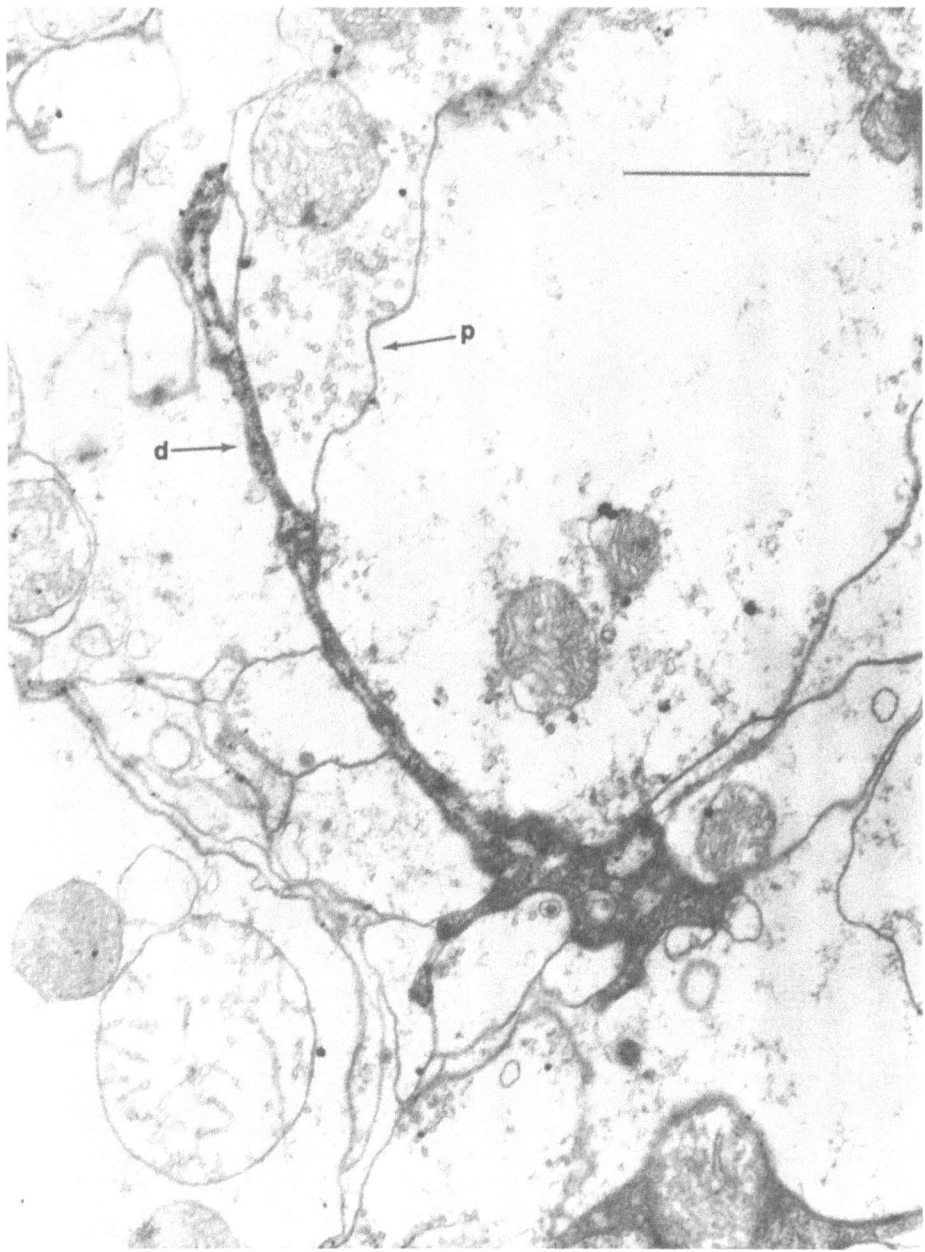

Fig. 7. An electron micrograph through the neuropil of a metathoracic ganglion in which cell 28 was stained by intracellular cobalt injection. The long dark profile (d) is from a branch off the initial process within 100 μm of the cell body. A profile containing clear vesicles (p) makes close contact with the cobalt-stained process and may be a pre-synaptic structure. Calibration: 1 μm.

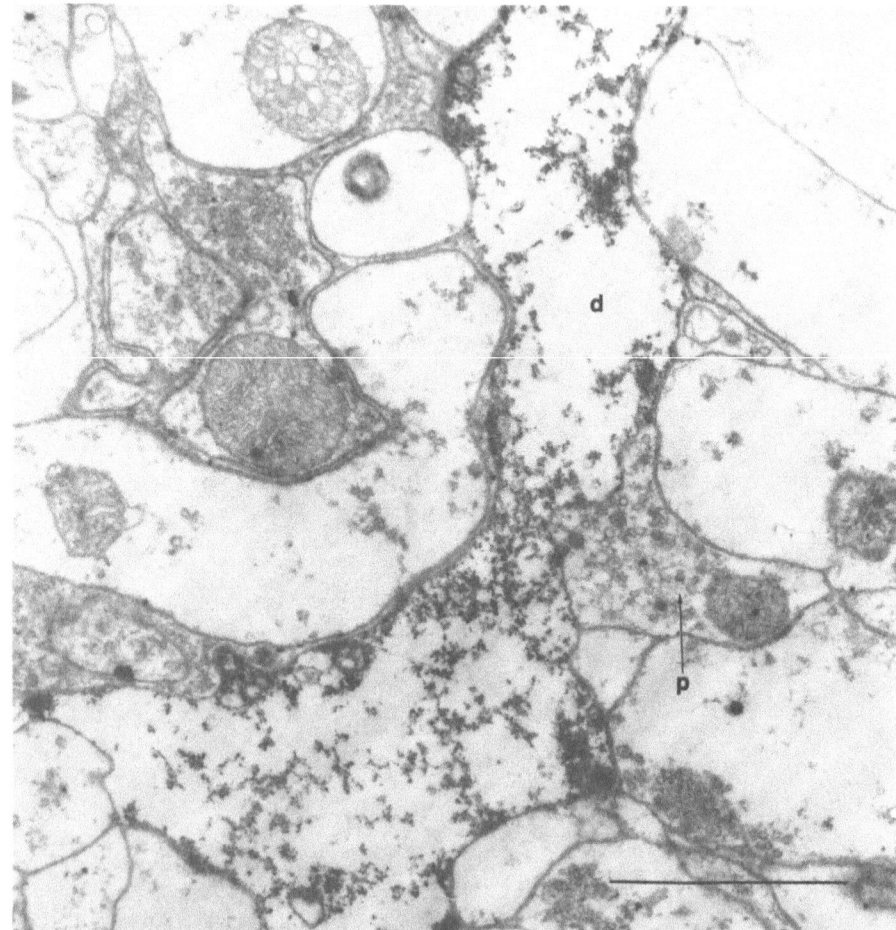

Fig. 8. An electron micrograph from a preparation and region of the ganglion similar to that described in Fig. 7. The long profile (d) shows the most dispersed state of the cobalt precipitate seen with the electron microscope. The structure (p) contains a variety of vesicles and is in close apposition to the cobalt-filled dendrite. It may represent a presynaptic element. Calibration: 1 μm.

long and stand out at right angles to a main dendritic trunk. The major trunks within the neuropil have been previously pictured in Methylene Blue studies as long smooth processes (Zawarzin, 1924). These fine branches have not yet been precisely studied with the electron microscope but they are strongly suggestive of synaptic loci. If this proves to be true, the density of synaptic input onto these motoneurons may be much greater than previously believed.

Branching Near the Soma

The increased ability to visualize fine branches with the cobalt technique has shown that the initial process leaving the soma gives off small branches much closer to the soma than previously thought. Examples of this in cell 28 are seen in Fig. 6, where short branches are present on the initial process within 100 μm of the soma. The massive dendritic "bush" close to the soma of cell 3 is another example (Fig. 3). The delicate brushlike structures along the initial process of the giant neuron shown in Fig. 4 also illustrate this point. These structures are in all likelihood, sites of synaptic activity. They may either receive information or perhaps even transmit signals via a form of dendro-dendritic interaction which has been found to be more prevalent in

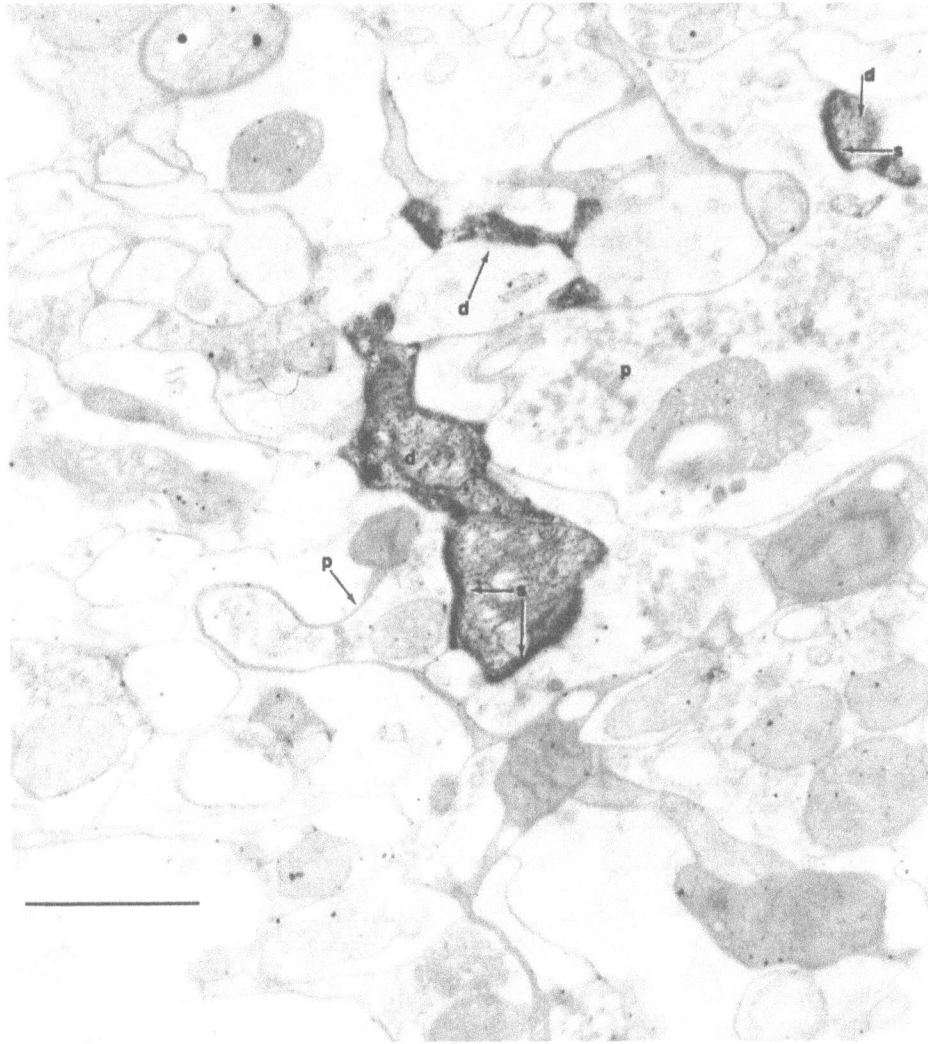

Fig. 9. An electron micrograph from a preparation and region of the ganglion similar to that described in Figs. 7 and 8. The dark profiles (d) are dendritic branches off the initial process within 100 μm of the soma. Structures containing vesicles (p) make close contact with the cobalt-filled profiles. The dense areas (s) along the membrane of the cobalt-stained structures may represent synaptic densities. Calibration: 1 μm.

the vertebrate central nervous system than previously assumed (Shepherd, 1970). The presence of these branches so close to the cell body is interesting in relation to the general difficulty of recording electrical activity from the soma of normal adult insect central neurons (Hoyle, 1970). Recent studies indicate more variability in the electrical responsiveness of the soma and initial process than previously supposed. This includes overshooting spikes in specific dorsal cells (Kerkut et al., 1969), small spike-like activity in some motoneurons (Hoyle, 1970), and experimental alterations in excitability induced by axotomy and the drug colchicine (Pitman et al., 1972b). The increased anatomical and physiological resolution of the properties of the insect soma-dendritic membrane may make this material especially suitable for examining the relationship between electrical activity and metabolism of excitable cells.

Consistency of Branching Pattern

The ability to view the entire dendritic complex of identified neurons has allowed us to examine the dendritic tree of identified cells in great detail. We are impressed by the similarity of the branching pattern from one animal to the next for any given cell thus far examined. This is apparent even with cursory examination of the two views of cell 28 taken from different preparations and shown in Figs. 2A and 6A. We are now in the process of analyzing the three-dimensional geometry of branching patterns using computer techniques common to x-ray crystallography and high energy physics. This should give precise quantitative information about the range of normal dendritic variation and provide a background for studies of factors that influence neuronal branching patterns during development.

Acknowledgments. Supported by PHS Research Grant 5 TO1 NS 08996 to M. J. Cohen, and PHS Postdoctoral Fellowship 2-F02-NS-47, 017-02 to C. D. Tweedle.

References

Blackstad, T. W.: Electronmicroscopy of Golgi preparations for the study of neuronal relations. In: Contemporary Research Methods in Neuroanatomy. Ed. W. J. H. Nauta and S. O. E. Ebbesson. pp. 186–216. New York: Springer-Verlag, 1970.

Cohen, M. J.: A comparison of invertebrate and vertebrate central neurons. In: The Neurosciences: Second Study Program. Ed. F. O. Schmitt, pp. 798–812. New York: Rockefeller University Press, 1970.

——, and J. W. Jacklet: The functional organization of motor neurons in an insect ganglion. Phil. Trans. R. Soc., Lond. B *252*, 561–572 (1967).

Hoyle, G.: Cellular mechanisms underlying behavior. In: Advances in Insect Physiology. Ed. J. W. L. Beament, J. E. Treherne and V. B. Wigglesworth. pp. 349–444. London: Academic Press, 1970.

Hughes, G. M., and L. Tauc: Aspects of the organization of central nervous pathways in *Aplysia depilans.* J. exp. Biol. *39*, 45–69 (1962).

Iles, J. F., and B. Mulloney: Procion yellow staining of cockroach motor neurones without the use of microelectrodes. Brain Res. *30*, 397–400 (1971).

Karnovsky, M. J.: A formaldehyde-glutaraldehyde fixation of high osmolality for use in electron microscopy. J. Cell. Biol. *27*, 137A (1965).

Kerkut, G. A., R. M. Pitman, and R. J. Walker: Iontophoretic application of acetylcholine and GABA onto insect central neurones. Comp. Biochem. Physiol. *31*, 611–633 (1969).

Nicholls, J. G. and D. A. Baylor: Specific modalities and receptive fields of sensory neurons in CNS of the leech. J. Neurophysiol. *31*, 740–756 (1968).

Otsuka, M., E. A. Kravitz, and D. D. Potter: Physiological and chemical architecture of a lobster ganglion with particular reference to gamma-aminobutyrate and glutamate. J. Neurophysiol. *30*, 725–752 (1967).

Pitman, R. M., C. D. Tweedle, and M. J. Cohen: Branching of central neurons: intracellular cobalt injection for light and electron microscopy. Science *176*, 412–414 (1972a).

————— ————— —————: Electrical responses of insect central neurons: augmentation by nerve section or colchicine. Science *178*, 507–509 (1972b).

Shepherd, G. M.: The olfactory bulb as a simple cortical system: experimental analysis and functional implications. In: The Neurosciences: Second Study Program. Ed. F. O. Schmitt, pp. 539–552. New York: The Rockefeller University Press, 1970.

Spurr, A. R.: A low-viscosity epoxy resin embedding medium for electron microscopy. J. Ultrastruct. Res. *26*, 31–43 (1969).

Stretton, A. O. W., and E. A. Kravitz: Neuronal geometry: determination with a technique of intracellular dye injection. Science *162*, 132–134 (1968).

Trujillo-Cenoz, O., and J. Melamed: Light and electronmicroscope study of one of the systems of centrifugal fibers found in the lamina of muscoid flies. Z. Zellforsch. mikrosk. Anat. *110*, 336–349 (1970).

Zawarzin, A.: Zur Morphologie der Nervenzentren. Das Bauchmark der Insekten. Ein Beitrag zur vergleichenden Histologie (Histologische Studien über Insekten VI). Z. wiss. Zool. *122*, 323–424 (1924).

Discussion

DR. KATER (Iowa): In my experience there has been a major difference between cobalt and Procion yellow.

Procion yellow, when it fills, does so very uniformly while cobalt results in clumping, that is, cobalt in some areas and not in others.

I wonder whether you have observed this clumping and whether it has hindered you in your electron microscopic studies?

DR. TWEEDLE (Michigan State): There are no extensive spaces where there is no cobalt; there may be variations in the density of the cobalt precipitate.

DR. KATER: And this offers no problem in your analysis? You couldn't cut through a neurite and fail to see enough cobalt to be sure it was the filled neurite?

DR. TWEEDLE: If you have a well-filled cell there is no problem. If the stain is diffuse within the cell, you may find a branch with a light cobalt precipitate arising from a more darkly stained branch. It is my impression that there is only rarely not enough cobalt precipitate to identify branches of the injected cell.

CHAPTER 7

Microelectrode Injection, Axonal Iontophoresis, and the Structure of Neurons

Brian Mulloney

This paper is a critique of two methods of filling identified neurons: microelectrode injection and axonal iontophoresis. Fig. 1 illustrates the differences between these two methods. Since most of the work reviewed in this book has been done with microelectrode injection, I am restricting my comments on this approach to generalities and concentrating on several applications of axonal iontophoresis. These demonstrate its limitations and advantages and introduce the methodological variations which have achieved the best results in specific systems. The microelectrode injection approach to filling neurons has some limitations. First, you must be able to penetrate and hold a physiologically-identified cell long enough to fill it. Iontophoresis may require an hour or more, and to hold a small cell this long often demands technical heroics (however, see Chapter 15 for a description of the "one-shot" technique). Second, each neuron must be filled individually. For most of us, that means one at a time. This may not be a problem, since the structure of a particular cell or the points of contact between two cells is frequently the information sought. But when one wants information about the structure and distribution of groups of cells, it can be very tedious to get using micropipettes. The alternative of filling the neurons by iontophoresing stain into their cut axons (Iles and Mulloney, 1971) can provide a solution in some cases to each of these difficulties. I shall limit my discussion to invertebrate preparations, since Chapter 15 and the discussion following it detail axonal iontophoresis in vertebrates.

Applications of Axonal Iontophoresis

Cockroach Leg Motoneurons

J. Iles first used axonal iontophoresis to fill neurons with Procion yellow (Iles and Mulloney, 1971). His apparatus consisted of two rings of petroleum jelly on a microscope slide. One ring contained a pool of saline and the other a solution of Procion yellow. A battery, a current-limiting resistor, and two silver wires for electrodes formed the current-generating circuit (Fig. 2A). Since he was interested in the control of walking in insects, he used this new method to discover the structure of the motoneurons which innervate the levator and depressor muscles of the metathoracic coxa of

Brian Mulloney

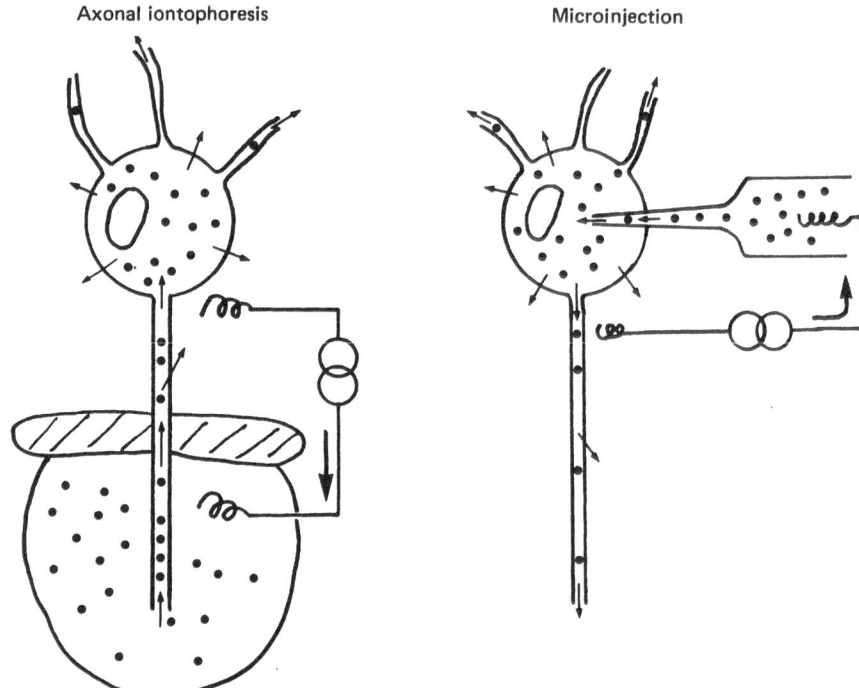

Fig. 1. This figure shows the current paths in the two different methods of filling neurons with stain. In microelectrode injection, a micropipette filled with a charged stain (dots) carries current into cell. The stain is then left in the cell as other ionic species move the current across the cell membrane (arrows). In axonal iontophoresis, the open end of a cut axon is insulated electrically from the rest of the neuron by some high resistance barrier. When a voltage drop is imposed across the insulating barrier, current flows in the axon and across the cell membrane (arrows). The current is carried into the open end of the axon by the ionic stain (dots), which fills the cell and is left inside as other ionic species carry the current across the cell membrane.

Periplaneta americana. These motoneurons are located in the metathoracic ganglion and had already been mapped by Cohen and Jacklet (1967).

The axons of the elevator motoneurons run in nerve 6br4, and the depressor neurons in nerve 5rl. Nerve 6br4 contains 12 motor axons, ranging in diameter from 3 to 25 μm. Iontophoresis of dye into 6br4, revealed two clusters of somata, one of which corresponded to a group of cells described by Cohen and Jacklet (1967). It required several preparations to fill all 12 neurons.

Iontophoresis of dye into nerve 5rl filled two cell bodies. This nerve contains five large axons which can be distinguished physiologically and anatomically. The axon of the fast coxal depressor neuron (D_f) is about 25 μm in diameter, and that of the slow coxal depressor motoneuron (D_s) is about 15 μm. The other three axons in nerve 5rl are axons of three common inhibitor neurons. These axons are each about 5 μm in diameter, and branch repeatedly from nerve 5 to innervate other muscles as well as the coxal depressors. Since these other branches are cut, they short circuit the iontophoretic current in these axons, and so the neurons do not fill with dye within

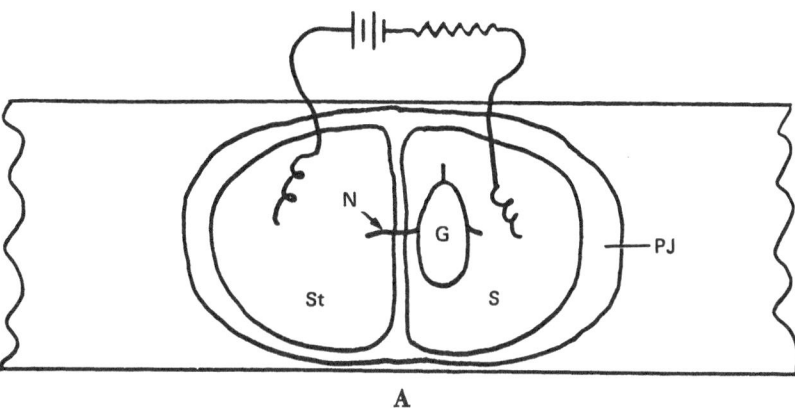

A

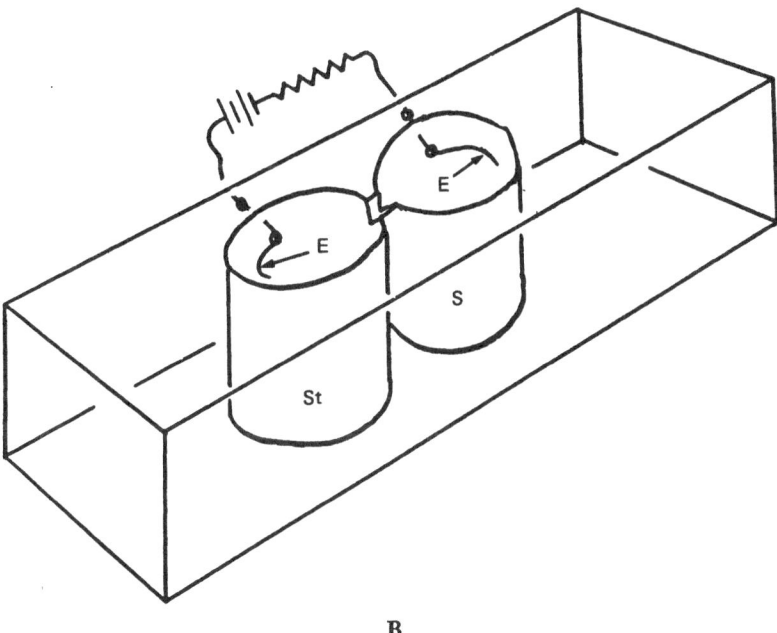

B

Fig. 2A and B. A shows the simplest arrangement for iontophoresing stain into cut axons. Two rings of petroleum jelly (PJ) on a microscope slide separate stain (St) and physiological saline (S) solutions. The ganglion (G) is in the saline. The nerve (N) to be filled spans the petroleum jelly barrier. Current generated by the battery flows in the nerve and moves the stain. B is a perspective diagram of a container for axonal iontophoresis (Iles and Mulloney, 1971) based on the system shown in A. Two wells are drilled in a lucite block and a notch cut to connect the two. Electrodes (E) enter each well through holes in the lucite. The electrodes are glued in place. One well is filled with saline (S), the other with stain (St). The nerve is led through the notch connecting the two wells, and sealed in place with petroleum jelly. The wells are covered with a coverslip during iontophoresis to control evaporation.

the time allowed to fill the two excitatory neurons. Iles identified D_f and D_s by tracing the neurite of each neuron from its soma to the root through which its axon left the ganglion. Since he knew the relative sizes of the axons in the periphery, he could identify the cells by the relative diameter of the filled axon. He reconstructed the injected cells from serial paraffin sections using a graphic reconstruction method (Pantin, 1960; Pusey, 1939) which is both accurate and comparatively fast. This method allows those of us who are inept artists to reconstruct the detailed processes of individual cells using only a compound microscope.

Iles (1972) used the new information about the structure of D_f to record from this neuron in the neuropil, and to show that D_f receives synaptic input from the smaller fibers which run from the abdomen into the thoracic ganglia but not from the largest giant fibers.

Several points are worth noting about Iles' work. First, his initial apparatus was very simple and worked well. The special chamber for iontophoresis (Iles and Mulloney, 1971; Fig. 2B) was developed so that preparations could be moved about, refrigerated, and covered to control evaporation, not because the initial system did not work. Second, it took many trials to get an optimum current density and time for iontophoresis. Third, migration of dye in the extracellular space inside the nerve sheath was a problem if iontophoresis continued too long. This unwanted dye formed a cloud of fluorescence in the ganglion and obscured the processes of the neurons. Fourth, not all the axons in a nerve filled each time the nerve was iontophoresed. Different axons seal each time the nerve is cut, and only axons with open ends will fill by iontophoresis. Fifth, the identification of D_f and D_s was based on preexisting knowledge of the system. If the axonal diameters of the two had not been conveniently different, the identification would have been more difficult, and impossible using anatomical information alone. Sixth, to get the structural information about D_f and D_s would have been very difficult by conventional methods. Procion staining with microelectrode injection would have required finding the soma of each cell, identifying it, and then filling it.

Embryonic Cricket Neurons

Field crickets have a set of songs (stridulation) which they produce by scraping one of their forewings over a rasp-like series of ridges on the other forewing (Huber, 1962). The muscles which move the forewings during stridulation are the same ones which move them during flight. These muscles are innervated by a small group of motoneurons, located in the thoracic ganglia, which can fire in the stereotyped patterns needed to generate either flight or stridulation. A map of the somata of all the meso-thoracic flight motoneurons, and some of their major processes has been published for a similar insect, the locust *Schistocerca gregaria* (Bentley, 1970). Crickets develop hemimetabolously, that is, they go through a series of 9–11 larval stages (instars) each of which is larger than its predecessor and more nearly resembles the adult. All the flight motoneurons of the adult are present in the thoracic ganglia at hatching (Gymer and Edwards, 1967), but neither flight nor stridulation occurs until the emergence of the adult cricket at the final moult.

Bentley and Hoy (1970) recorded indirectly the activity patterns of the motoneurons which innervate the subalar and remotor muscles of the mesothoracic wings during induced stridulation in *Teleogryllus commodus*. They found that the full adult motor pattern was present in the last larval instar. Furthermore, they were able to elicit elements of the adult flight motor pattern in instars which had as many as four moults to complete before reaching adulthood. As the crickets matured, the flight patterns they generated included more neurons, became more perfect in their temporal structure, and lasted for longer periods. Several alternative mechanisms might produce these results: one of the most interesting would be the sequential development of synaptic connections between the motoneurons at each moult. Bentley (University of California, Berkeley, personal communication) has recently begun to examine this possibility by iontophoresing Co^{++} into the cut axons of selected flight motoneurons to see if the neuropilar branches of these neurons in the adult are present in the earlier larval stages. Using an ingenious adaptation of the standard axonal iontophoresis system, he has filled motoneurons which innervate muscles 81 (the mesothoracic dorsolongitudinal muscles) and 112 (the metathoracic dorsolongitudinal muscles), the large stretch receptor axon which runs in the same nerve as the motor axons, and a previously unknown neuron which sends its axon bilaterally out to muscles 112 in the larval instars and disappears at the adult moult. The central question about the possible ordered sequence of formation of synaptic connections is not answerable yet, but a few points can be made (Fig. 3). First, the, map of the somata of the flight motoneurons in the adult thoracic ganglia of *T. commodus* are very similar to the published map of these somata in *S. gregaria*. This encourages the hope that the structure of some identifiable neurons will be generalized in related groups of animals. Second, the structure of dorsolongitudinal motoneurons is complete in the last larval instar, at a time when the adult motor pattern can be elicited in the subalar and remotor neurons. Third, the somata of these neurons, which are tightly clustered in the adult ganglia, are much more loosely grouped in the early larval stages. Fourth, many of the major branches of each neuron which can be seen in the adult are already present in the third instar.

Bentley's method of filling the neurons requires that the thorax of the cricket be opened dorsally and the nerve containing the desired axons be dissected off the surface of the muscle they innervate. Then he removes the unwanted muscles and gut, builds a bowl of petroleum jelly on top of the thoracic ganglia, and leads the nerve to be filled through the bottom of the petroleum jelly bowl. All four of the nerves innervating the meso- and metathoracic dorsolongitudinal muscles can be filled simultaneously with this method. He first fills the bowl with distilled water for 3 min to open the cut ends of the axons, then with a 250 mM $CoCl_2$ solution in distilled water and passes current between the Co^{++} pool and the cricket's haemolymph. The appropriate currents decrease with the size of the animal; for adults he uses 10^{-7} A for several hr, and for very small larvae, 10^{-11} A for similar periods. After iontophoresis is finished, the Co^{++} is precipitated as CoS (Pitman et al., 1972) with $(NH_4)_2S$; the animal is fixed in Carnoy's fixative for 15 min and dehydrated in ethanol; the thoracic ganglia are dissected out, cleared, and stored in methyl benzoate, yielding good whole mounts. During the whole iontophoresis, the cricket is alive and ventilating, and some of the tracheal system is still intact, so the thoracic ganglia remain in good condition.

Bentley has filled motoneurons in third instar crickets (3 to 4 mm long); their thoracic ganglia are about 100 μm in diameter. It should be possible using this approach to fill flight motoneurons in flies and other small but well-studied insects.

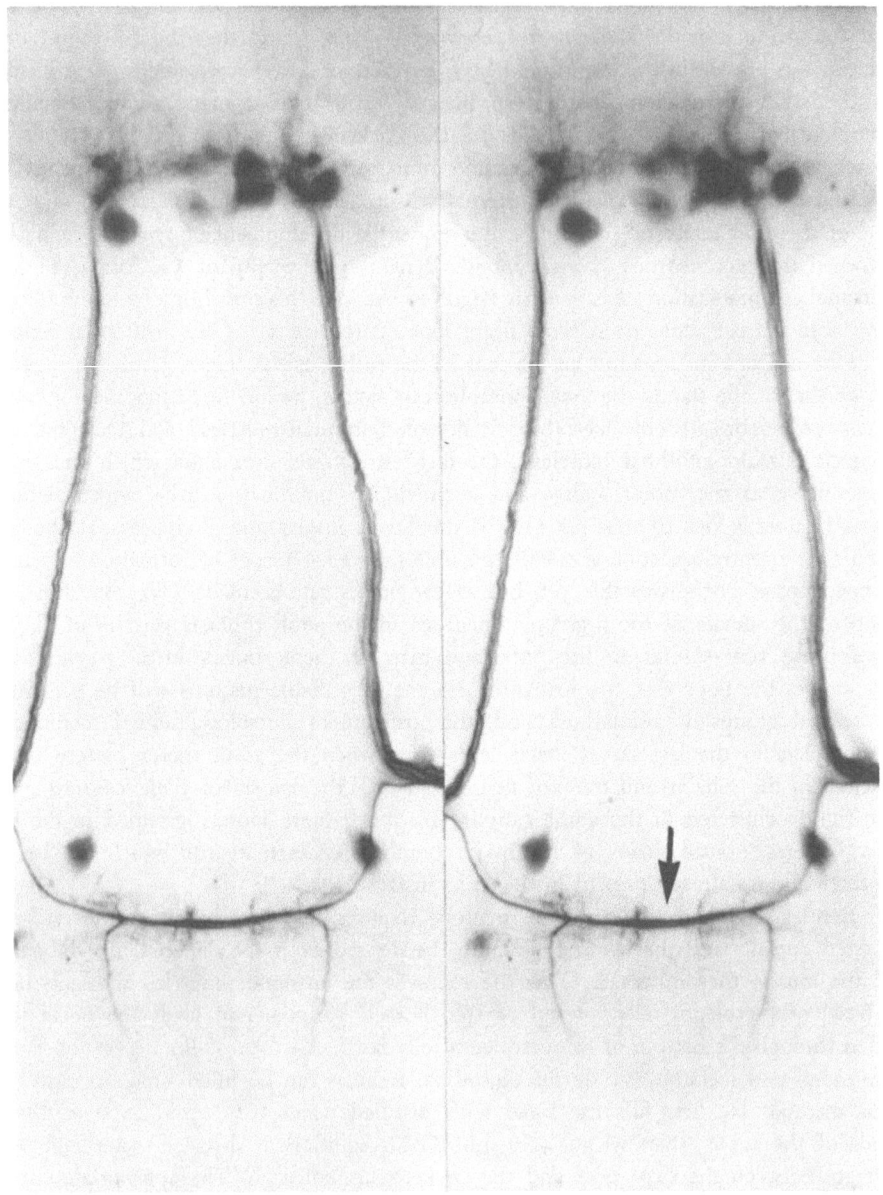

Fig. 3. Whole mount of the meso- and metathoracic ganglia of an adult male polynesian field cricket, *Teleogryllus oceanicus*. The mesothoracic ganglion is at the top of the figure, and is connected to the metathoracic ganglion by the paired connectives of the ventral nerve cord. The motoneurons innervating muscles 112 (the metathoracic dorsolongitudinal muscles) were filled with Co^{++} by axonal iontophoresis through the recurrent nerve on each side of the ganglion. The four somata in each hemisphere of the mesothoracic ganglion are ipsilateral to the muscle they innervate. The somata in the metathoracic ganglion are contralateral to the muscles they innervate. The neurites of these two metathoracic neurons approach each other closely (arrow) as they cross the midline of the ganglion. The branching structures of each of these neurons is ipsilateral to the muscle it innervates. Each ganglion is about 1 mm wide. Histological procedures are discussed in the text. This is a stereo pair. (Photo courtesy Bentley, unpublished).

The Descending Contralateral Movement Detector

The descending contralateral movement detector (DCMD) is an interneuron found in at least three superfamilies of Orthopteran insects: The Acridoidea, Grylloidea, and Tettigonioidea. It receives visual input from the contralateral eye and sends its axon down the ventral nerve cord from the tritocerebrum through the subesophageal, prothoracic, and mesothoracic ganglia to end in the metathoracic ganglion. The neuron responds to the movement of small objects in the visual field but does not respond when the animal moves or to any stimulus which affects all the visual field simultaneously. Although the DCMD has been studied for years, its function is not known (Rowell, 1971a,b).

M. O'Shea and C. H. F. Rowell (University of California, Berkeley, personal communication) have succeeded in filling the DCMD in the sub- and supraesophageal ganglia and in each of the thoracic ganglia. They filled the cell with Co^{++} by axonal iontophoresis through the cut connective. The cell body of the DCMD is located in the dorsal protocerebrum on the side contralateral to the axon (Fig. 4). A neurite

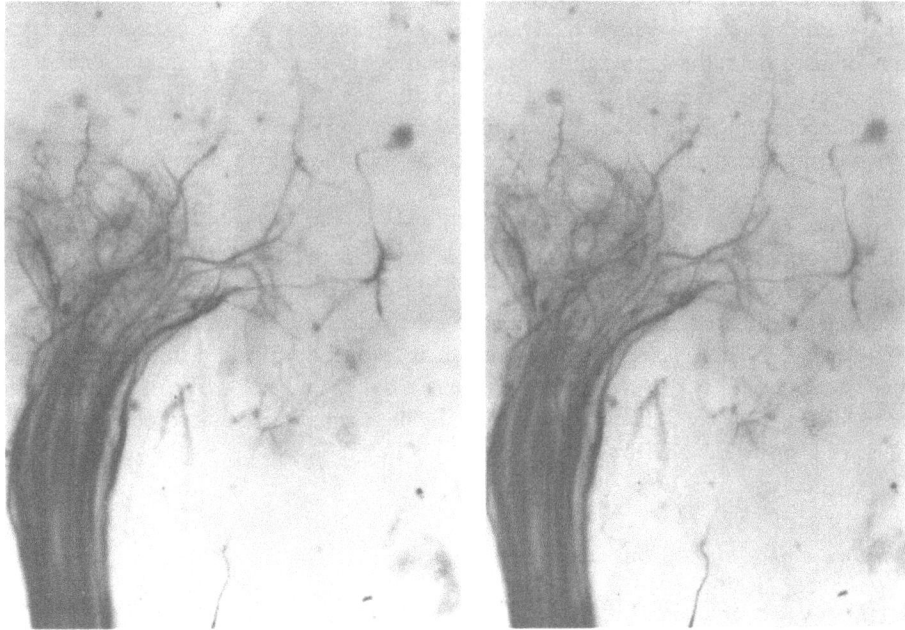

Fig. 4. Posterior view of the central area of the supraesophageal ganglion of the grasshopper *Schistocerca vaga* Scudder. Axonal iontophoresis of Co^{++} from the left hand side cervical connective. Most of the filled axons terminate at dendrites or cell bodies on the ipsilateral side of the brain, and are mainly out of the plane of focus of this picture. The large neuron which crosses over the midline of the brain to a thickened integrating segment, bearing dendrites and a neurite running to a large dorsal and posterior soma, is the descending contralateral movement detector (DCMD), a well known visual interneuron which runs from the protocerebrum to the metathorax. This is a stereo pair and should be viewed with a viewer set for an interocular distance of approximately 65 mm. (Photo courtesy O'Shea and Rowell, unpublished).

descends from the soma to an integrating segment, which has, among others, a branch leading towards but not reaching the optic lobe. The neurite crosses the midline and widens to form the axon which descends the contralateral connective. The axon passes through the subesophageal ganglion without branching, through the prothoracic and mesothoracic ganglia with limited branching in each, but branches extensively in the metathoracic ganglion. The seeming suitability of the DCMD for a warning function has led Rowell to suggest that it may alter the sensitivity of thoracic motor systems to cerebral input. The complex branching of the DCMD in the metathoracic ganglion is consistent with this hypothesis, because the metathoracic ganglion contains most of the motoneurons involved in jumping. The DCMD may also act as a synergist in the activation of the jump.

The methods developed by O'Shea and Rowell are worth considering further. The DCMD is among the largest axons (12–15 μm) in the connectives between the ganglia, so stain would be expected to migrate rapidly in it. But each connective contains several hundred axons and considerable extracellular space. Therefore, to overcome the migration of stain in the extracellular space and in unwanted smaller axons, they began the iontophoresis one ganglion away from the ganglion in which they wanted to examine the DCMD. Using 10^{-6} to 10^{-7} A for several hours, only the largest axons fill through to the supraesophageal ganglion. These are clearly visible because all unwanted stain has diffused into the extracellular space of the subesophageal ganglion. O'Shea and Rowell place the cut end of the connective in distilled water for 3 min before beginning iontophoresis and find that the axons fill much more reliably after this treatment. Apparently the sealed ends of some of the axons pop open again because of the osmotic stress.

Earlier attempts to fill the DCMD with Procion yellow by injection into the axon did not work very well because the Procion migrated too slowly. The greater mobility of Co^{++} allowed O'Shea and Rowell to fill these neurons in each of the five ganglia within a short period of time. Once they knew the location of the soma of the DCMD, Rowell and O'Shea were able to penetrate it with Co^{++}-filled microelectrodes placed stereotactically. They have recorded synaptic activity and spikes from the soma, and have filled the cell by iontophoretic injection. The anatomical results so obtained confirm those of axonal iontophoresis.

Sensory and Motoneurons in Flies

Flies detect food in part by sensory hairs on their mouthparts. If a fly is starved long enough, stimulation of a single hair with a suitable food substance will elicit a feeding response from the animals (Getting, 1971). Little is known about the central processing of the sensory information or about the central interactions of sensory and motoneurons. These neurons are located in the subesophageal ganglion, which contains thousands of neurons, all very small. Recently, M. Nelson (Tufts University, personal communication) has begun to fill the axons and central ramifications of the primary sensory neurons in the labial nerve with Procion yellow. S. Fredman (University of California, Berkeley, personal communication) has filled the three motoneurons which innervate the extensor muscles of the haustellum in Calliphorids, using the same method as O'Shea and Rowell. The three somata are very small and are located near the base of the labial nerve.

Fredman identified the neurons by cutting their axons at the surface of the extensor muscles, so that only these axons were filled. This procedure also reduced the problem of migration of Co^{++} in the extracellular space. This work is at too early a stage to be able to test any of the hypothetical central mechanisms (Getting, 1971) against the anatomical evidence.

The Retrocerebral Neuroendocrine Complex of Locusts

The retrocerebral neuroendocrine complex consists of several clusters of neurosecretory cells in the supraesophageal ganglion. Their axons leave the ganglion via two pairs of nerves to terminate in a neurohemal organ called the corpus cardiacum and an endocrine gland called the corpus allatum. These neurosecretory cells have been studied extensively because they are involved in the control of growth and moulting in insects. Attempts to describe their structure using standard histological methods have not been very successful, and their branching structure within the supraesophageal ganglion is largely unknown (Adiyodi and Bern, 1968).

C. A. Mason (University of California, Berkeley, personal communication) has filled three sets of neurosecretory cells in the locust *Schistocerca vaga* by iontophoresing Co^{++} in the nervi coporis cardiaci I and II (NCC I and II) and the nervi corporis allati (NCA I) using the methods of O'Shea and Rowell. The NCC I leaves the posterior side of the supraesophageal ganglion and ends in the corpus cardiacum. It contains the axons of the median group of neurosecretory cells. Iontophoresis of Co^{++} into NCC I filled two groups of neurons: the medial neurosecretory cells and a new group. The medial neurosecretory cells occur in a pyramidal cluster in the protocerebrum. The new group consists of about 10 cells of unknown function in the tritocerebrum.

The NCC II leaves the supraesophageal ganglion just dorsal to NCC I and contains the axons of the lateral group of neurosecretory cells which terminate in the corpus cardiacum. Iontophoresing stain into the cut NCC II fills the lateral neurosecretory cells. The somata of these cells form a curving stripe on the anterior margin of the supraesophageal ganglion. Their axons decussate without branching and exit via NCC II. Since axonal iontophoresis of Co^{++} demonstrated very fine branches in the new group of cells in the tritocerebrum, the apparent absence of branches from the neurities and axons of the lateral neurosecretory cells is probably correct.

The NCA I connects the corpora allata and the corpus cardiacum. The lateral neurosecretory cells were thought to innervate the corpora allata (Strong, 1965), as well as the corpus cardiacum. Iontophoresing stain toward the corpus cardiacum and the supraesophageal ganglion fills some but not all the lateral neurosecretory cells. Other neurons in the supraesophageal ganglion and the corpus cardiacum fill via NCA I, so the innervation of this gland appears to be complex.

Oculomotor Neurons in Crabs

A. Steinacker (Smith Kettewell Institute, personal communication) has filled the eyestalk withdrawal motoneurons and optokinetic motoneurons in the crab, *Callinectes sapidus,* as part of an intracellular analysis of the oculomotor system. She used a piece of micropipette glass with a bowl blown in the tip to form a pool of Procion dye solution. Using a micromanipulator to position the bowl, she placed the cut end of the nerve in the bowl, and sealed it in place with silicone stopcock grease. This allowed her to bring the dye closer to the ganglion than would be possible with the vaseline bridge or plexiglas dish methods.

The nerve in which the optokinetic and withdrawal motoneurons run contains many sensory axons as well. These sensory axons are much smaller in diameter than the motor axons. Steinacker controlled the migration of dye in the extracellular space of the nerve and in the unwanted sensory axons by simply cutting the nerve again at the boundary of dye in the extracellular space once iontophoresis was complete. In this system, dye migrated faster and farther in the motoneurons than in the sensory axons and extracellular space, so cutting off the distal part of the nerve left dye only in the motoneurons. Using these methods, Steinacker has filled the axons and some branches of these neurons. The somata of these neurons did not fill.

Slow Flexor Neurons in Crayfish

The slow flexor motoneurons of *Procambarus clarkii* are part of the system controlling abdominal posture in these crayfish (reviewed by Kennedy et al., 1966). There are six bilateral pairs of these neurons in each abdominal ganglion. A member of each pair innervates the slow flexor muscle on each side of the abdomen. The central and reflexive connections of these neurons have been widely studied using extracellular recording methods (Page and Sokolove, 1972). However, they have not been amenable to intracellular work because their axons and central processes are relatively small and their somata are located beneath the outermost layer of cell bodies and so are not visible. The axons of these neurons are too small to use the axonal microelectrode injection method (Selverston and Remler, 1972), and their cell bodies are electrically silent.

Jay Mittenthal (Stanford University, personal communication) has filled some of the neuropilar processes of these cells using axonal iontophoresis of Procion yellow. He first tried to fill the cells using the petroleum jelly dam method and found that dye moved faster in the extracellular space than it did in the axons, and that even intense currents and very long iontophoresis would not fill the axons as far as the ganglion. Mittenthal then developed an oil bath method where the whole excised ganglion was immersed in light oil. A large suction electrode filled with Procion solution contacted the cut end of the nerve. He completed the circuit by contacting the opposite side of the ganglion with a large suction electrode filled with saline. The oil bath severely restricted the extracellular space around the axons so that migration of dye inside the sheath was reduced. Several slow flexor neurons filled well into the ganglion, past several branch points. The somata of these neurons filled only after prolonged iontophoresis.

Gastric Mill Neurons in Lobsters

Each of the four gastric mill (GM) neurons of the stomatogastric ganglion of *Panulirus interruptus* has one spike initiating zone, but the axon which leaves this zone then branches four times to innervate three bilaterally paired sets of muscles at opposite ends of the stomach (D. Hartline, University of California, San Diego, personal communication). These axons leave the ganglion by three different nerves. The synaptic and electrotonic connections which these neurons make are known (Selverston and Mulloney, 1973), and their structure is known qualitatively from iontophoretic injection of Procion yellow into their somata (Selverston and Mulloney, University of California, San Diego, personal communication). Because of their complex structure, these neurons seemed an ideal system in which to test the equivalence of the results achieved by axonal iontophoresis and microelectrode injection of Procion yellow and Co++ in the same cells.

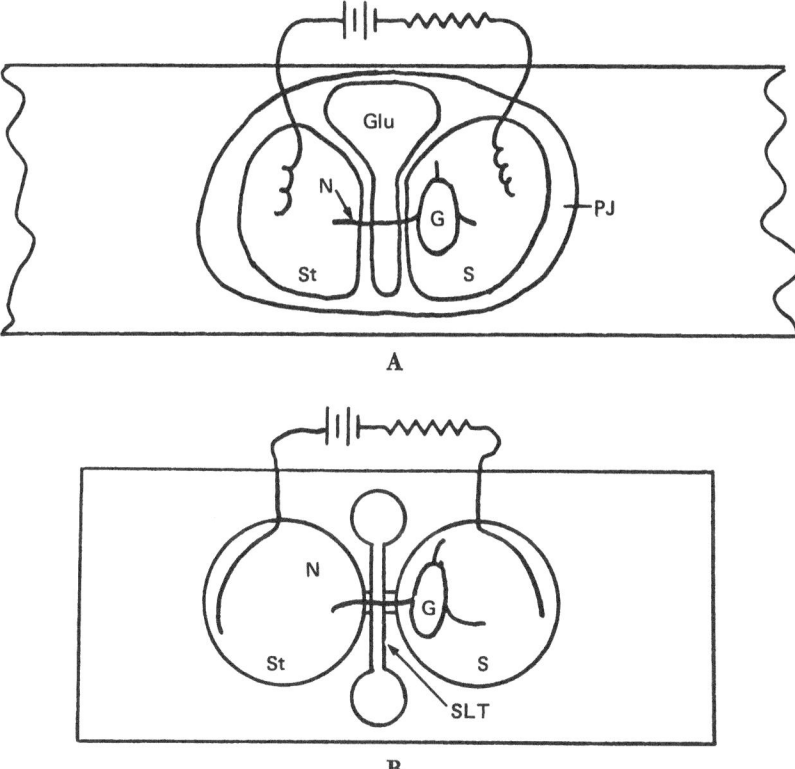

Fig. 5A and B. A shows a simple arrangement for iontophoresing stain into a nerve across a glucose gap. Three enclosures of petroleum jelly (PJ) on a glass microscope slide separate the stain (St), glucose (Glu), and physiological saline (S) solutions. The nerve (N) to be filled spans the two barriers of petroleum jelly. Current generated by the battery then moves preferentially in axons, because the sucrose solution reduces the conductivity of the extracellular fluids. B is a top view of a container for axonal iontophoresis which creates a glucose gap along the nerve. Four wells are drilled in a lucite block. The two smallest are connected by a slot (SLT) which is as deep as the wells themselves. Notches connect the three compartments. Electrodes enter the stain (St) and saline (S) wells through holes in the lucite. The nerve (N) spans the sucrose gap through the notches and is sealed in place with petroleum jelly. The wells are covered during iontophoresis.

The first attempts to fill these neurons with Co^{++} or Procion failed because migration of stain in the extracellular space inside the sheath obscured the neural processes and stain moved preferentially in the axons. Current flowed through the ganglion out the other axons and not into the soma or neuropilar processes. Two additions to the system solved these problems. First, a barrier of slightly hypotonic glucose solution along the nerve blocked extracellular movement of stain. This barrier was formed by putting a second petroleum jelly dam parallel to the first one which separated the pools of saline and stain (Fig. 5A). This gap was filled with glucose solution, and the nerve to be iontophoresed was led across the two dams to connect the stain and saline pools. The glucose solution removed the electrolytes in the extracellular fluid of a section of nerve. This greatly increased the extracellular resistance and eliminated the unwanted stain migration. Second, each of the nerves which contained axons of the GM neurons and which were not being iontophoresed were ligated with fibers of surgical silk. This eliminated the paths of low resistance out of each GM cell. When these two modifications were used, both the neuropilar branches and the somata of the GM neurons filled well. The glucose gap method should be especially useful for molluscan systems.

A variety of configurations have been used to fill molluscan and crustacean neurons by axonal iontophoresis because their axons are more loosely packed inside the nerve sheath than are those of insects. Since there is relatively more extracellular space in these nerves, the conductivity of this space is greater than in insects. Therefore, stain moves faster and farther in the extracellular space. Chapter 10 of this volume discusses the application of the axonal iontophoretic technique to molluscan preparations.

Photographing Whole Mounts

Many of the neurons discussed in this review have been studied in whole mounts—cleared but unsectioned blocks of tissue, often whole ganglia. Producing quality photographs of these neurons can be difficult because they extend beyond the plane of focus of most conventional photo-micrographic systems. The problem is usually depth of field, not magnification. Larger apertures and greater magnifications cause a decreased depth of field. Most compound microscope lenses, and almost all stereo-microscopes have no provision for adjusting the aperture of the lens.

The best way to photograph whole mounts is with a large format camera with a bellows extension, and a good quality photo–micrographic lens. With such a system, one gets negatives which can be contact-printed. The magnification can be controlled continuously through a wide range with the bellows, and the depth of field can be controlled with the aperture diaphragm in the lens (Blaker, 1965, pp. 38–44). The useful magnification of this system is limited to about $60\times$. The next best system is a small format camera with a bellows and a photo-micrographic lens. It requires two separate negatives to produce a stereo-pair. With a large format camera, one can make a stereo-pair in one negative by making a double exposure with the specimen first on the left and then on the right of the field of view (Rowell, University of California, Berkeley, personal communication). A third alternative is to get the magnification of a bellows extension by replacing the objective lens of one channel of a stereo-microscope with a photo-micrographic lens. These systems do not produce photographs of the quality of the bellows systems because the reduced aperture necessary to get good depth of field affects the resolution of the final photograph even more adversely than in the bellows system.

Conclusions

These examples of the application of axonal iontophoresis show both its utility and limitations. Having criticised microelectrode injection methods, this paper would not be complete without a similar critique of axonal iontophoresis. First, migration of stain in the extracellular space within a nerve can be a severe problem requiring oil bath or glucose gap methods to control it. Second, the identity of the neurons filled can be difficult or impossible to determine from the anatomical evidence alone. Third, it may take prolonged iontophoresis to fill complete neurons, particularly the somata of monopolar neurons. This seems to be a problem particularly in some crustacean nervous systems. Fourth, it is difficult to fill reproducibly particular individual neurons when their axons are part of a nerve containing many axons. Five, axonal iontophoresis should drive stain selectively across low-resistance electrotonic junctions between cells and so complicate the study of individual cells. I have seen no conclusive evidence that this occurs, but Procion dyes do cross some electrotonic junctions (Payton et al., 1969; Mulloney, 1970; Chapter 8), and axonal iontophoresis should accentuate the problem.

Nonetheless, the examples reviewed here show that axonal iontophoresis provides a way to fill groups of cells, cells whose small size rules out injection through micro-pipettes, and cells whose position is unknown. The elegant results obtained with this technique confirm our initial enthusiasm.

Acknowledgments. NIH fellowship #2 FO2 NS22310-02 from the National Institute of Neurological Diseases and Stroke and the Alfred P. Sloan Foundation supported the author's research.

I thank all the people who told me about and gave me permission to review results which they had not yet published. They observed and strengthened a tradition of frankness which should be characteristic of scientists. I thank Eve Marder, David Bentley, Hugh Rowell, and Jay Mittenthal for criticizing drafts of this paper, and Diane Newsome for typing the text from the countless illegible pages of manuscript.

References

Adiyodi, K. G., and H . A. Bern: Neuronal appearance of neurosecretory cells in the pars intercerebralis of *Periplaneta americana* L. Gen. Comp. Endocrinol. *11,* 88–91 (1968).

Bentley, D. R.: A topological map of the locust flight system motor neurons. J. Insect Physiol. *16,* 905–918 (1970).

———, and R. R. Hoy: Post embryonic development of adult motor patterns in crickets: a neural analysis. Science *170,* 1409–1411 (1970).

Blaker, A. E.: Photography for scientific publication: A handbook. San Francisco: W. H. Freeman, 1965.

Cohen, M. J., and J. W. Jacklet: The functional organization of motor neurons in an insect ganglion. Phil. Trans. R. Soc., Lond. B *252*, 561–572 (1967).

Getting, P. A.: The sensory control of motor output in fly proboscis extension. Z. vergl. Physiol. *74*, 103–120 (1971).

Gymer, A., and J. S. Edwards: The development of the insect central nervous system. I. An analysis of post embryonic growth in the terminal ganglion of *Acheta domesticus.* J. Morph. *123*, 191–197 (1967).

Huber, F.: Central nervous control of sound production in crickets and some speculations on its evolution. Evolution *16*, 429–442 (1962).

Iles, J. F.: Structure and synaptic activation of the fast coxal depressor motor neurone of the cockroach, *Periplaneta americana.* J. exp. Biol. *56*, 647–656 (1972).

———, and B. Mulloney: Procion yellow staining of cockroach motor neurones without the use of microelectrodes. Brain Res. *30*, 397–400 (1971).

Mulloney, B.: Structure of the giant fibers of earthworms. Science *168*, 994–996 (1970).

Page, C. H., and P. G. Sokolove: Crayfish muscle receptor organ: role in regulation of postural flexion. Science *175*, 647–650 (1972).

Pantin, C. F. A.: Notes on microscopical technique for zoologists. Cambridge: University Press, 1960.

Payton, B. W., M. V. L. Bennett, and G. D. Pappas: Permeability and structure of junctional membranes at an electronic synapse. Science *166*, 1641–1643 (1969).

Pitman, R. M., C. D. Tweedle, and M. J. Cohen: Branching of central neurons: intracellular cobalt injection for light and electron microscopy. Science *176*, 412–414 (1972).

Pusey, H. K.: Methods of reconstruction from microscope sections. J. Roy Microsc. Soc. *59*, 232–244 (1939).

Rowell, C. H. F.: The orthopteran descending movement detector (DMD) neurones: a characterization and review. Z. vergl. Physiol. *73*, 167–194 (1971a).

———: Variable responsiveness of a visual interneurone in the free-moving locust, and its relation to behavior and arousal. J. exp. Biol. *55*, 727–747 (1971b).

Selverston, A. I., and M. P. Remler: Neural geometry and activation of crayfish fast flexor motor neurons. J. Neurophysiol. *35*, 797–814 (1972).

Discussion

DR. BENNETT (Albert Einstein): It seems to me that you really aren't going to be iontophoresing stain for very long distances. The kind of setup shown suggests that you are just using this as a way of loading stain into axons. The stain is then going on and diffusing or traveling by axonal transport along the axon.

DR. MULLONEY (University of California, San Diego): That is probably true. I am not convinced, however, that it is simply a diffusional process which is carrying either cobalt or Procion beyond three or four space constants from the point at which you are injecting current. The precise mechanisms responsible for stain migration are unclear (see also Chapter 20).

DR. LLINÁS (Iowa): That is a very elegant technique, Dr. Mulloney. There is an interesting variation of it that I have tried recently with Stan Kater. The dye was injected intracellularly in the soma of a snail neuron and current was passed between the intracellular electrode and the cut end of the axon of the same cell in the periphery. With this approach we were able to fill the soma and axon of the cell in a few seconds.

ANONYMOUS: Is there any limitation in the length of the axon with the axonal iontophoretic method?

DR. MULLONEY: I know of no theoretical limit on the length to which one could iontophorese either cobalt or Procion. What you observe in practice is that cobalt migrates farther and faster than Procion and the farthest I know that anyone has moved Procion is about a centimeter.

Dr. KRAVITZ (Harvard): We have seen dye move at least 3 cm. Using some of the large lobster ganglion cells which have filled quite well, you can see dye going from the ganglion, through the connective and out into the third root.

We don't know whether dye would have gone further as our preparations were generally cut at this point.

Dr. BENNETT: What was the post-filling incubation time?

Dr. KRAVITZ: Our usual time was 16 hr.

Dr. ROWELL: How far has anybody moved cobalt?

Dr. COHEN: We have moved it 3 cm, the length of a good-sized *Periplaneta.*

CHAPTER 8

Permeability and Structure of Electrotonic Junctions and Intercellular Movements of Tracers

M. V. L. Bennett

In this paper I shall present a brief review of the data that led to our concept of the structure of the electrotonic junction or synapse. Some find the arguments persuasive, to a variable degree, and certainly the idea is at least hinted at in the work of many who have studied these junctions. In the second part of the paper I shall deal with some effects of histological fixation on junctional properties and the implications for interpretation of tracer measurements. Much of this work is in progress, and I am indebted to my colleagues for permission to use and refer to unpublished findings of collaborative work.

The key point in our formulation is that there are small cytoplasmic channels connecting coupled cells and that these channels are responsible for cell to cell movement of small molecules and electric current. The evidence indicates that the channels are localized in "gap junctions" which are one specific kind of close apposition between cells. The first level of argument is an inductive one, namely regions of close apposition are found between cells that are electrotonically coupled and the incidence correlates roughly with the closeness of coupling (cf. Bennett, 1972; Bennett et al., 1967c; Pappas et al., 1971). The inductive evidence may extend to experimental treatments that block coupling and disassociate the junctions, although it should be admitted that coupling may also be decreased without as yet resolvable change in junctional structure (cf. Pappas et al., 1971; Rose, 1971). It now appears that these close appositions are identical to those called gap junctions or nexuses. In the next level of argument we deduce that the coupling pathway requires close apposition of cell membranes. At this point we apply calculations of electrical coupling in a model system and measurements of permeability to various substances. Finally, we take the great leap forward to the proposal of a channel connecting cells, which provides the simplest explanation of the entire body of morphological and physiological findings.

The Nature of Electrotonic Coupling

I shall simply assert here that many cells behave as if their cytoplasms are connected by a resistive pathway. This pathway is in general voltage-independent over a wide range, that is, the resistance is constant. (We shall not be concerned with the few rectifying electrotonic synapses from which there are inadequate morphological data in any case, cf. Bennett, 1972.) The electrical coupling between cells can in general be shown to require a specialized relation between them, for it is too great to arise merely from the cells lying in the same volume conductor even where they have a large area of unspecialized membrane in contact.

The Morphological Basis of Electrotonic Coupling

I shall not review here the extensive data correlating electrotonic coupling and close membrane appositions. Suffice it to say that the correlation is observed widely in the nervous system and in many other tissues, both excitable and inexcitable, as well as in every embryo that has been examined. Wherever the data are unambiguous as to class of close apposition, that is gap junction or occluding junction, the differing structures of which are discussed below, it is the gap junction that is observed (Brightman and Reese, 1969; Gilula et al., 1972; Johnson and Sheridan, 1971; McNutt and Weinstein, 1970; Pappas et al., 1971).

> The term "tight junction" formerly was used as a general term for all close appositions, those arranged in zonules, *zonulae occludentes,* and those found as spots or maculae. Of course many other terms were used as well. Now that one distinguishes between the two junctional types, it seems desirable for clarity's sake to denote each type by a previously unused term. Occluding junction seems preferable to (truly) tight junction because one is always apt to drop the truly. Gap junction is all right and may well be with us to stay although it is misleading in that structures do bridge the gap. Nexus would probably be better except that, speaking for myself, it has such an awkward plural.

In a number of tissues studied before this distinction was made, the occurrence of structural periodicities indicates that these close appositions also belong to the class of gap junctions (e.g., Robertson, 1963; Bennett et al., 1967a, b). In all tissues but epithelia, close appositions between coupled cells are exclusively gap junctions (excluding occasional clearly artifactual cases of extensive fusion, cf. Karlson and Schultz, 1965; Brightman and Reese, 1969). There are no data indicating that occluding junctions mediate coupling because in no tissue have coupled cells been found to be joined exclusively by occluding junctions without there being any gap junctions. There are a few indications that occluding junctions do not mediate coupling. A number of epithelial receptor cells are joined to neighboring cells by occluding junctions, but the physiological conditions suggest that they are not coupled to them for a loss of sensitivity and specificity would result (cf. Bennett et al., 1967c; Bennett, 1971; Brightman and Reese, 1969).

A similar argument applies to the septate desmosome which has been proposed as a site of electrotonic coupling. Cells found to be joined by septate desmosomes, including one type known to be coupled, have subsequently been found to be joined by gap junctions as well (Gilula and Satir, 1972; Rose, 1971), and there is no evidence that the gap junctions do not mediate the observed coupling.

The gap junction or nexus is characterized in perpendicular sections not by fusion or contact but by a separation of unit membranes of 20–40 Å (Fig. 1A; Revel and

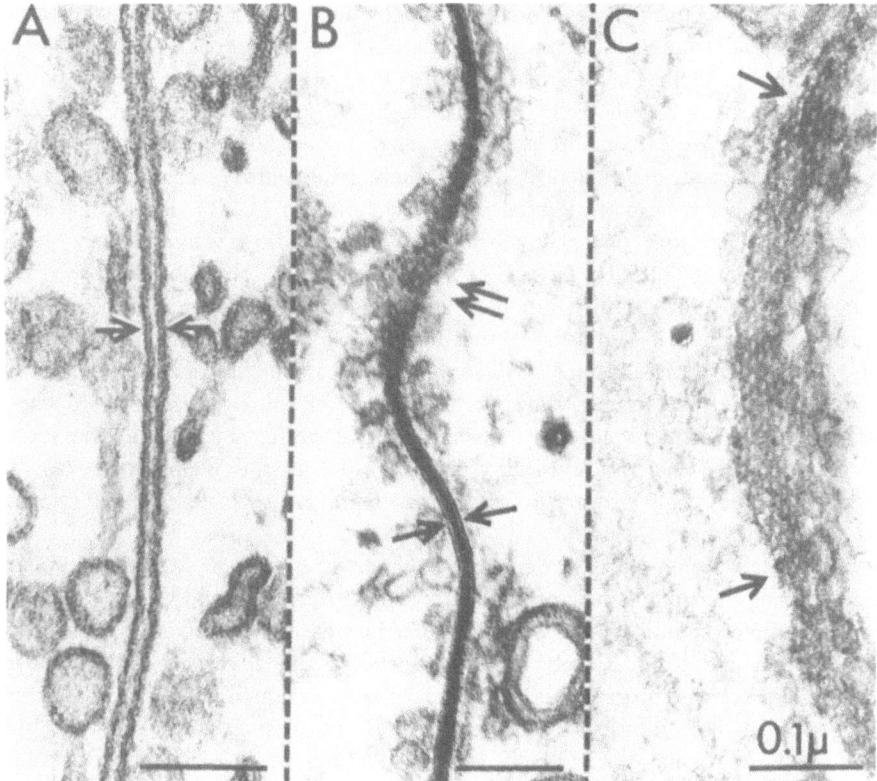

Fig. 1A–C. Electron micrographs of electrotonic synapses at septa of the crayfish septate axon. A. A thin section perpendicular to the closely apposed membranes. The overall thickness of the junction is about 180 Å at the arrows. There is a central light area between the two unit membranes which is 30–40 Å wide. B. Similar to A but the extracellular space between membranes is filled with a dense deposit of lanthanum that was present during fixation. The overall thickness of the junction is about 165 Å at the arrows. At the double arrow the membranes are cut somewhat tangentially and the lanthanum deposit appears not to be a uniform sheet. C. An approximately tangential section through the synaptic region following fixation in the presence of lanthanum. The lanthanum deposit (between arrows) forms a roughly hexagonal network outlining an array of clear regions. All calibrations 0.1 μm. (From Pappas et al., 1971).

Karnovsky, 1967; Brightman and Reese, 1969). The space between membranes with conventional staining is essentially unstructured and of low density, but in many cases the presence of material within the gap would be difficult to see because the gap is so narrow that very slight tilt or bending of the junction within the section would obscure the gap. The patency of the gap is emphasized by fixation in the presence of La(OH)$_3$. The La permeates the gap and forms an electron dense line that extends across the middle of the junction from the ordinary unspecialized regions of intercellular cleft at either side (Fig. 1B). However, in tangential sections it becomes clear that the gap is not uniform; rather there is an hexagonal array of structures bridging the gap and the extracellular space permeated by La is actually a hexagonal lattice of channels surrounding these bridging structures (Fig. 1C). A hexagonal subunit structure was first clearly defined in tangential sections by Robertson (1963) in permanganate-stained material from goldfish medulla. We observed indications of it in osmic acid fixed material in another neural tissue (Bennett et al., 1967a).

Our proposal is that the bridging structure has in its center a channel that connects the cytoplasms of the joined cells. There are a number of morphological indications of a specialization in this region of the junctions. In Robertson's original permanganate fixed preparations there was a central dot in the hexagons (1963). In isolated junctions negatively stained with phosphotungstic acid there is also a central spot (Benedetti and Emmelot, 1968). Even $La(OH)_3$ fixed preparations can show the spot (Revel and Karnovsky, 1967); which can be interpreted as a small degree of penetration by ionic La. Since the pH is adjusted so that $La(OH)_3$ is just at the point of precipitation, it is reasonable that there is some small concentration of ionic La remaining in solution.

Freeze cleave preparations show further aspects of the structure of gap junctions. Each membrane contains an array of particles that presumably overlie the structures bridging the gap (Chalcroft and Bullivant, 1970; McNutt and Weinstein, 1970). These particles show a small central pit which may represent a portion of the cytoplasmic channels connecting the cells.

Freeze cleave techniques also demonstrate important differences between gap junctions and occluding junctions (Chalcroft and Bullivant, 1970; Friend and Gilula, 1972; Goodenough and Stoeckenius, 1972). Occluding junctions have linear regions of actual contact between cells that form a web around (generally) the apical margins of the cells. The contact region apparently is made up of a closely packed row of particles. In most epithelia these junctions completely occlude the space between cells and prevent material from crossing the epithelia by way of extracellular space. In perpendicular section the junctions are seen as points of contact (sometimes called focal tight junctions) which are actually sections of the linear regions. Between these contacts membranes may be separated by 20–40 Å, but the structure is not to be confused with that of gap junctions. However, freeze cleave preparations show that gap junctions may sometimes be included within the web of occluding junctions where it would not be possible to demonstrate their occurrence by the La fixation method (Friend and Gilula, 1972; Goodenough and Stoeckenius, 1972). Tracers are sometimes seen to have crossed a contact of an occluding junction in section; presumably they have crossed out of the plane of section (Brightman and Reese, 1969).

Permeability Measurements

The electrical measurements only indicate that one or more available ions cross the junctions. While K^+ is the obvious candidate, the large conductance changes associated with Cl-mediated IPSPs and the large outward Na^+ currents obtained with large inside positive clamping voltages demonstrate that these ions are also present in sufficient concentration to carry significant transjunctional currents. And other ions might be involved as well. Permeability measurements indicate that many small molecules can cross the junctions in different tissues. In the heart, K^+ passes freely along the tissue, indicating high K^+ permeability of the intercellular junctions (Weidmann, 1966). In the septate axon, Na^+, K^+, Cl^-, and SO_4^{--} all cross the septum as shown by injection of isotopes and subsequent isolation (Bennett et al., 1967d). Co^{++} crosses the septum as shown by subsequent precipitation with sulfide (Politoff et al., 1972). However, the rate of injection must be kept low or decoupling of the junctions will occur. Labeled sucrose (M.W. 342) also crosses (Fig. 2; Bennett and Dunham, 1970). Dye injection, either iontophoretically or by pressure, can allow direct visual observation of transjunctional fluxes. Fluorescein (ion M.W. 330), neutral red (ion M.W. 252), and Procion yellow (ion M.W. about 630) all cross the junctions of the septate axon (Fig. 3D; Bennett et al., 1967; Pappas and Bennett, 1966; Payton et al., 1969). Fluorescein, dansyl-L-glutamate, dansyl-DL-asparate, and Procion yellow pass between coupled cells of a cultured hepatoma line (Johnson and Sheridan, 1971). Fluorescein passes between a number of kinds of coupled cells in tissue culture (Furshpan and Potter, 1968). Fluorescein and Procion yellow pass between cells of dipteran salivary glands (Rose, 1971). Procion yellow has been reported to pass between horizontal cells of the skate retina, but movement after fixation is a possibility (Kaneko, 1971 and Chapter 11).

There are a few data indicating that junctions between embryonic cells are less permeable than junctions in adult tissues. The most reliable results are from isolated and reaggregated blastomeres of the teleost, *Fundulus* (Bennett et al., 1972a, b). Large cells about 200 μm in diameter can be reassembled in pairs, and electrotonic coupling develops between them in a matter of 10 or so minutes (Bennett et al., 1972a). No transjunctional movement of fluorescein is observed, even when the cells are more closely

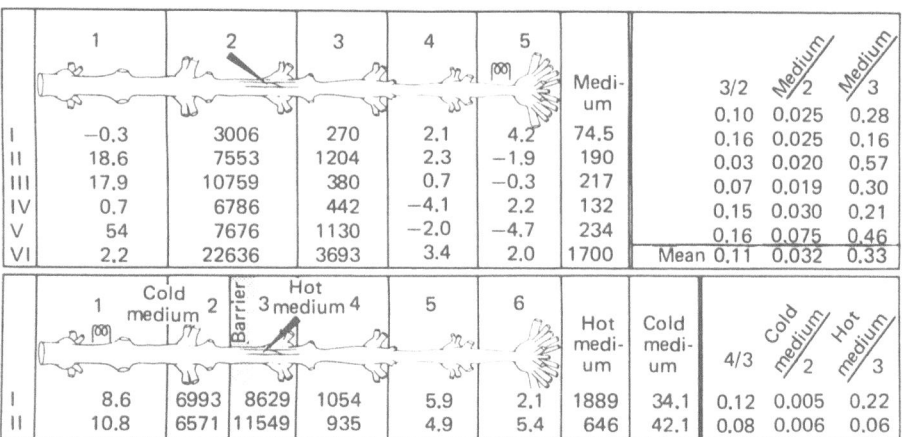

Fig. 2. Transsynaptic passage of sucrose across the septum of the septate axon of the crayfish. Tritium-labeled sucrose was injected just anterior to the septum in the third abdominal ganglion as indicated by the black diagrammatic electrode. After about 30 min the nerve cord was removed from the bathing medium and rapidly frozen. It was then cut as indicated by the vertical lines. The fragments were extracted and tritium in them and in the medium were determined by a liquid scintillation counter. The numbers indicate counts minus background. In the average of the six experiments, sucrose in the next posterior segment was 11% of that in the injected segment (upper part of the figure). Most if not all of the material in the medium probably escaped the injected axon around the site of injection, for if a vaseline barrier divided the bath close to the injection site, the medium over the unimpaled axon contained a much smaller fraction of that in the underlying axon (two experiments, lower part of the figure). (Unpublished work in collaboration with P. B. Dunham, see Bennett and Dunham, 1970).

coupled than segments of the septate axon in which such transjunctional movement is readily demonstrated. Fluorescein does not pass between coupled cells of intact cleavage stages and blastulae of *Xenopus* (Slack and Palmer, 1969), although it does cross in reaggregated cells from later stages (Sheridan, 1971). In a small series of experiments, fluorescein and Procion yellow were not observed to pass between echinoderm embryonic cells at a stage where coupling was developing (Tupper and Saunders, 1970). The dye Chicago Sky Blue (ion M.W., 897) was seen to pass between occasional neighboring cells in the squid embryo (Potter et al., 1966). The significance of this observation is questionable, however, since the cells are rapidly dividing and the actual site of dye movement may be a mid-body or cytoplasmic bridge remaining from a division that has not gone to completion (cf. Bennett et al., 1972a).

Another if Unlikely Possibility

From the electrical measurements, one can calculate a junctional resistance, and, from junctional area, one then obtains the areal resistivity of the junctional membrane. One can then ask what would happen if this membrane were closely apposed but actually separated by, say, 200 Å. It turns out that if one assumes the low resistance membrane is circular in shape and separated by a cleft that is open around the rim so that the potential can be taken as zero at this point, one can compute the form of potential in the cleft as a function of the potential in two cells (Bennett and Auerbach, 1969; Heppner and Plonsey, 1970). If one makes assumptions as to the nonappositional resistances, one can go backwards to obtain the potential in one cell for current applied in the other and also determine the leakage out of the cleft. It turns out that if low resistance membrane is restricted to the region of apposition, significant coupling can be obtained without there being excessive leakage out of the edges of the appositional region. Furthermore, the electrical measurements obtained with a pair of electrodes in each of the coupled cells do not distinguish whether current crosses nonappositional membrane or leaks out of the appositional region between them. An example of the potential within such a model apposition is diagrammed in Fig. 4. The calculation assumes that the nonappositional resistance is very high compared to the appositional resistance, but this does not require unreasonable values in relation to real cells. The arrows show that current flows into the postappositional cell over the central region of apposition while current flows out of it near the edge of the apposition.

A surprising result of our calculations was that, as far as the electrical measurements were concerned, coupling observed at many sites could be mediated across gaps of the order of several hundred Å. The closeness of the gap at gap junctions could allow even closer electrical coupling. Only where coupling was very close, as in cardiac muscle, was it necessary to assume that the hypothetical gap between cells was much smaller than that at gap junctions. One could then conclude that the pathway between cells did not involve a gap large enough to let ions pass out to the surrounding medium (cf. Bennett and Trinkaus, 1970, for a similar calculation as to the possible gap at occluding junctions).

By analogy to this electrical model one can ask whether small ions and dyes such as fluorescein leak out of one cell into the extracellular space and then enter the neighbor-

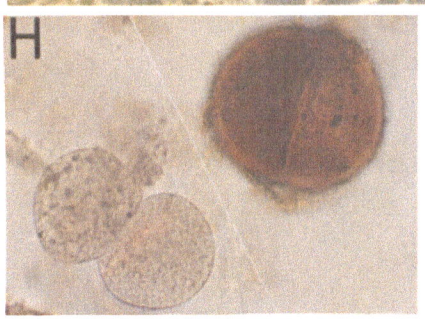

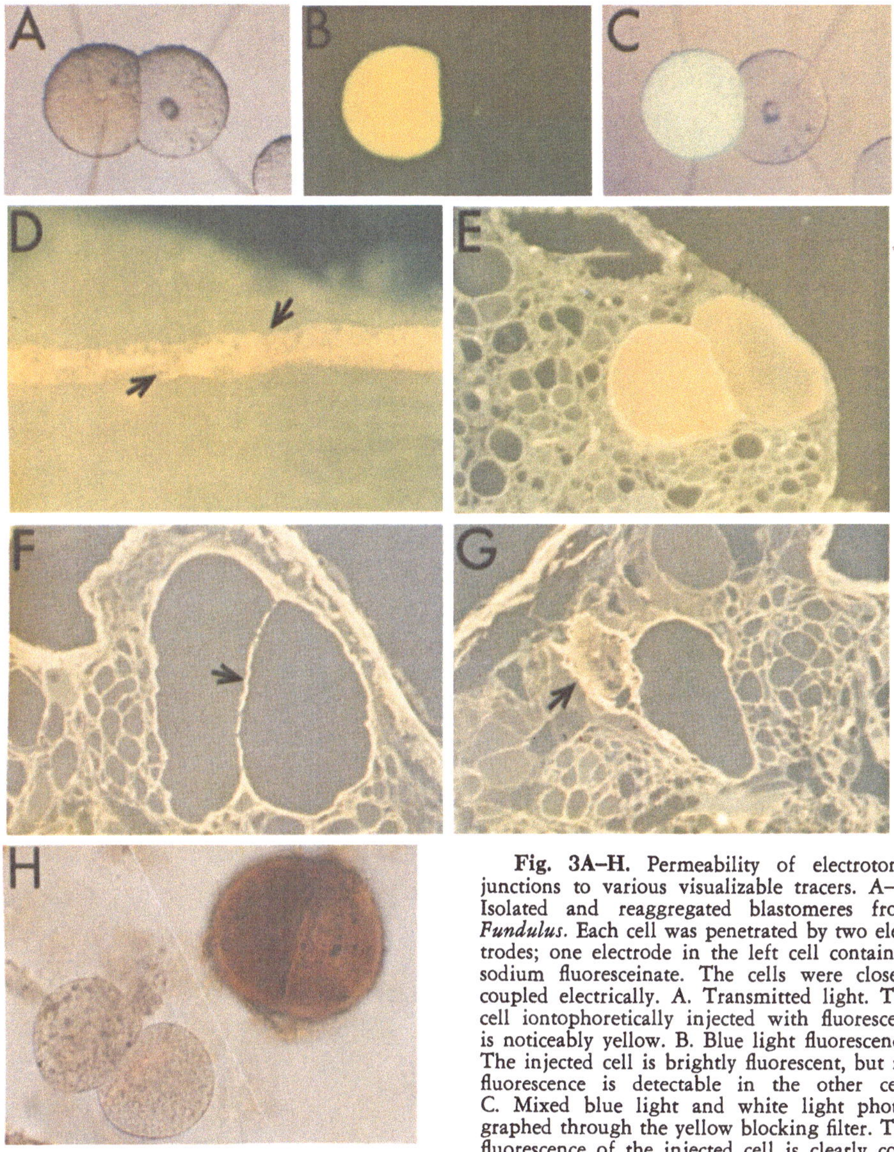

Fig. 3A–H. Permeability of electrotonic junctions to various visualizable tracers. A–C. Isolated and reaggregated blastomeres from *Fundulus*. Each cell was penetrated by two electrodes; one electrode in the left cell contained sodium fluoresceinate. The cells were closely coupled electrically. A. Transmitted light. The cell iontophoretically injected with fluorescein is noticeably yellow. B. Blue light fluorescence. The injected cell is brightly fluorescent, but no fluorescence is detectable in the other cell. C. Mixed blue light and white light photographed through the yellow blocking filter. The fluorescence of the injected cell is clearly confined to that cell. The cells are about 200 μm in diameter (unpublished findings obtained in collaboration with M. E. Spira and G. D. Pappas). D–G. The septate axon of the crayfish. D. Fluorescence micrograph of an intact preparation. Procion yellow was injected in the anterior axon segment (to the right in the figure) and the preparation was allowed to stand overnight. The concentration of dye on the two sides of the septum (arrows) was approximately equal. The background fluorescence shows the outline of the ganglion and nerve cord. E. A fluorescence micrograph of a cross section through the septum of a similar preparation embedded in Epon. F. A fluorescence micrograph of a cross section through the septum in a nerve cord that had been soaked overnight in a 1.0% solution of Procion yellow. The axonal sheaths and connective tissue are well stained as is the septum (arrow). However, no dye has penetrated the septate axon, and the core of the axon has the same dark appearance as the Epon of the section outside of the nerve cord. G. From the same ganglion as F but on the contralateral side. The anterior axon (arrow) had been opened to admit the dye which stained its axoplasm but failed to cross the septum, because the injury caused decoupling. The axons are 75–125 μm in diameter. H. Transmitted light micrograph of *Fundulus* blastomeres. The larger, nearly spherical pair of cells had been shown to be coupled: the left cell had then been injected with peroxidase and finally glutaraldehyde plus La(OH)₃ was applied for fixation. The two cell pairs lay in the same field but at different depths and the micrograph is a montage. Although only the left cell of the large pair had been injected with peroxidase (by pressure) and co-injected fluorescein had failed to pass between cells, following fixation and development of reaction product, the right cell was almost as intensely stained. The difference between this pair of cells and the uninjected pair is so great that no question of false positives is raised. Presumably peroxidase crossed as a result of new large channels opened during the process of fixation in the presence of La (from unpublished work in collaboration with M. E. Spira and G. D. Pappas; see Bennett et al., 1972a,b).

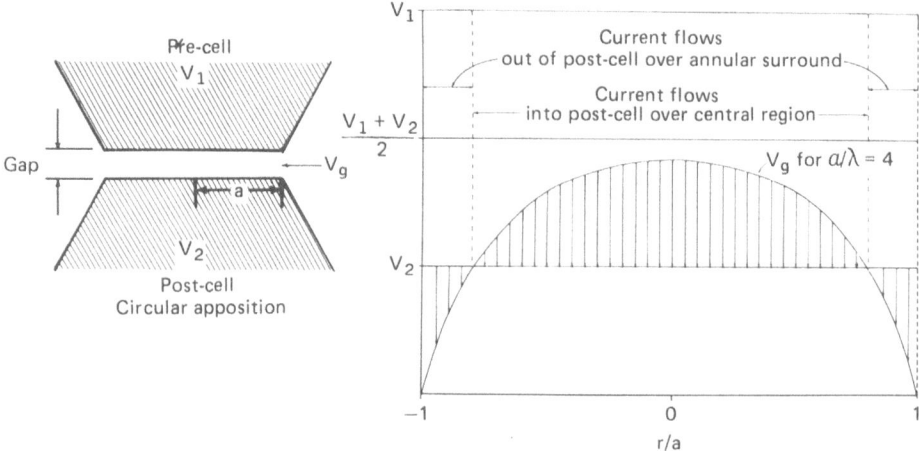

Fig. 4. Electrical coupling between cells that have a circular region of low resistance membranes closely apposed. Diagram on the left, potential in the gap V_g on the right. Because the gap widens at the edge of the apposition V_g is assumed to go to zero at this point. V_1 and V_2 are the potentials in pre- and postappositional cells respectively. Current is applied in the pre–cell setting up V_1; the input resistance of nonappositional membrane in the post-cell is assumed to be high relative to the resistance of the apposition membrane. With these assumptions and an apposition with a radius of four space constants defined in the usual way, the curved line in the graph shows V_g as a function of radial position in the gap. The coupling coefficient, V_2/V_1, is approximately ⅓ (unpublished work in collaboration with A. A. Auerbach; see Bennett and Auerbach, 1969).

ing cell. It is possible that intercellular movement of tracer is by way of extracellular space comparable to the electrical coupling by way of extracellular space. This latter explanation is certainly plausible for the small ions with the possible exception of sulfate to which not many cells are permeable. Sucrose also is a standard for determination of extracellular space because of its usual impermeability. However, it is difficult to know the properties of the membrane in the region of coupling. It might well be permeable to large molecules in correlation with its lower resistivity.

Evidence that the Coupling Pathway between Cells Does Not Involve Extracellular Space

We found a dye that apparently does not cross by way of extracellular space, Procion yellow (Payton et al., 1969). It does pass from cell to cell in the living axon (Fig. 3D). In the better experiments it can be seen to have crossed the septum in the intact axon within an hour of injection. The preparation in Fig. 3D was left overnight, and the amount of dye on the two sides of the septum had become about equal. There was some coagulation of the axoplasm which accounts for the uneven distribution of the dye. The transsynaptic movement of the dye can also be seen in fixed and sectioned material (Fig. 3E).

The failure to see dye outside the junctional region when one cell is injected with dye does not exclude movement from cell to cell by way of extracellular space, for the extracellular volume rapidly increases with distance from the edge of the apposition and the concentration of dye would rapidly fall. A better approach is to apply dye extracellularly in a concentration comparable to that in the injected cell, for that is the level it could approach in the cleft between cells, if it were crossing by way of this space. Procion yellow fails to enter the axon when applied externally (Fig. 3F). In the illustrated experiment the nerve cord was soaked overnight in a 1.0% (w/v) solution of the dye, which is an intensely colored solution. Following fixation, embedding, and sectioning, essentially no fluorescence is seen in the core of the septate axon. The background of the Epon section outside the nerve cord is indistinguishable from the axoplasm itself. The surfaces and connective tissues all are well stained, however. Apparently the dye is quite capable of binding to the external surfaces of cells (Payton et al., 1969; Payton, 1970). Most significant for the present argument, the dye filled the septum right to the center (as does La, as illustrated in Fig. 1C). There are small interruptions which may represent the gap junctions. We were able to envisage but not satisfactorily photograph a faint fluorescence continuing across these regions, which perhaps represented dye that had penetrated into the gap itself. At any rate the dye certainly had access very close to the gap junctions and would have been expected to enter the cells if the pathway between them was open to the exterior.

In the same experiment the anterior axon segment on the other side was cut open a short distance anterior to the septal region. This causes electrical uncoupling of the axons and separation of the junctional membranes (Asada and Bennett, 1971; Pappas et al., 1971). Thus, when the nerve cord was immersed in Procion yellow solution, the dye penetrated the opened axon up to the septal region, but failed to cross into the intact axon segment (Fig. 3G). The axoplasm of the injected cells was coagulated in this experiment as in the experiment in Fig. 3D, and the illustrated section was through a site of relatively greater staining.

From data of this kind we draw two important conclusions. First, the pathway between cells does not involve extracellular space and, second, the pathway between cells is permeable to larger molecules than is the nonjunctional membrane. We can say without qualification that the junctional membrane or pathway is more permeable to large molecules than the nonjunctional membrane, for we are interested in the qualitative difference without worrying about precise ratios of permeability. Since electrical resistance of the septum and nonjunctional membrane of an axon segment are comparable and much more dye crosses the junctional membrane than the nonjunctional membrane, the two membranes must have markedly different permeability properties. There is an important quantitative question with respect to movement by way of extracellular space. If the pathway between cells were by way of extracellular space, would enough externally applied dye get in to be detectable?

A limiting case for electrical coupling by way of extracellular space is that the input resistance is entirely due to current leaking out of the gap. It follows easily from our calculations (Bennett and Auerbach, 1969 and unpublished) that for the septate axon this would involve a comparable leakage from extracellular space back into the coupled cells. The same consideration would apply for dye movement provided the (hypothetical) extracellular leakage path were at least as permeable to the dye as the (hypothetical) apposed but separated membranes. Thus, for the similar exposure times to transsynaptic movement (Fig. 3D, E) and external application (Fig. 3F) one would expect comparable degrees of staining. (We assume comparable concentrations of dye in the injected cell and externally, an assumption which appears reasonable in view of the similarity of the degree of staining in the section of the injected axon and the section of the opened axon into which the dye had penetrated, Fig. 3E, G). As noted above, our calculations indicate that observed coupling at many sites could be mediated across significant gaps as far as the electrical measurements were concerned, a gap of several hundred Å for the septate axon, in the worst case assumption that all the input resistance represents leakage out of the cleft. The dye measurements require a significant narrowing of the cleft, which is also required by the fact that current flows down the axon away from the septum as well as into the junctions. Thus, we can reliably conclude that at least the major portions of dye and current passing between cells cross a markedly reduced extracellular space between them.

The accuracy of present data for the septate axon does not eliminate the possibility of dye movement across a narrowed cleft of the dimensions of the gap of a gap junction. More careful quantification of transjunctional fluxes and rates of entry may allow one to narrow the hypothetical gap by another order of magnitude, which would be small enough, compared to molecular dimensions, that one could reject it as a significant pathway. By analogy to the electrical measurements, a transjunctional flux 100 times the leakage would be adequate to narrow the hypothetical uniform gap to the point of absurdity.

We can make another argument. When one axon segment is injured, it decouples from the adjacent segment and the junctional membranes separate and increase markedly in resistance (Asada and Bennett, 1971; Pappas et al., 1971). However, the input resistance of the adjacent segment changes by little if at all, no more than a few percent. Thus, either there was no leakage out of the cleft prior to injury or the separated junctional membranes increase in resistance just to the point where they compensate for the missing leakage component; the former seems more likely.

Let us question whether the dye and electric current move by the same pathway. The dye certainly is very hydrophilic as presumably are the current carrying ions. By what pathway might the dye move that current would not? Although there are vesicles at the gap junctions in the septate axon (Fig. 1A), there is no sign of exo- and endo-pinocytosis, which would be an analogue of carrier-mediated transport. We are left with channels between cells. Such channels, if permeable to Procion yellow, would certainly be expected to let small ions through and would also account for the permeability to the other dyes and sucrose. Indeed, the existence of hydrophilic channels is supported by the other observations; namely, the low and fixed electrical resistance and the nonspecific permeability to a number of small ions, dyes, and other tracer molecules. The molecular dimensions of fluorescein, sucrose, and Procion yellow require channels of about 10 Å in diameter. Although the data are not yet entirely compelling, it seems likely that the channels do not open to extracellular space but run directly between cytoplasms of the joined cells. In the septate axon the only actual contact between cells bridging extracellular space is at the gap junctions, which occupy a very large fraction of the actual appositions in the septum where connective tissue does not separate the axon segments. If the channels are to have walls they must be located near the center of the bridging structures of the gap junctions. Thus we come to our hypothetical structures which is diagrammed in Fig. 5.

A

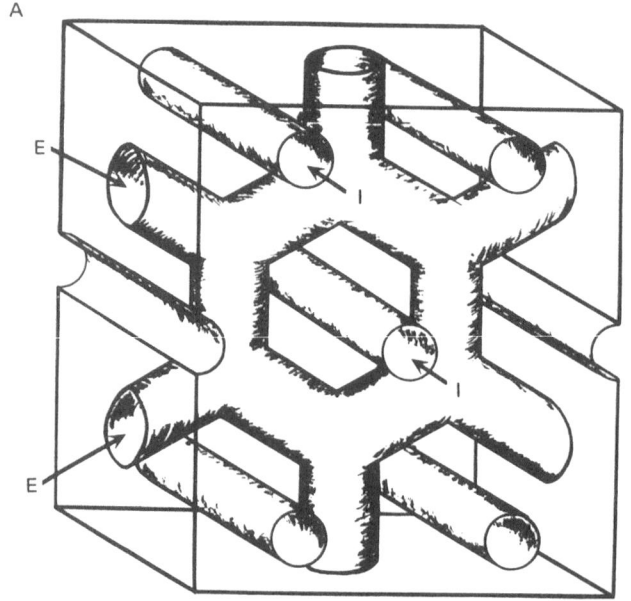

B

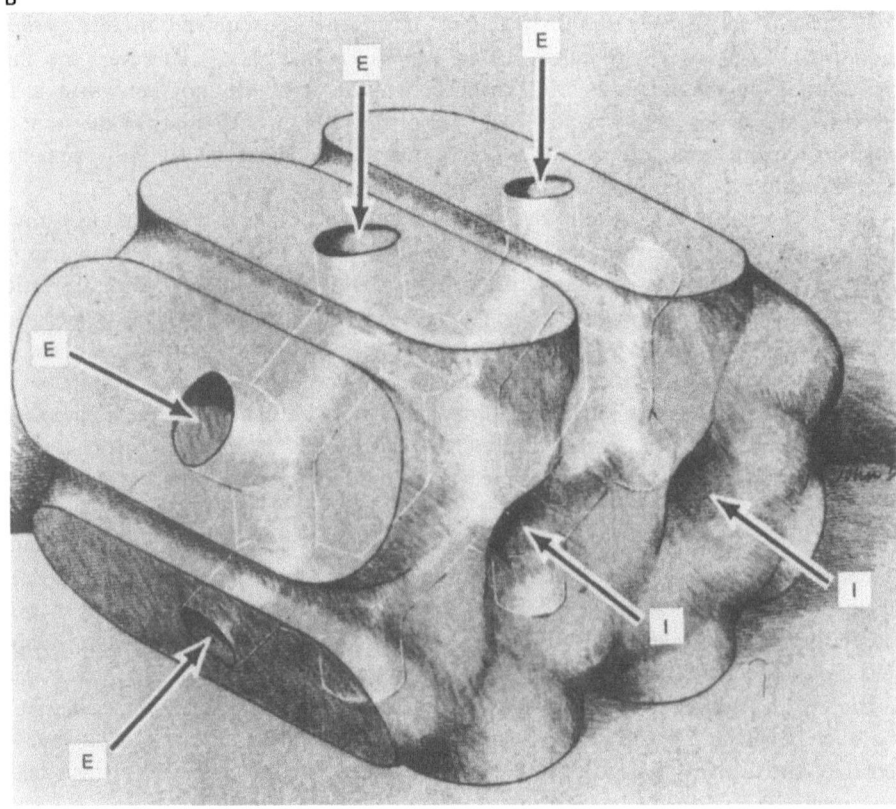

The model presents certain structural problems. In the 90–100Å repeat period (center to center spacing) of the hexagons there is the diameter of one cytoplasmic channel, about 10 Å, the diameter of the extracellular channel, about 20 Å, and two walls separating them which leaves about 30 Å for each wall. This dimension is well below that of unit membranes, but is not unreasonable for a bimolecular lipid membrane which would have the necessary impermeability. Isolation of gap junctions and subsequent chemical and x-ray diffraction analysis provide great promise for further understanding (Goodenough and Stoeckenius, 1972).

An important point remains that will come up later, namely, we do not know the largest size molecule that will cross the junctions. It is unlikely to be very much larger than what has already been observed to cross, because a larger cytoplasmic channel is difficult to fit into the proposed junctional structure. If the model is correct, of course, most cytoplasmic constituents won't fit into what we are calling a cytoplasmic channel. Our expectation is that the embryonic junctions have narrower channels than the adult junctions, but that the overall structure is the same (cf. Bennett et al., 1970a). We think it unlikely, based solely on speculations about structure, that the lower permeability of the embryonic junctions represents an earlier stage of adult junctions. Rather we suspect that the embryonic junctions are "designed" to restrict larger molecules. This would allow embryonic fields to be set up in terms of concentrations of small ions (most likely cations) while individual cells retained larger molecules that operate in control of cell processes such as division and differentiation (Bennett et al., 1972a).

Granted that dye moves by way of cytoplasmic channels and that current utilizes the same pathway, one may then ask whether current uses this pathway exclusively. One approach is to estimate the electrical resistance of the channels. If we assume the specific resistivity in the channels is 100 Ω cm and that the channels are 150 Å long and comprise 1% of the area of the junctions, we arrive at a figure of 0.015 Ω cm^2. (Calculated access resistance becomes comparable to junctional resistance only for junctions greater than about 1 μm in diameter.). Thus, the proposed structure of gap junctions easily provides a low enough resistance, for measurements of gap junction resistivity are generally around 1 Ω cm^2. Actually none of these resistivity measurements are terribly accurate and an order of magnitude lower value would be reasonable. Similarly we could easily assume a higher value of resistance for the cytoplasmic channels and have the calculated and measured values meet in the middle of where they now lie.

Another check on the model is to assume aqueous channels and see if the resistance and permeability values are consistent. The resistance R will be given by

$$R = \frac{\rho l}{A}$$

Fig. 5A and B. Proposed structure of gap junctions, diagrammatic (A) and less diagrammatic views (B). The arrows labeled E indicate the channels within the gap continuous with extracellular space. The arrows labeled I indicate the cytoplasmic channels running between cells in the bridges crossing the gap. The cytoplasmic faces are parallel to the plane of the page in A and to the lower right and back left in B. The overall thickness measured from cytoplasm to cytoplasm is about 150 Å. The center to center spacing of the I channels is about 100 Å.

where ρ is the specific resistivity in the channels, A is the total area of channels, and l is their length.

For permeability, the flux or substance moving per unit time $\dfrac{\Delta Q}{\Delta t}$ will be approximated by:

$$\frac{\Delta Q}{\Delta t} = D \left(\frac{Q_1}{V_1} - \frac{Q_2}{V_2} \right) \frac{A}{l}$$

here D is the diffusion constant, $\dfrac{Q_1}{V_1}$ and $\dfrac{Q_2}{V_2}$ are concentrations on the two sides, and A and l are as before. These equations can be combined to eliminate A and l whence

$$\frac{\Delta Q}{\Delta t} = D \left(\frac{Q_1}{V_1} - \frac{Q_2}{V_2} \right) \frac{\rho}{R}$$

An approximation of the permeability valve can be obtained from the sucrose measurements. For sucrose $D = 0.5 \times 10^{-5}$ cm^2/sec. We may express ΔQ, Q_1, and Q_2 in radioactive disintegrations per minute and V_1 and V_2 in cm^3. As a rough approximation for these experiments, we may take Q_2 as zero, the volume V_1 as 10^{-5} cm^3, $\rho = 100\ \Omega$ cm R as $10^5\ \Omega$ and $t = 10^3$ sec. Whence

$$\frac{\Delta Q}{Q_1} = 0.5 \times 10^{-5} \times 10^5 \times \frac{100}{10^5} \times 10^3 = 0.5$$

This figure is to be compared to the ratio of counts in injected and uninjected segments in Fig. 2, which averaged 0.11 with a maximum of 0.17. This difference is in the right direction, for material in the intraganglionic portion of the uninjected axon was included with the injected sample. There are enough approximations in the calculation, particularly in respect to constant concentration on the injected side, zero concentration on the uninjected side, and junctional resistance, that no great confidence can be placed in the computed value. Nevertheless the absence of any great discrepancy supports the conclusion that small ions and larger molecules cross the junctions by the same pathway.

An Artifact of Fixation?

An obvious goal is to demonstrate the cytoplasmic channels of gap junctions directly by electron microscopy. We injected the septate axon with peroxidases which can be used as tracers localizable both electron and light microscopically (Brightman and Reese, 1969; Feder, 1970; Karnovsky, 1967). In the peroxidase procedure, the enzyme treated tissue is developed with hydrogen peroxide and diamino-benzidine. A reddish brown precipitate is formed which is visible light microscopically and which reacts strongly with osmic acid to be visible electron microscopically. Because of the multiplicative effect of enzymatic action, the peroxidases are sensitive tracers that may yet have usefulness in marking single cells. Horse radish peroxidase has been used to demonstrate axonal uptake and somatopetal axoplasmic transport of proteins, a use that must require considerable sensitivity (Kristensson et al., 1971; LaVail and LaVail, 1972).

Our experiments revealed a quite different aspect that may be of some general significance. In an initial study we found that horse radish peroxidase (M.W. 40,000) did not appear to cross the septum but that a smaller synthetic molecule "microperoxidase" (M.W. c. 1800, cf. Feder, 1970) did cross (Reese et al., 1971). These conclusions cannot be accepted at this time, for we have subsequently found that not all of injected horse radish peroxidase is immobilized by the fixative. After fixation, some peroxidase can diffuse for long distances in the axoplasm, and, most important, after fixation it can cross the septa between axon segments (Bennett et al., 1973). We have not resolved the extent to which the microperoxidase behaves similarly. It is not yet certain that horse radish peroxidase crosses the septum only after fixation but its molecular diameter would be about 50 A if it is spherical, a size which would not fit through the channels in the model of Fig. 5. We are currently carrying out experiments using fluorescein-labeled albumin, which is visible prior to fixation, to determine if such large molecules can cross the junctions in life or only after fixation. Further experiments are planned using differently sized tracers to delimit accurately the size of molecules that will pass through the junctions in their normal state.

Electrophysiological measurements provide further data as to the effects of fixation on junctional membranes of the septate axon. Application of glutaraldehyde in the standard fixation conditions first causes a pronounced and rapid increase in junctional resistance (Fig. 6). After a few minutes, however, junctional resistance gradually declines again to somewhere near its prefixation level. In contrast, fixation initially causes a much smaller increase in nonjunctional resistance, which may be partly a result of removal of the usual current carrying ions. Nonjunctional resistance gradually declines over an hour or more after application of fixative, presumably just due to general disruption of membrane structure.

We believe the fixative somehow occludes the cytoplasmic channels between cells to cause the initial rise in junctional resistance. Our explanation of the reversal of the rise in junctional resistance is that there are microscopic breaks occurring in the junctional membranes and that these breaks constitute new channels. The postfixation permeability of the junctions to a molecule as large as horse radish peroxidase is consistent with this viewpoint. Breaks in the junctional membranes that may represent such channels can be seen electron microscopically (Fig. 7), but of course we do not know if these breaks occur between cells during fixation or in dehydration and subsequent embedding.

Somewhat different results are obtained with electrotonically coupled blastomeres of *Fundulus*. Paraformaldehyde or glutaraldehyde fixation causes a rapid increase in junctional resistance which is stable for well over an hour (Fig. 8; Bennett et al., 1972a, b). Nonjunctional resistance also increases to a new relatively stable value, but the change is much smaller than that in junctional membrane. Buffers do not produce the effect nor are they necessary for it to occur. When glutaraldehyde fixation is carried out in the presence of $La(OH)_3$ (Bennett et al., 1972b), the junctional resistance rises as before, but, after a variable period from a few minutes to an hour, it falls again. Now fluorescein is able to pass between the cells (Fig. 9), although it rarely crosses between cells fixed in glutaraldehyde alone. Again, it appears that a new, more permeable intercellular pathway has been formed. We believe that it is also due to breakage, for junctional resistance can be lowered in sudden steps which are precipitable by slight mechanical shock to the experimental table. As an extension of the experiments with

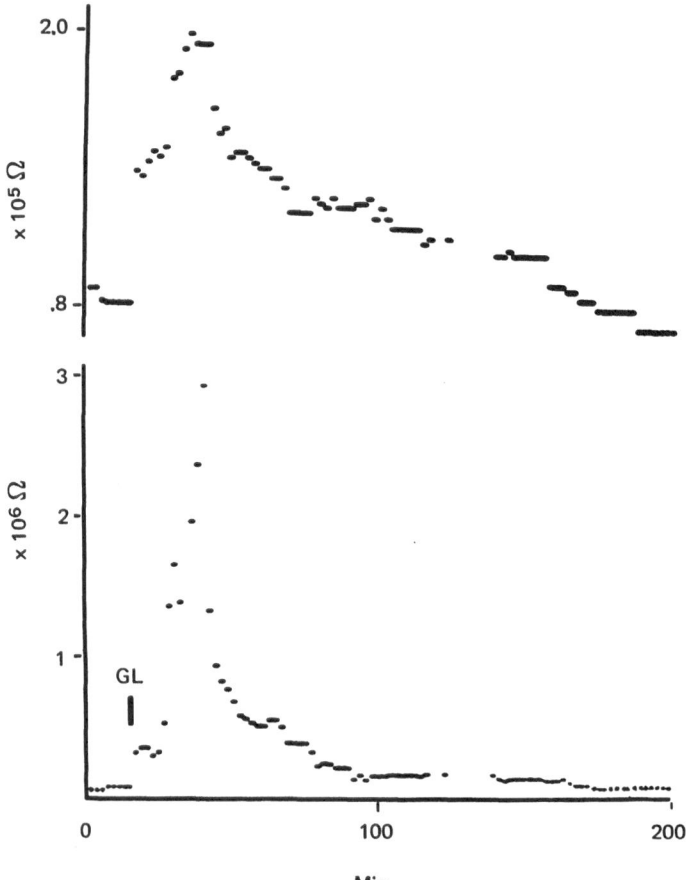

Fig. 6. Change in junctional and nonjunctional resistances at the septum of the crayfish septate axon. Glutaraldehyde fixation with phosphate buffer. The junctional resistance (lower graph) was first increased more than 30-fold by application of the fixative, but the resistance gradually fell again to near its prefixation value. The nonjunctional resistance increased but to a much smaller extent and then also decreased again. (Methods as in Asada and Bennett, 1971, and Pappas et al., 1971. From unpublished work in collaboration with A. Politoff and G. D. Pappas; see Politoff and Pappas, 1972.)

the septate axon, we found that horse radish peroxidase passes between blastomeres fixed in glutaraldehyde plus La (Fig. 3H). However, we have yet to show that peroxidase does not pass between cells fixed simply in aldehydes, in which junctional resistance remains high. Nonetheless, the result is suggestive of formation of relatively large channels as a result of fixation.

Having shown the effects of buffer alone and of buffer plus fixative with and without La it seemed almost unnecessary to try La plus buffer, that is, to omit the aldehyde from the mixture. To our surprise we found that La plus buffer treatment led to cell fusion. Although electron microscopy has not yet been carried out, one can see large particles moving between cells and there is no indication of a barrier to movement

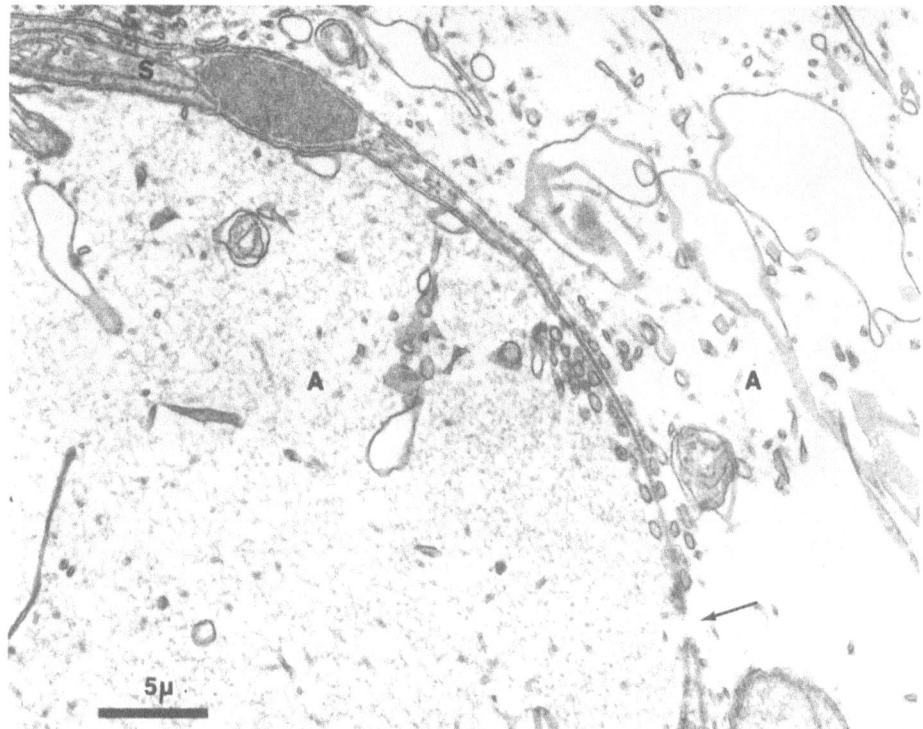

Fig. 7. Electron micrograph of a hole in a gap junction in a septum of the crayfish septate axon (A). The arrow indicates a small hole connecting cytoplasms of the two axon segments. Glutaraldehyde fixation. (Methods as in Pappas et al., 1971. From unpublished work in collaboration with G. D. Pappas and A. Politoff.)

of electric current or fluorescein between them. The concentration of La in this procedure is 25 mM, but because the pH is adjusted to the point of precipitation of the hydroxide, La ion activity must be much less. Conceivably the mechanism of fusion is similar to the mechanism whereby La augments transmitter release at the neuromuscular junction (Heuser and Miledi, 1971). In the latter process, La^{+++} may facilitate fusion of presynaptic vesicles with the external membrane by reducing surface charges. It is not clear whether cells must be electrotonically coupled prior to La treatment for fusion to occur. It is reasonable that formation of channels during fixation in the presence of La has the same mechanism as is operative with La plus buffer alone. The failure of fusion to go to completion with fixative present could merely be a result of stiffening, or fixation, of the general cell structure. On the other hand the presumed formation of new channels in the septate axon requires only the aldehyde, and the mechanism in this case could well differ from that for blastomeres treated with La plus buffer alone.

An important implication of these experiments is that tracers localized after fixation may not have been there prior to fixation. Of particular concern for neuroanatomical studies is the postfixation movement between cells that are joined by electrotonic synapses. Actually, although caution is indicated, not all hope is lost. Provided it can be shown

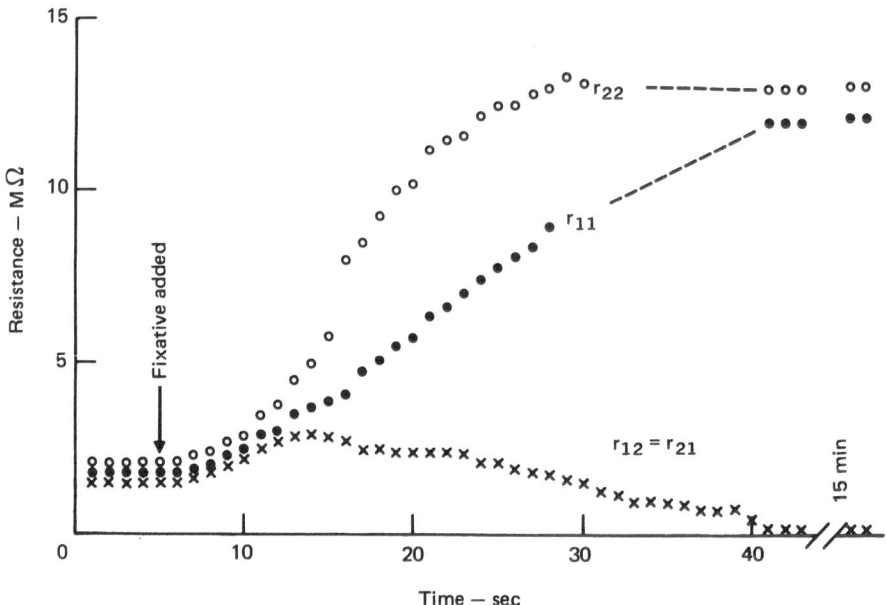

Fig. 8. Time course of changes in coupling of a pair of *Fundulus* blastomeres during glutaraldehyde fixation. Abscissa shows time in seconds; a 15 min break is indicated near the end. Ordinate shows input resistances r_{11} and r_{22}, of the two cells and transfer resistances r_{12} and r_{21}, which were approximately equal in the two directions. Input resistances of the two cells rose rapidly after addition of fixative, but the spread of current between them rapidly decreased indicating a very large increase in junctional resistance, by calculation well over 400-fold. (From Bennett et al., 1972a).

that whatever moves after fixation will keep on moving, then with adequate washing it should be possible to get a reasonably accurate picture of the prefixation distribution. Along this line of argument, it is unlikely that Procion yellow will be found to cross electrotonic synapses after fixation and become fixed there on the far side from the injection site. Since aldehyde fixation would rapidly tie up all the free amino groups with which the Procion yellow would combine more slowly, any dye crossing after fixation would be expected to remain unbound and to be lost in subsequent histological procedures. Soluble Procion yellow labeled protein that had reacted with fixative might also cross the synapses, and this would probably be more difficult to wash out.

Conclusions and Summary

Gap junctions are in general the morphological basis of electrotonic coupling between cells. Although coupling by way of an extracellular gap between cells is at least a theoretical possibility, evidence is adduced that at gap junctions small channels that are walled off from extracellular space connect the cell cytoplasms. From morphological and permeability data these channels are presumed to lie in the center of the structures bridging the extracellular space of the gap. The channels are electrically linear, and

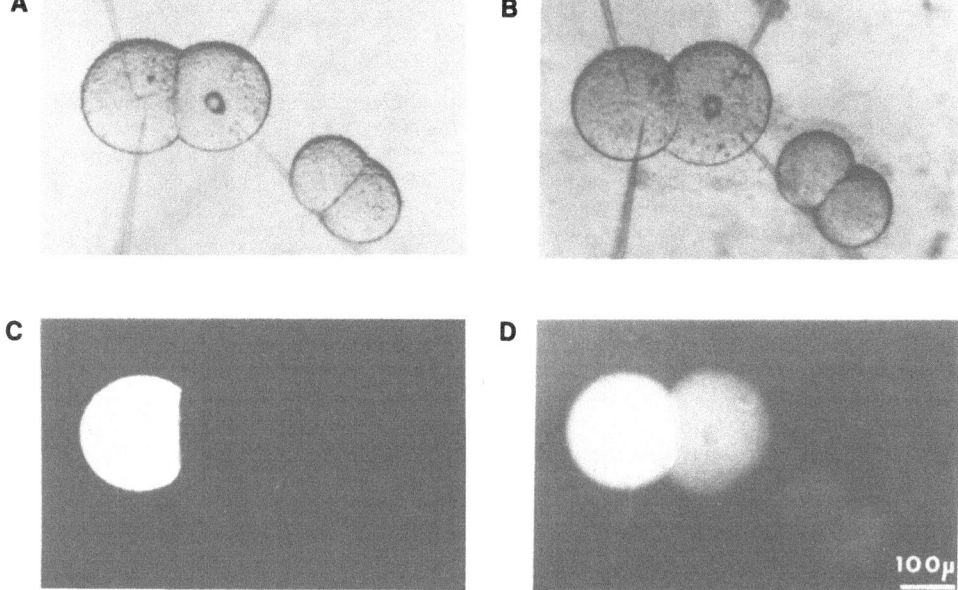

Fig. 9A–D. Effect of glutaraldehyde fixation in the presence of La(OH)₃ on permeability of junctions between *Fundulus* blastomeres. A, B. Prior to fixation viewed in transmitted and blue light fluorescence. One cell pair is impaled by four electrodes, one of which contains fluorescein. A second cell pair not impaled by electrodes serves as a control for autofluorescence. The impaled cells were closely coupled electrotonically. The left cell was injected with fluorescein iontophoretically and was highly fluorescent, but no fluorescein spread to the adjacent cell. C, D. The same cells after fixation in the presence of La(OH)₃ and after junctional resistance had increased and then decreased. Some fluorescein spread to the uninjected cell of the coupled pair. There was also some autofluorescence as shown by the unimpaled pair of cells, but this is insufficient to account for the degree of fluorescence in the uninjected cell in the impaled pair. The same experiment as in Fig. 3A–C. (From unpublished work with M. E. Spira and G. D. Pappas; see Bennett et al., 1972b).

the gap junctions have a low areal resistivity that can be put in reasonable agreement with the model. Calculations of permeability to sucrose and electrical resistance are consistent with passage of each by the same channels.

Fixation by the usual techniques causes an increase in junctional resistance, which may be then followed by a decrease in resistance. The increase in resistance indicates that the impermeability of junctions after various preparative treatments may arise as a result of the procedures themselves. Conversely, the subsequent decrease in resistance is associated with increased permeability to large molecules and may simply represent breaking of the junctional membranes between cells. Thus, at least some large tracers can move from cell to cell after fixation, and distributions observed under these conditions must be carefully controlled for postfixation movement.

Acknowledgments. This work was supported in part by grants from the National Institutes of Health (NB-07512 and HD-04248) and the Alfred P. Sloan Foundation. M. V. L. Bennett was a Kennedy Scholar.

References

Asada, Y., and M. V. L. Bennett: Experimental alteration of coupling resistance at an electrotonic synapse. J. Cell Biol. *49,* 159–172 (1971).

Benedetti, E. L., and P. Emmelot: Hexagonal array of subunits in tight junctions separated from isolated rat liver plasma membranes. J. Cell Biol. *38,* 15–24 (1968).

Bennett, M. V. L.: Electroreception. In: Fish Physiology. Ed. W. S. Hoar and D. S. Randall. pp. 493–574. New York: Academic Press, 1971.

———: A comparison of electrically and chemically mediated transmission. In: Structure and Function of Synapses. Ed. G. D. Pappas and D. P. Purpura. pp. 221–256. New York: Raven Press, 1972.

———, and A. A. Auerbach: Calculation of electrical coupling of cells separated by a gap. Anat. Rec. *163,* 152 (1969).

———, and P. B. Dunham: Sucrose permeability of junctional membrane at an electrotonic synapse. Biophys. J. *10,* 117a (1970).

———, and J. P. Trinkaus: Electrical coupling between embryonic cells by way of extracellular space and specialized junctions. J. Cell Biol. *44,* 592–610 (1970).

———, G. D. Pappas, E. Aljure, and Y. Nakajima: Physiology and ultrastructure of electrotonic junctions. II. Spinal and medullary electromotor nuclei in Morymyrid fish. J. Neurophysiol. *30,* 180–208 (1967a).

———, Y. Nakajima, and G. D. Pappas: Physiology and ultrastructure of electronic junctions. III. Giant electromotor neurons of *Malapterurus electricus.* J. Neurophysiol. *30,* 209–235 (1967b).

———, G. D. Pappas, M. Gimenez, and Y. Nakajima: Physiology and ultrastructure of electrotonic junctions. IV. Medullary electromotor nuclei in gymnotid fish. J. Neurophysiol. *30,* 236–301 (1967c).

———, P. B. Dunham, and G. D. Pappas: Ion fluxes through a "tight junction." J. gen. Physiol. *50,* 1094 (1967d).

———, M. E. Spira, and G. D. Pappas: Properties of electrotonic junctions between embryonic cells of *Fundulus.* Developmental Biol. *29,* 419–435 (1972a).

——— ——— ———: Effect of fixatives for electron microscopy on properties of electrotonic junctions between embryonic cells. J. Cell Biol. *55,* 17a (1972b).

———, N. Feder, T. S. Reese, and W. Stewart: Movement during fixation of peroxidases injected into the crayfish septate axon. J. gen. Physiol. (in the press).

Brightman, M. W., and T. S. Reese: Junctions between intimately apposed cell membranes in the vertebrate brain. J. Cell Biol. *40,* 648–677 (1969).

Chalcroft, J. P., and S. Bullivant: An interpretation of liver cell membrane and junction structure based on observation of freeze-fracture replicas of both sides of the fracture. J. Cell Biol. *47,* 49–60 (1970).

Feder, N.: Microperoxidase. An ultrastructural tracer of low molecular weight. J. Cell Biol. *51,* 339–343 (1970).

Friend, D. S., and N. B. Gilula: Variations in tight and gap junctions in mammalian tissues. J. Cell Biol. *53,* 758–776 (1972).

Furshpan, E. J., and D. D. Potter: Low-resistance junctions between cells in embryos and tissue culture. In: Current Topics in Developmental Biology. Ed. A. A. Moscona, and A. Monroy. Vol. 3, pp. 95–127, 1968.

Gilula, N. B., O. R. Reeves, and A. Steinbach: Metabolic coupling, ionic coupling and cell contacts. Nature, Lond. *235,* 262–265 (1972).

———, and P. Satir: Septate and gap junctions in molluscan gill epithelium. J. Cell Biol. *51,* 869–872 (1971).

Goodenough, D. A., and W. Stoeckenius: The isolation of mouse hepatocyte gap junctions. Preliminary chemical characterization and x-ray diffraction. J. Cell Biol. *54,* 646–656 (1972).

Heppner, D. B., and R. Plonsey: Simulation of electrical interaction of cardiac cells. Biophys. J. *10*, 1057–1075 (1970).

Heuser, J. E. and R. Miledi: Effect of lanthanum ions on function and structure of frog neuromuscular junctions. Proc. Roy. Soc. (Lond.) *B179*, 247–260 (1971).

Johnson, R. G., and J. D. Sheridan: Junctions between cancer cells in culture: ultrastructure and permeability. Science *174*, 717–719 (1971).

Kaneko, A.: Electrical connexions between horizontal cells in the dogfish retina. J. Physiol., Lond. *213*, 95–105 (1971a).

Karlsson, U., and R. L. Schultz: Fixation of the central nervous system for electron microscopy by aldehyde perfusion. J. Ultrastruct. Res. *12*, 160–206 (1965).

Karnovsky, M. J.: The ultrastructural basis of capillary permeability studied with peroxidase as a tracer. J. Cell Biol. *35*, 213–236 (1967).

Kristensson, K., Y. Olsson, and J. Sjostrand: Axonal uptake and retrograde transport of exogenous proteins in the hypoglossal nerve. Brain Res. *32*, 399–406 (1971).

LaVail, J. H., and M. M. LaVail: Retrograde axonal transport in the central nervous system. Science *16*, 1416–1417 (1972).

McNutt, N. S., and R. S. Weinstein: The ultrastructure of the nexus. A correlated thin section and freeze cleave study. J. Cell Biol. *47*, 666–688 (1970).

Pappas, G. D., and M. V. L. Bennett: Specialized junctions involved in electrical transmission between neurons. Ann. N.Y. Acad. Sci. *131*, 495–508 (1966).

——, Y. Asada, and M. V. L. Bennett: Morphological correlates of increased coupling resistance at an electrotonic synapse. J. Cell Biol. *49*, 173–188 (1971).

Payton, B. W.: Histological staining properties of Procion yellow. J. Cell Biol. *45*, 659–662 (1970).

——, M. V. L. Bennett, and G. D. Pappas: Permeability and structures of junctional membranes at an electrotonic synapse. Science *166*, 1641–1643 (1969).

Politoff, A., and G. D. Pappas: Mechanisms of increase in coupling resistance at electrotonic synapses of the crayfish septate axon. Anat. Rec. *172*, 384–385 (1972).

—— ——, and M. V. L. Bennett: Cobalt: a tracer for light and electron microscopy that can cross an electrotonic synapse. J. Cell Biol. *55*, 204a (1972).

Potter, D. D., E. J. Furshpan, and E. S. Lennox: Connections between cells of the developing squid as revealed by electrophysiological methods. Proc. natn. Acad. Sci. U.S.A. *55*, 328–336 (1966).

Reese, T. S., M. V. L. Bennett, and N. Feder: Cell-to-cell movement of peroxidases injected into the septate axon of crayfish. Anat. Rec. *169*, 409 (1971).

Revel, J. P., and M. J. Karnovsky: Hexagonal array of subunits in intercellular junctions of the mouse heart and liver. J. Cell Biol. *33*, C7–C12 (1967).

Robertson, J. D.: The occurrence of a subunit pattern in the unit membranes of club endings in Mauthner cell synapses in goldfish brains. J. Cell Biol. *19*, 201–221 (1963).

Rose, B.: Intercellular communication and some structural aspects of membrane junctions in a simple cell system. J. Membrane Biol. *5*, 1–9 (1971).

Sheridan, J. D.: Dye movement and low resistance junctions between reaggregated embryonic cells. Devl. Biol. *26*, 627–636 (1971).

Slack, C., and J. F. Palmer: The permeability of intercellular junctions in the early embryo of *Xenopus laevis*, studied with a fluorescent tracer. Expl Cell Res. *55*, 416–419 (1969).

Tupper, J. T., and J. W. Saunders: Intercellular permeability in the early *Asterias* embryo. Devl Biol. *27*, 546–554 (1972).

Weidmann, S.: The diffusion of radiopotassium across intercalated disks of mammalian cardiac muscle. J. Physiol., Lond. *187*, 323–342 (1966).

Discussion

MR. GILLETTE (Toronto): In my work I inject dye into *Aplysia* neurons. I have put one color dye into an interneuron and another color into a follower neuron.

With this method I can see the zones of apposition of these neurons within the neuropil of the ganglion. Quite often I can find what seems to be an exchange of colors between the two neurons that extends beyond the zone of apposition up to 100 μm. These are chemically coupled neurons.

There are electrically coupled neurons in *Aplysia* which could account for another result I have seen. In that case I was able to trace a neuron that I had not acutally injected with dye. Its axon apposed that of the dye-filled interneuron for a 170 μm length and had acquired enough yellow fluorescence to enable me to trace this axon, in serial section, 1300 μm across a commissure to its cell body.

It seems to me if these phenomena could be brought under control in some way, this would be a valuable way to identify and trace postsynaptic cells.

DR. BENNETT, (Albert Einstein): Yes, I would certainly agree if it could be brought under control.

Are you sure that there is not an electrical component at these "chemical" synapses?

MR. GILLETTE: No, I am not sure.

DR. R. LLINÁS (Iowa): A word about blasting dye across chemical synapses, Charles Nicholson and I were quite surprised to see, on several occasions, Procion dye going across a chemical synapse. This happens occasionally in the squid giant synapse if excess pressure is utilized in injecting the dye pre- or postsynaptically and may be an artifact to be aware of when studying the passage of dye between alleged electrotonically coupled cells. The interesting point here is that the subsynaptic membrane appears to be more easily ruptured than the nonsynaptic membrane and the dye would readily move into the next cell across the synapse rather than into the extracellular medium. We have similar observations with intracellular oil injection but have never had this problem when the dye was injected iontophoretically.

DR. NICHOLSON (Iowa): In a paper in which you were a co-author (Payton, et al., 1969; and see also Payton, 1970) it was stated that Procion acted as a local anesthetic when applied extracellularly. Would you amplify this point?

DR. BENNETT: I think that is true. Our impression, and it is purely an impression, is that it acts as an anesthetic outside but not inside the cell, but we really haven't done that with proper dosages (see also Chapter 15).

DR. PURVES (University College, London): Has anyone looked at stain migration across rectifying electrical junctions?

DR. BENNETT: I haven't. Do you have a comment Dr. Kravitz?

DR. KRAVITZ (Harvard): We never saw it go across the motor-giant synapse of the lobster.

DR. BARRETT (Colorado): Globus, Lux and Schubert (Society for Neuroscience Abstracts, 1972, p. 254) report that injections of tritiated glycine into one leech Retzius cell produces labeling of both Retzius cells in the ganglion. When puromycin, an inhibitor of protein synthesis, is injected into the contralateral Retzius cell, this cell no longer appears labeled in microautoradiographs. These data suggest that the labeled glycine crosses the inter-Retzius gap junction as the free amino acid, rather than in synthesized proteins. This finding complements Dr. Bennett's observation that relatively small molecules, such as Procion yellow, do cross at least some gap junctions, whereas molecules as large as horseradish peroxidase do not.

Dr. Bennett, does extracellularly applied Procion yellow eliminate electrical coupling at the septate junction?

DR. BENNETT: No.

CHAPTER 9

Characterization of an Insect Neuron which cannot be Visualized *in situ*

R. K. Murphey

The inability to visualize a target is a problem common to intracellular recording in many nervous systems. This problem can be mitigated by the injection of stain into the neurons of interest. By determining the geometry of a neuron and correlating that geometry with respect to external landmarks, it is possible to record routinely from the dendrites and somata of single, physiologically identifiable but "invisible" neurons in the central ganglia of crustacea (Zucker et al., 1971; Zucker, 1972a, b, c) and insects (illustrated here). In the insects, injection of stain into "blindly" impaled neurons can provide unique morphological identification of those neurons. By combining intracellular recording and staining techniques, we can now obtain meaningful data on the integrative processes underlying behavior in this well-studied group of organisms. In addition, intracellular staining techniques should facilitate analysis of the geometry of neurons which develop under various experimental conditions.

Neuronal Morphology as an Aid to Intracellular Recording in the Insect Nervous System

In contrast to the Crustaceans, where the integrative events underlying a number of behaviors have been analyzed (e.g., Sandeman, 1969; Zucker et al., 1971; Zucker, 1972a, b, c), intracellular recording of events underlying behavior in insects has proceded very slowly. With a few notable exceptions (Hagiwara and Watanabe, 1956; Narahashi, 1963), the insect nervous system remained refractory to routine intracellular recording until 1969 (Bentley, 1969a, b; Kerkut et al., 1969a, b). One of the major problems with insect nervous systems is the difficulty encountered in trying to visualize the somata of insect neurons. If one removes the ganglionic sheath in order to improve visibility and facilitate penetration with micropipettes the insect somata are damaged. When activity is recorded from such somata, it is of poor quality (Bentley, 1969a; Hoyle, 1970; Hoyle and Burrows, 1970). Hoyle and Burrows (1970) demonstrated that recordings obtained from insect somata were improved by driving electrodes through the intact ganglionic sheath. By stabilizing the ganglion on a platform [a technique used extensively in molluscs (Willows, 1967)] Hoyle and Burrows (1970) were able to obtain somatic recordings from a large number of motoneurons and interneurons in the metathoracic ganglia of the locust *Schistocerca*. I have extended these techniques and now record routinely from, and inject stain into, the large dendritic branches as well as the somata of the giant interneurons in the terminal ganglion of the cricket *Acheta domesticus*.

Intracellular staining techniques have facilitated intracellular studies in insects in several ways. First, the injection of a stain demonstrates that the electrode is actually in a cell rather than in some quasi-intracellular recording situation. Second, it can provide information about which region of the neuron has been penetrated. Third, knowledge of the anatomy of the neurons makes it possible to selectively impale different regions of the neuron (Selverston and Kennedy, 1969). Fourth, intracellular injection of stain can confirm that recordings are being obtained from the "same" neuron in different preparations.

Regional Characterization of an Insect Neuron

The single neuron which I have studied most extensively can be identified anatomically by: (1) the position of its large axon in the connective (Fig. 2), (2) the position of its large soma (Fig. 3), and (3) a characteristic branching pattern of its dendritic field (Fig. 3). The most reliable of these characteristics is the axon position in the connective, and the cell will be called the medial giant interneuron (MGI) on this basis.

Methodology

The semi-intact preparation used throughout these experiments is illustrated in Fig. 1. The cricket was induced to autotomize the metathoracic legs and then restrained dorsal side up on a dissecting dish. The abdominal nervous system was exposed by removing the dorsal cuticle of all of the abdominal segments, and the terminal ganglion was lifted onto a micromanipulated platform in order to stabilize it for intracellular recording. A recording micropipette, prepared by the fiberglass method of Tasaki et al. (1968), filled with 5% (w/v) Procion yellow (M4RS or M4RAN) and having a resistance of 100–150 MΩ was then positioned over the ganglion. The micropipette was driven through the sheath in a vertical path through the ganglion. When a recording was obtained, from what appeared on physiological grounds to be a cell of interest, dye was injected by passing 5×10^{-9} to 5×10^{-8} A in 800 msec pulses at 1 Hz for 10–30 min for a neuropilar injection site or 30–90 min for a soma injection. The stain was allowed to migrate for 5–8 hr. (Longer migration times resulted in severe degeneration of the material.) Standard histological procedures were then carried out on the tissue and the resulting sectioned material photographed (Milburn and Bentley, 1971).

Locus of Stain Injection

The anatomical region of the insect neuron into which stain is injected significantly influences the anatomical results. Injecting the stain through one of the main dendritic branches gives the most complete, intense, and reproducible fill of the dendritic field.

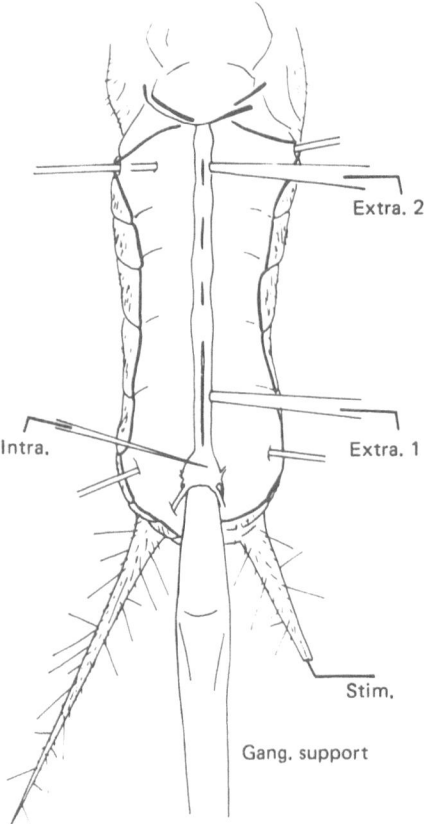

Fig. 1. A. Sketch of the dorsal dissection of *Acheta* showing the arrangement of stimulating and recording electrodes. *Extra 1* and *Extra 2* are *en passant* suction electrodes recording activity at different points on one connective. *Intra* is a recording micropipette D.C.-coupled via a cathode follower to an oscilloscope. *Stim* is a stimulating lead placed in the cercus for direct stimulation of cercal afferents.

The soma is the most stable recording site, but, as in crustacean neurons (Remler et al., 1968; Davis, 1970), the long thin neurite acts as a barrier to migration. Therefore, even very intense 1.5 hr injections of the MGI provide relatively poor dendritic detail. A similar problem exists for axonal injections; the thin region between axon and dendrites acts as a migration barrier. Thus longer migration times (greater than 8 hr) are required for axonal or somatic injections, and the histologically prepared material is less well preserved. The axonal injection sites were also the least stable of the three possibilities in the semi-intact preparation used here.

Recording from the Neuropil

Records from large dendritic branches of this cell were obtained regularly in a series of exploratory recordings. The neuropilar recording site is near the confluence of all the major cell branches, as indicated by the arrow in Fig. 4. This estimate of electrode position is based on the gross anatomical position at which the recording pipette penetrates the sheath and on the very intense staining of the cell in this region when compared with other regions of the dendritic field. This neuron has also been impaled in more posterior regions of the two main dendritic branches and the site of impalement is often detectable as a region of localized swelling, presumably due to osmotic pressure induced by the high concentration of dye near the electrode tip. The relatively high probability (approx. 0.50, N = 30 preparations) of impaling the unit in a given preparation is undoubtedly due to the large area of the terminal ganglion occupied by branches of this cell (Fig. 3). (The criterion for this estimate of success is simply that a large unit, as recorded extracellularly, is impaled.) However, the high probability of impaling the cell is counteracted by the difficulty of getting a stable penetration which does not depolarize the cell and kill it. In this series of experiments, approximately 15% (N = 30 preparations) of the neuropilar penetrations were stable enough to allow recording of physiological data as well as a 10–20 min injection of stain.

Synapses in arthropod ganglia are restricted to neuropil, and therefore neuropilar recording sites should be optimal for recording synaptic activity (Takeda and Kennedy, 1964; Sandeman, 1969; Zucker et al., 1971). This is very clearly the case in the MGI (e.g., Figs. 4 and 6). However, the correlation of unitary potentials with the firing of presynaptic neurons, like that elegantly demonstrated with an extracellular method in cockroach giant interneurons (Callec et al., 1971), has yet to be carried out.

The axonal spikes, which are monitored extracellularly by *en passant* suction electrodes, are seen as 20–30 mV deflections when recorded in the neuropil. They have a time course of 2.5 — 5 msec and never overshoot zero potential (resting potential = 50 — 60 mV), indicating that the spike is conducted electrotonically from a distant spike initiating zone.

When the frequency of firing is high, as at the beginning of the burst in Fig. 4B, spikes appear to be initiated on the falling phase of preceding spikes. In extreme

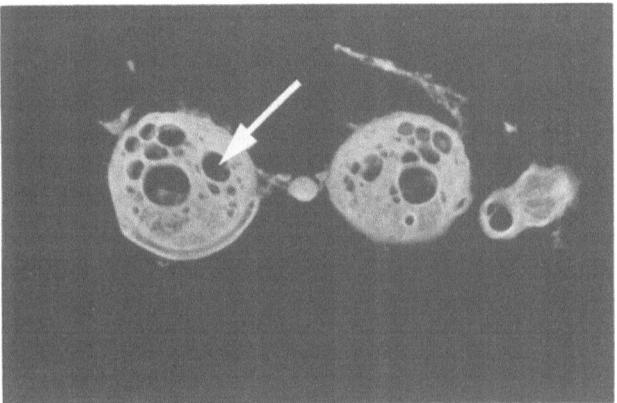

Fig. 2. A transverse section of the connective of *Acheta domesticus* just anterior to the terminal abdominal ganglion. The medial giant interneuron is indicated by the arrow.

examples, there is a mere notch at the top of a depolarization, the notch separating two axonal spikes as recorded by the suction electrodes. This is consistent with results from a crab motoneuron, in which increased resistive and capacitive properties of the main dendritic branches (compared with the thin axon leading to this region) caused a prolongation of the repolarization phase. This was demonstrated by moving the recording site further from the spike initiating zone. The anatomical and electrophysiological similarities between MGI and the crab eye withdrawal motoneuron suggest that the two large dendritic branches of the MGI function as "integrating segments" as defined by Sandeman (1969).

Recording from the Soma

The soma of the MGI was found in the anterolateral region of the ganglion by injecting a process of this neuron located in the neuropil (Fig. 3). It is one of the largest somata in the entire ganglion (diameter = 30–40 μm) and is significantly larger than the other somata in this area. Therefore, the soma of MGI is an easy target to hit. Once its position was known, it could be impaled 85% of the time (N = 20 preparations). Somatic recordings are maintained routinely for 30–60 min in the semi-intact preparation, and, when compared with axonal or dendritic recordings, are extremely stable.

The electrical events recorded in the soma are greatly affected by the long neurite which separates the soma from the integrating segment. Since the neurite is of fairly uniform diameter and its length is known, it is possible to calculate the approximate attenuation factors for events of known time course taking place at the distal end of the neurite and measured in the soma. Using the equations given by Falk and Fatt (1964) and by Katz and Miledi (1965) for the frequency dependence of the space constant and the values given by Narahashi (1963) for cockroach giant axons for the space (λ = 1 mm) and time (τ = 4.2 msec) constants for a 50 μm axon it can be shown that the frequency-dependent space constant for the spike (frequency of about 330 Hz) seen in the neuropil will be 90–144 μm for a 2–5 μm neurite. Thus a 25 mV spike recorded in the integrating segment should be recorded in the soma (227 μm away as measured in Fig. 8) as a potential of 2–5 mV. This agrees well with the observed spike sizes of 1–5 mV (e.g., Figs. 4, 5, and 7).

The constants used in the above calculation are in agreement with the observations. I therefore made similar calculations for a 100 Hz polarization to provide an estimate of the attenuation factor for a synaptic potential of 10 msec duration (Callec et al., 1971) as recorded in the soma. A 5 mV unitary potential of 10 msec duration appearing at the distal end of the neurite would appear as a 1–1.5 mV potential in the soma. Thus spike-like events in the neuropil are attenuated to 10–20% of their original size and chemical EPSPs are attenuated to 20–30% of their original size.

Due to these filtering characteristics of the neurite, the electrical events recorded from the soma can be difficult to interpret. The spikes are often difficult to pick out on the intracellular record. For instance, in response to optimal sound stimulation (400 Hz) a mere notch may be seen separating two axonal spikes, or the initial spikes in a barrage may be completely absent from the intracellular record (Fig. 5). Voltage changes in response to natural stimuli, such as sound or wind puffs, are recorded as smooth de-

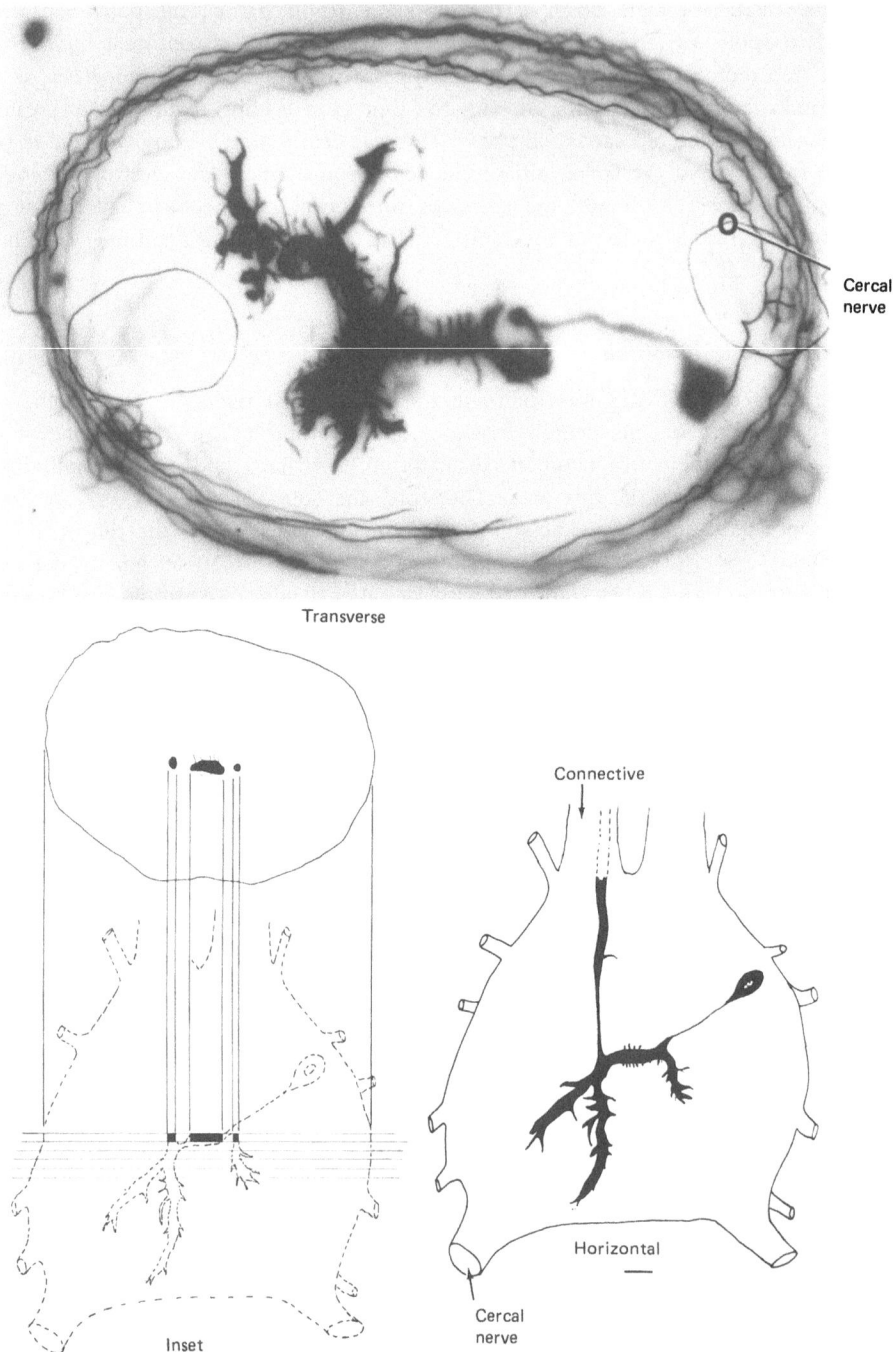

Transverse

Connective

Cercal nerve

Horizontal

Cercal nerve

Inset

Fig. 3. Horizontal and transverse reconstruction of the medial giant interneuron of the cricket *Acheta domesticus.* The original material was sectioned in the transverse plane, and the reconstruction in this plane is the most accurate. The horizontal reconstruction was obtained by projecting the cross sections onto a grid with spacings scaled to the thickness of the cross sections, as shown in the inset.

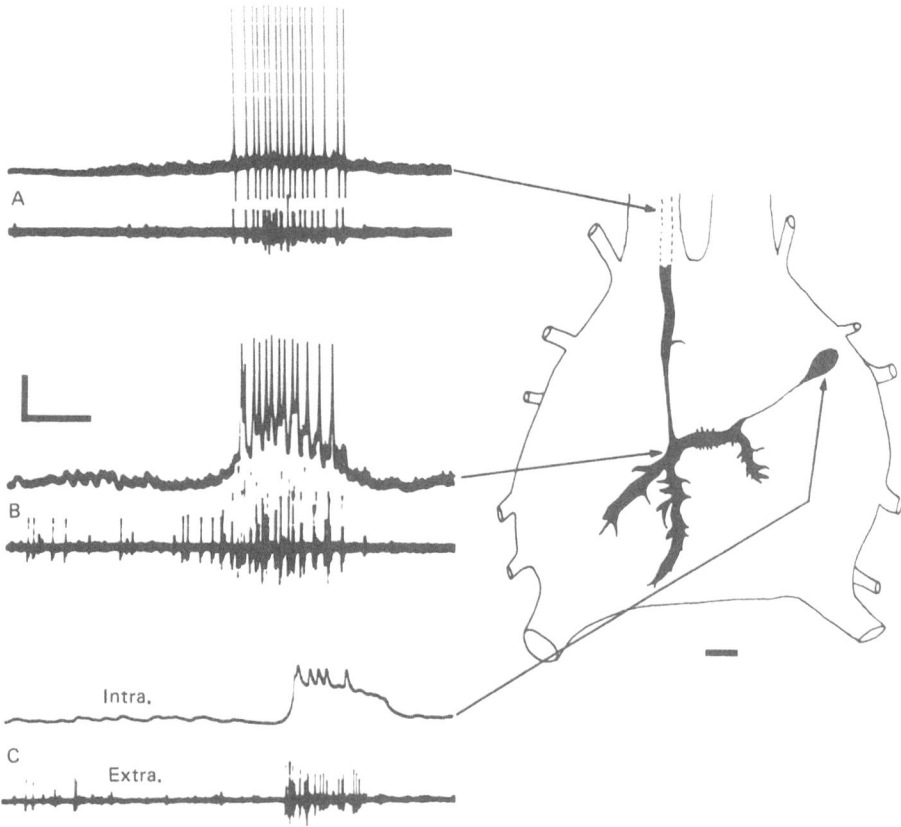

Fig. 4A–C. Recordings from different regions of the medial giant interneuron (MGI) of *Acheta.* Anatomical calibration bar 50 μm. The upper trace in each record is an intracellular recording from the MGI; the lower trace is an extracellular recording from the connective as it leaves the ganglion. In A, the extracellular spike appears to lead the intracellular one, but this is an artifact of alignment of the two beams. The true relationship is shown in Fig. 5. The axonal recordings are thought to be from this cell on the basis of electrode position and physiological response, but anatomical confirmation was not obtained. The other two recording sites were identified unequivocally by dye injection. Calibration: vertical, upper beam only, A, 8 mV; B and C, 10 mV; horizontal, 50 msec).

polarizations with no evidence for unitary synaptic potentials. When no stimulation is applied, the baseline shows low amplitude, short time course fluctuations which may be unitary PSPs, but these have not been unequivocally identified as such. Inhibitory potentials caused by stimulation of the contralateral cercus (Callec et al., 1971; Palka, personal communication) could not be detected in the soma; only a smooth but depressed wave of depolarization could be recorded under the appropriate stimulus conditions.

Recording from the Axon

The axon was found to be the least stable recording site of the three used in these semi-intact preparations. Overshooting spikes (50–80 mV) were often observed

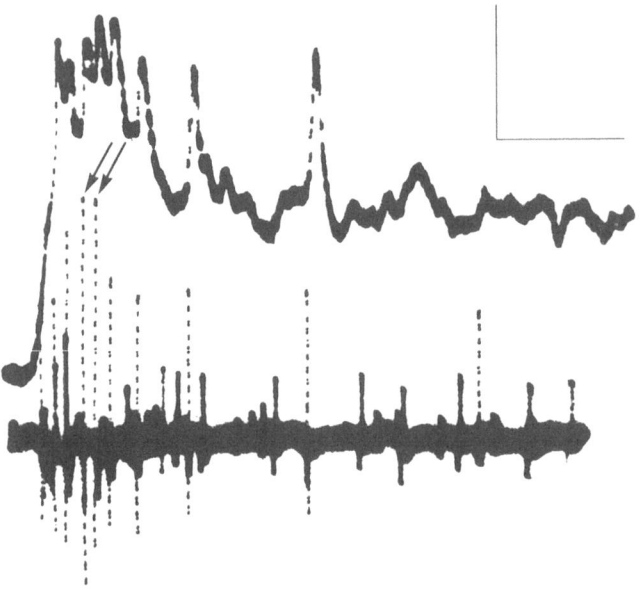

Fig. 5. Recordings from the soma of the medial giant interneuron of *Acheta*. Note the poorly defined spikes in the intracellular record at the onset of stimulation. The arrows indicate
summation of MGI with other units in the extracellular record.
The stimulus was a 100 msec tone pulse (400 Hz at an intensity
of 85 dB). Calibration: vertical, 2 mV; horizontal, 20 msec.

(Fig. 4). The recordings were seldom maintained for more than a few minutes; the
overshoot and postspike undershoot usually disappeared very quickly, leaving a mono-
phasic, short time course spike which was propagated along the axon. These results
bracket the spike initiating region since spikes are not overshooting in dendrites or
soma. As in the case of a crustacean motoneuron (Sandeman, 1969), spike initiation
must be occurring at some point along the narrow stalk between axon and integrating
segment. Very little of the postsynaptic activity in the neuropil can be recorded from
the axonal recording site (Fig. 4).

An unusual morphological feature of this cell is its small dendritic tree on the
axon stalk (at *a* in Fig. 3). This may represent the segmental branches associated with
the primitive ganglion now fused with others into the terminal ganglion. Such branching
patterns in other anatomically distinct segmental abdominal ganglia as well as in the
fused terminal ganglion have been demonstrated by Harris and Smyth (1971) in the
cockroach. Whatever its origin, if this branch receives synaptic input, synapses at this site
are likely to have a disproportionate affect on spike initiation due to its proximity
to the site of spike initiation.

Synaptic Input to the Medial Giant Interneuron

To date, my studies have concentrated on the input side of this neuron. The main
dendritic field of the MGI is asymmetric, with the majority of its branches in the

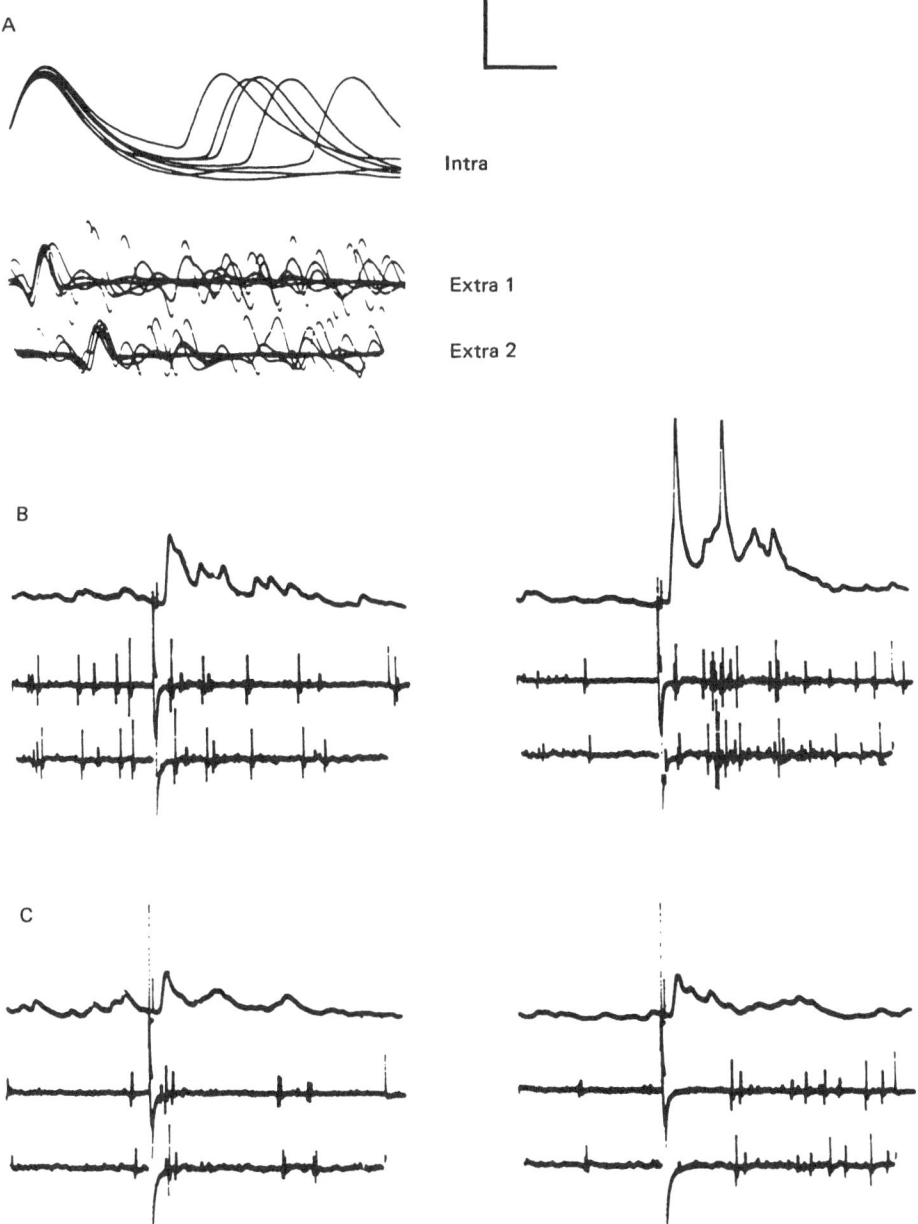

Fig. 6A–C. The asymmetry of cercal afferents to the MGI of *Acheta*. A. Demonstration that the cell has an axon in the connective ipsilateral to the intracellular recording (labels are the same as those in Fig. 2). The distance between extracellular electrodes was 6 mm. B. Response to ipsilateral cercal nerve stimulation. C. Response to contralateral cercal stimulation. Calibration: vertical, A–C, 20 mV; horizontal, A, 1 msec; B and C, 20 msec.

neuropil ipsilateral to the axon (Fig. 3). A smaller dendritic branch is located contralateral to the axon, and a major midline process connects the branches on the two sides of the ganglion. The soma is contralateral to the axon just behind Nerve 8 and is connected via the thin neurite to the large midline branch (Figs. 3 and 8). This particular neuron is sensitive to sounds of frequencies of less than 1000 Hz delivered to the animal. The responses to 100 msec, 80 dB tone pulses are shown in Fig. 7 and are plotted in the accompanying graph. The response is maximal at 300 Hz. At very low frequencies (e.g., 60 Hz) the afferents provide a synchronous barrage of PSPs such that the membrane potential of the interneuron follows the sound pressure.

This interneuron receives most of its input from the cercus ipsilateral to the axon field. Direct electrical stimulation of cercal afferents demonstrated that the ipsilateral cercal nerve (i.e., ipsilateral with respect to the main portion of the interneuron's dendritic field and its axon) triggered a barrage of interneuron spikes while stimulation of the contralateral cercal nerve elicited only a very small depolarization (Fig. 6). Similar results were obtained in response to touching the tip of the cercus.

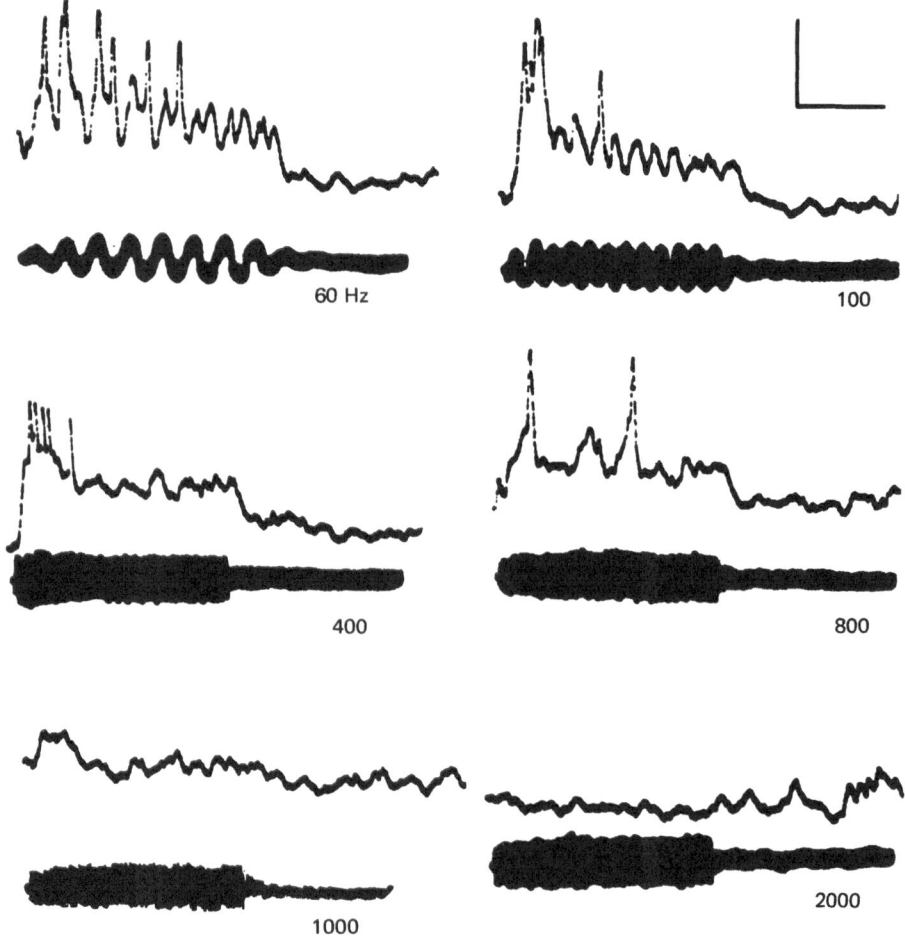

Fig. 7A.

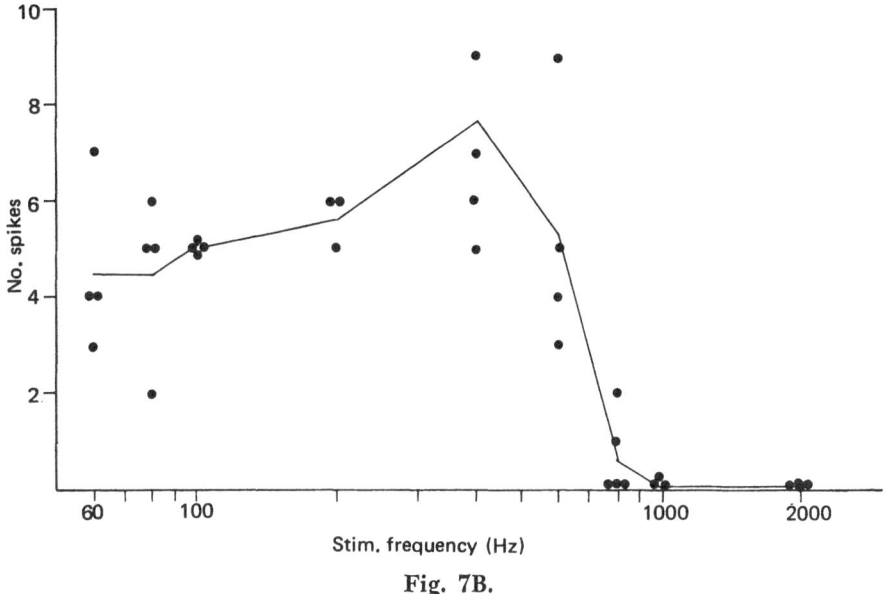

Fig. 7B.

Fig. 7A and B. Sensitivity of the medial giant interneuron of *Acheta* to tones. The sound pulse was 100 msec long and had an intersity of 80 dB. A. Sample of the recordings obtained from the soma. B. Response as a function of stimulus frequency for the preparation sampled in A. Calibration: vertical, 4 mV; horizontal, 40 msec.

The asymmetric input to the neuron was confirmed by crushing the cercal nerve ipsilateral to the interneuron and then recording the response to tones. It was impossible, at the same sound intensities (80 dB), to obtain a frequency-response curve: even the loudest tones (95 dB) elicited only a single spike.

The asymmetric dendritic field of this giant interneuron and the asymmetric response to stimulation of the two cerci is correlated with results from degeneration experiments carried out by John Edwards (personal communication) using the technique of Lamperter et al. (1969). His experiments show that the projection of cercal afferents is completely ipsilateral to the cercus.

There are other interneurons with smaller axons which possess symmetric dendritic fields. Their main branches wrap around the entering cercal nerve on both sides of the ganglion rather than only one side as is the case for the MGI. At the moment nothing is known about the responsiveness of these cells but it is very likely that they are nondirectional and receive input as strongly from one cercus as from the other. On the basis of the following criteria, the medial giant interneuron is one of the large, 'driven' cells recorded from by Edwards and Palka (1971): (1) it is one of the largest potentials, although never the largest, in extracellular recordings; (2) it possesses the second largest axon in the connective; (3) it usually shows no background discharge; (4) ipsilateral input is vastly more effective than contralateral input, and (5) it responds to tones and other stimuli in the same way.

Procion yellow injected into the integrating segment region of one MGI crossed to its bilateral mate in one out of seven injections (Fig. 8). This evidence suggests that an electrotonic junction exists between the two cells (Remler et al., 1968; Payton et al., 1969; also see Chapter 8). A similar suggestion has been made by Milburn and Bentley (1971) for cockroach giant interneurons on strictly anatomical grounds—there were no instances of stain passing from one cell to another. The fact that the uninjected neuron was as intensely stained as the injected one is unusual and may indicate that the electrotonic junction was injured. The result is similar to that shown by Remler et al. (1968), although they were using pressure injection and felt that the junction between the lateral giant interneuron and the giant motoneuron of the crayfish may have been injured by excess pressure. Since dye was injected into the MGI of the cricket electrophoretically, direct pressure was not applied to the proposed junction. An osmotically-induced pressure is a possible explanation for this result since regions near the injection sites often appear enlarged, probably due to water flow into the neuron induced by the high concentration of dye near the injection site. Whatever the reason, the "weakness" is confined to the region between the bilaterally symmetric cells since dye has never been observed to leak into other neurons nor has it been observed to leak out into the extracellular space. If the proposed electrical junction can be confirmed electrophysiologically then the contralateral excitatory input (Fig. 6) to the MGI must be due in part to this junction.

Developmental Aspects of the Form and Function of the MGI

One of the major questions one can begin to answer with the stain injection technique is: how is the geometry of an identified neuron's dendritic field controlled? More specifically, how much of the dendritic field is subject to alteration due to experimentally induced changes in afferents of a developing neuron?

When a cricket grows from hatching to adulthood deprived of one of its cerci, the gross morphology of the terminal ganglion is altered and the asymmetric responses of the giant fibers to cercal stimulation are altered (Edwards and Palka, 1971). The axon of the MGI is present in its usual position in the connective ipsilateral to the "lesion" but the volume of the neuropil ipsilateral to the lesion is decreased significantly (Edwards, personal communication). These changes immediately suggest the possibility that a change in the geometry of the neuron may account for new connections which have been made and for the alterations in the volume of the neuropil.

In order to answer these questions an estimate of the reproducibility of the anatomical technique is required. Two examples provide a qualitative estimate of the reproducibility inherent in the stain injection techniques in the insect nervous system. First, a fast motoneuron of the bug *Gerris* has been filled from a neuropilar recording site in two preparations; and the dendritic field which lies almost completely in one 45 μm cross sectional plane of the ganglion was remarkably similar for the two preparations, down to the finest resolvable branches (Fig. 9). Second, the general shape and position of the main dendritic branches of the cricket MGI were always the same. Smaller branches were also very similar from preparation to preparation. In order to demonstrate this in more detail, a portion of the dendritic field was reconstructed. This selected area of dendrite was easily recognized and readily reconstructed in three cases. In a fourth case the field was recognized but a continuous series of branches could not be reconstructed.

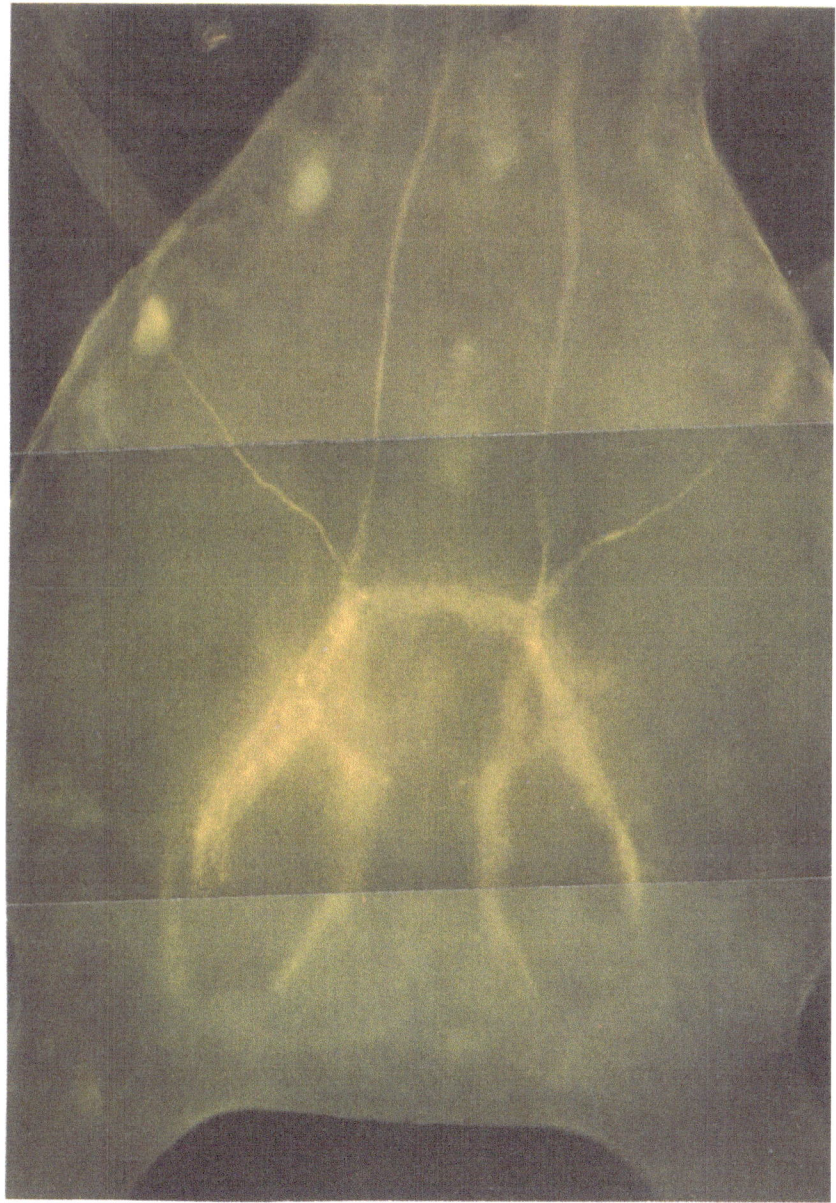

Fig. 8. The paired MGIs seen in a whole mount of the terminal ganglion of *Acheta*. One medial giant interneuron was injected from the thick integrating segment on only one side. Dye spread across a presumed electrotonic junction and filled the bilaterally symmetric cell as well as the one from which the injection was made.

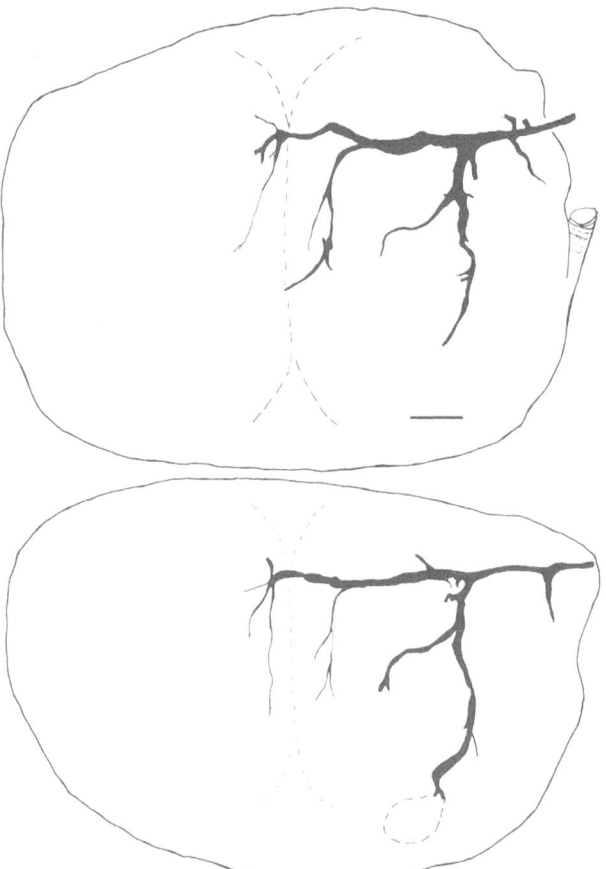

Fig. 9. Reproducibility in the branching pattern of a fast motoneuron in the bug *Gerris*. The neuron innervated mesothacic muscle 52 (nomenclature, Guthrie, 1961) and was injected from a dendritic branch. Calibration bar = 50 μm.

Two of the examples have been reproduced photographically by superimposing the negatives (Milburn and Bentley, 1971) from three successive 15 μm sections (Fig. 10A and B). There are very clear similarities between preparations. The general shape of the field is preserved from preparation to preparation but the details are not. For example, the antler-shaped field projecting ventrally possesses a Y-shaped branch near the midline in both cases (a), and a claw-shaped group of branches (b). The claw is made up of three to four branches but these are not strictly comparable in the two examples. Both the orientation and the size of the branches differ in detail. Similarly, in the dorsal branch (c) which projects toward the midline, there is a right-angle bend in the main branch in B but none in A.

I think these results demonstrate that it is possible to determine whether certain main areas of dendritic field are present or altered in an identified neuron; if we obtain

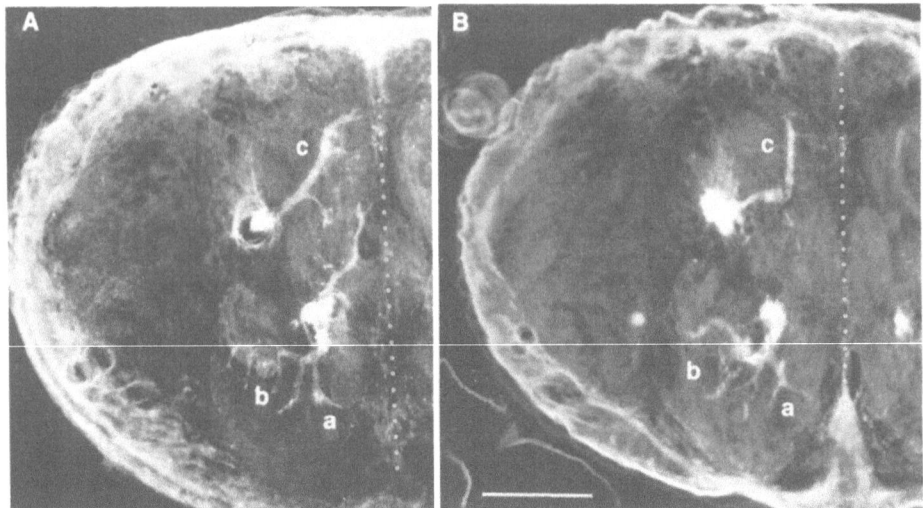

Fig. 10A and B. Reproducibility in the fine branches of one area of the dendritic field of MGI of *Acheta*. The photographs are 45 µm sections through ganglia from two different preparations obtained by superimposing negatives of three successive sections and printing them as a single photograph. The midline is dotted white in both cases. Calibration: 100 µm.

large changes in morphology we should be able to detect them. If the results are very subtle we have as yet no quantitative method of handling them.

Medial giant interneurons of adult *Acheta* which have been "deprived" (by cutting off the cercus ipsilateral to the neuron being studied, at hatching) have surprisingly normal dendritic fields. The data available do not allow a detailed analysis at this time but it is clear that the main dendritic branches are still present and the portion of dendritic field shown in Fig. 10 could be identified in these preparations. John Edwards, John Palka, and I are continuing these experiments and intend to present a more detailed analysis of differences as well as similarities between lesioned and normal interneurons.

Conclusion

The material presented here demonstrates some aspects of the physiology and anatomy of neurons in the insect nervous system. The problems which had hindered the use of insects as neurophysiological preparations are described and the resolution of these difficulties demonstrated. It is suggested that results such as those reported here might be most profitably followed up by developmental studies of neurons which can be identified from preparation to preparation. In this regard, intracellular staining techniques will certainly play a major role.

Acknowledgments. A number of people have contributed to this work. They know who they are and I thank each of them. Dr. David Bentley, Dr. John Palka, and Dr. John Edwards have kindly provided laboratory space at different times during the course of this work. This investigation was supported in part by Biomedical Sciences Support Grant FR-07035 from the General Research Support Branch, Division of Research Resources, Bureau of Health Professions Education and Manpower Training, National Institutes of Health.

References

Bentley, D. R.: Intracellular activity in cricket neurons during the generation of behavior patterns. J. Insect Physiol *15*, 677–699 (1969a).
——: Intracellular activity in cricket neurons during generation of song patterns. Z. vergl. Physiol. *62*, 267–283 (1969b).
Callec, J. J., J. C. Gullet, Y. Pichon, and J. Boistel: Further studies on synaptic transmission in insects. II. Relations between sensory information and its synaptic integration at the level of a single giant axon in the cockroach. J. exp. Biol. *55*, 123–149 (1971).
Davis, W. J.: Motoneuron morphology and synaptic contacts: determination by intracellular dye injection. Science *168*, 1358–1360 (1970).
Edwards, J. S., and J. Palka: Neural regeneration: delayed formation of central contacts by insect sensory cells. Science *172*, 591–594 (1971).
Falk, G. and P. Fatt: Linear electrical properties of striated muscle fibres observed with intracellular electrodes. Proc. R. Soc., Lond. B *160*, 69–123 (1964).
Guthrie, D. M.: The nervous system of *Gerris*. Phil. Trans. R. Soc., Lond. B *224*, 65–102 (1961).
Hagiwara, S., and A. Watanabe: Discharges in motoneurons of Cicada. J. cell comp. Physiol. *47*, 415–429 (1956).
Harris, C. L., and T. Smyth Jr.: Structural details of cockroach giant axons revealed by injected dye. Comp. Biochem. Physiol. *40A*, 295–303 (1971).
Hoyle, G.: Cellular mechanisms underlying behavior. In: Advances in Insect Physiology. Ed. J. W. L. Beament, J. E. Treherne and V. B. Wigglesworth. pp. 349–444. London: Academic Press, 1970.
——, and M. Burrows: Intracellular studies on identified neurons of insects. Fedn Am. Soc exp. Biol. *29*, 589 Abs. No. 1922 (1970).
Katz, B., and R. Miledi: Propagation of electric activity in motor nerve terminals. Proc. R. Soc., Lond. B *162*, 453–482 (1965).
Kerkut, G. A., R. Pitman and R. Walker: Sensitivity of neurons of the insect CNS to iontophoretically applied ACH or GABA. Nature, Lond. *222*, 1075 (1969a).
—— —— ——: Iontophoretic application of acetylcholine and GABA onto insect central neurones. Comp. Biochem. Physiol. *31*, 611–633 (1969b).
Lamparter, H. E., K. Akert, and C. Sandri: Localization of primary sensory afferents in the prothoracic ganglion of the wood ant (*Formica lugubris* Zett): A combined light and E.M. study of secondary degeneration. J. comp. Neurol. *137*, 367–376 (1969).
Milburn, N., and D. R. Bentley: On the dendritic topology and activation of cockroach giant interneurons. J. Insect Physiol. *17*, 607–623 (1971).
Narahashi, T.: The properties of insect axons. In: Advances in Insect Physiology. Ed. J. W. L. Beanent, J. E. Treherne, and V. B. Wiggleworth. pp. 175–256, 1963.
Payton, B. W., M. V. L. Bennett, and G. D. Pappas: Permeability and structure of junctional membranes at an electrotonic synapse. Science *166*, 1641–1643 (1969).
Remler, M. P., A. I. Selverston, and D. Kennedy: Lateral giant fibers of crayfish: location of somata by dye injection. Science *162*, 281–283 (1968).
Sandeman, D. C.: Integrative properties of a reflex motoneuron in the brain of the crab *Carcinus maenas*. Z. vergl. Physiol. *64*, 450–464 (1969).
Selverston, A. I., and D. Kennedy: Structure and function of identified nerve cells in the crayfish. Endeavour *28*, 107–113 (1969).
Takeda, K., and D. Kennedy: Soma potentials and modes of activation of crayfish motoneurons. J. cell comp. Physiol. *64*, 165–182 (1964).

Tasaki, I., Y. Tsukahara, S. Ito, M. J. Wayner, and W. Y. Yu: A simple, direct and
    rapid method for filling microelectrodes. Physiol. Behav. *3*, 1009–1010 (1968).
Willows, A. O. D.: Behavioral acts elicited by stimulation of single identifiable brain
    cells. Science *157*, 570–574 (1967).
Zucker, R. S.: Crayfish escape behavior and central synapses. I. Neural circuit exciting
    lateral giant fiber. J. Neurophysiol. *35*, 599–620 (1972a).
——: Crayfish escape behavior and central synapses. II. Physiological mechanisms under-
    lying behavioral habituation. J. Neurophysiol. *35*, 621–637 (1972b).
——: Crayfish escape behavior and central synapses. III. Electrical junctions and dendritic
    spikes in fast flexor motoneurons. J. Neurophysiol. *35*, 638–651 (1972c).
——, D. Kennedy, and A. I. Selverston: Neuronal circuit mediating escape responses
    in crayfish. Science *173*, 645–650 (1971).

Discussion

DR. BENNETT (Albert Einstein): A further point. I don't know what Dr. Murphey
would say, but if you press on a ganglion with its connective tissue intact, I wonder
how much you are deforming it? Is it possible that you form a channel that wasn't
there before?

DR. MURPHEY (Iowa): Mike, is there another possibility? In my particular case where
I am iontophoresing the dye maybe I am killing the cell or at least affecting the proposed
electrical junction, uncoupling it as I inject, thus blocking the movement of dye across
the junction. Occasionally, as in the case shown, this uncoupling occurs more slowly allowing
the dye to cross the junction.

DR. R. CHAPPELL (City University, New York): Regarding the problem of getting good
fills from different points of penetration, and in light of what Dr. Cohen told us this
morning about being able to record impulses from cell bodies in the roach four days
after colchicine is applied systemically, it might be interesting to try colchicine to see
if it facilitates stain migration from the cell body. Positive results might also provide
insight into the mechanism which accounts for the differences between the cell body and
the dendritic field in their staining and recording properties.

CHAPTER 10

Intracellular Staining Techniques in Gastropod Molluscs

C. R. S. Kaneko and S. B. Kater

Many laboratories, including ours, employ gastropod molluscs in the study of neural connectivity and the neuronal basis of specific behaviors. One of the primary reasons for selecting this group is that, by virtue of contrasting pigmentation, the cortically-arranged neuronal somata are easily seen and routinely identifiable. In this respect, gastropods differ from the arthropods and vertebrates, in which intracellular stain injection facilitates identification of individual neurons (see Chapters 6, 7, 9, 11, 12, 13, 14, 15, and 17). However, intracellular stain technology has valuable applications in addition to cell identification. In this chapter we will discuss the applications and methodology of intracellular stain injection in gastropod molluscs. These comments may be germane to other classes of animals, but, in our opinion, slight differences between animals can obviate direct transfer of methods.

Applications of Intracellular Staining Techniques

Intracellular staining techniques have been applied to gastropods to provide solutions to two sorts of problems. Microelectrode injection of stain has been used to obtain information about the neuronal architecture of visually identifiable (Kerkut et al., 1970; Sakharov and Salánki, 1971) and physiologically characterized neurons (Cottrell, 1970, 1971; Kater et al., 1971; Kater and Kaneko, 1972). Axonal iontophoresis has been used to locate the somata of both peripheral (Kater and Rowell, 1973) and central (C. R. S. Kaneko, unpublished) neurons whose axons are found in specific nerve trunks. These applications represent a small proportion of the potential uses of intracellular staining technology, and there is no reason why more sophisticated information, like that discussed in Chapters 16, 17, and 18 cannot be obtained in gastropods.

Our experience with the pulmonate snail *Helisoma trivolvis* suggests that any neuron that can be penetrated with a microelectrode can be filled successfully. The microanatomy of any penetrable neuron is therefore amenable to description. More important, however, is the complement that intracellular staining can provide to established electrophysiological procedures. For example, the course of axons of particular neurons cannot be established solely by recordings of electrical activity in the somata. An inactive zone between the soma and the axon may significantly distort somatically recorded potentials. Furthermore, the activity of axonal branches might not be recorded reliably from the soma because of low safety factors at branch points (cf. Mulloney and Selverston, 1972). Finally, endogenous activity of somatic origin tends to distort and/or obscure potentials arising in branches. In our opinion, electrophysiological properties do not affect the staining characteristics, and reliable identification of axonal processes is obtained readily by intracellular injection of stain.

Microelectrode techniques are not always required to reveal the geometry of specific molluscan neurons. There are many situations in which extracellular recording and stimulation can reveal the presence of neural elements of interest within a particular nerve trunk. In such cases the axonal iontophoretic technique (Iles and Mulloney, 1971; Chapter 7) can be employed to locate the somata and neurites of these axons. For example, in the course of an investigation on the neural control of feeding in *Helisoma*, extracellular recordings revealed the presence of a mechanoreceptor in the buccal mass. Physiological approaches failed to discern whether the somata of these receptors were located centrally or peripherally, and more important, whether our recordings were from primary sensory cell axons or from axons of second order neurons. The relatively simple technique of axonal iontophoresis of Procion yellow located the somata of primary sensory neurons in the periphery (Fig. 1; Kater and Rowell, 1973).

Integrating both microelectrode injection and axonal iontophoresis into our usual physiological approach to the study of neural interactions in *Helisoma* has saved immeasurable time. The best example of this is the increased efficiency of locating the somata of motoneurons. Previously, this required microelectrode penetration of a large number of neurons in the CNS with a low probability of finding the motoneurons of interest. We now employ the following procedure routinely. First, the nerve trunk containing axons of motoneurons is identified by extracellular stimulation of the distal

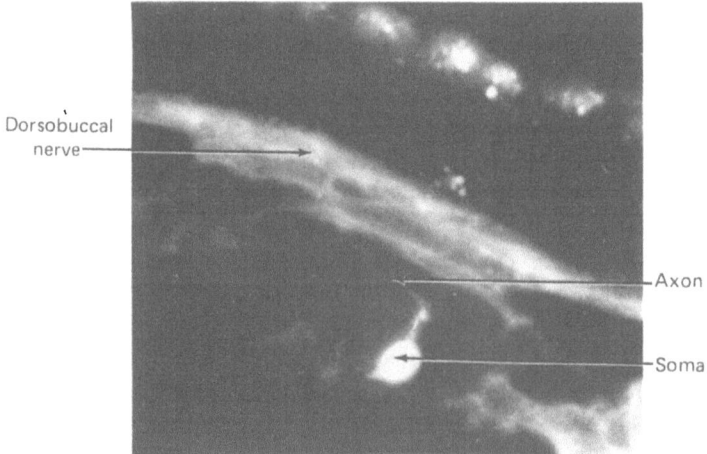

Fig. 1. The localization of mechanoreceptor somata in the periphery. The dorsobuccal nerve was cut at its junction with the buccal ganglia and the cut proximal end was subjected to axonal iontophoresis. The dye Procion yellow M4RAN was passed down the nerve while imposing a 5×10^{-9} A cathodal current for a period of 2 hr. Material was fixed in Carnoy's fixative, embedded in paraffin, and 10 μm sections were cut. This procedure revealed a number of candidate mechanoreceptor somata. One such soma is shown along with the dorsobuccal nerve and the faint axon arising from the soma and joining the dorsal buccal nerve. Calibration: 10 μm.

end of the cut nerve while monitoring the muscle for signs of activity. Once the appropriate trunk is located, the proximal end of the cut nerve is subjected to axonal iontophoresis (usually with cobalt, because of the ease of viewing whole mount preparations). This results in filling a number (10–30) of somata within the CNS. Not all of these somata are motoneurons since most nerve trunks are mixed nerves. However, the probability of locating motoneuronal somata is greatly increased over random penetration. Physiological criteria are applied to support the identification of motoneuron somata. Finally, somata are filled by microelectrode injection (usually with Procion yellow, because it has least effect on the electrical properties of the cell and because of the greater contrast with the heavily pigmented periphery). If the distance is not too great, one can expect the motoneuron to fill with stain all the way to its terminals on the muscle.

Technical Considerations

Intrasomatic Injection

As discussed in Chapter 1, intracellular staining techniques offer many advantages over standard histological techniques, the most obvious being speed and simplicity. For instance, with *Helisoma* we have used the "one shot" technique (Chapter 15) in which a massive current (tens of microamperes as compared with the normal tens of nanoamperes) is delivered through the stain-containing electrode in one extended pulse (of a few seconds duration). This technique, in combination with a rapid fixation, dehydration series, and clearing has allowed us to visualize the main axonal branches of a particular neuron within an hour of penetration by the stain-filled microelectrode. We have used this technique with both cobalt (Pitman, Tweedle, and Cohen, 1972) and Prussian Blue (Bultitude, 1958). This technique should be applicable to Procion yellow injection since it is our impression that the mobility of Procion yellow is greater than that of either of the former stains. Longer fill times, followed by a period of incubation, have allowed us to visualize some of the fine processes in whole mount preparations (see Fig. 2), thereby eliminating the necessity for sectioning and reconstruction. This has been done with both cobalt and Procion yellow.

A problem does arise if one is interested in the finest neurites and their branching patterns. Kerkut (personal communication), working on the land snail *Helix aspersa,* finds that Procion yellow often fails to invade the finer branches of the axons, especially when there is a change in the size of the axon due to branching. We have noted the same phenomenon using cobalt and Prussian Blue, and to a lesser extent using Procion yellow. Using the larger (i.e., greater molecular weight) dyes, Procion black H-GSA and Procion gray H-NS, we were unable to pass dye beyond the large proximal axonal segments. On the other hand, Strumwasser (personal communication) has had a very high rate of success in completely filling neurons of the sea hare *Aplysia californica* with Procion yellow. Strumwasser's success in filling the finer neurites might be due to one or both of two modifications of the original technique of Stretton and Kravitz (1968). First, he has quite successfully adapted, for use with Procion yellow, a technique perfected in his laboratory for introduction of various substances into neural somata by pressure injection. This technique is explained fully in Chapter 20 of this

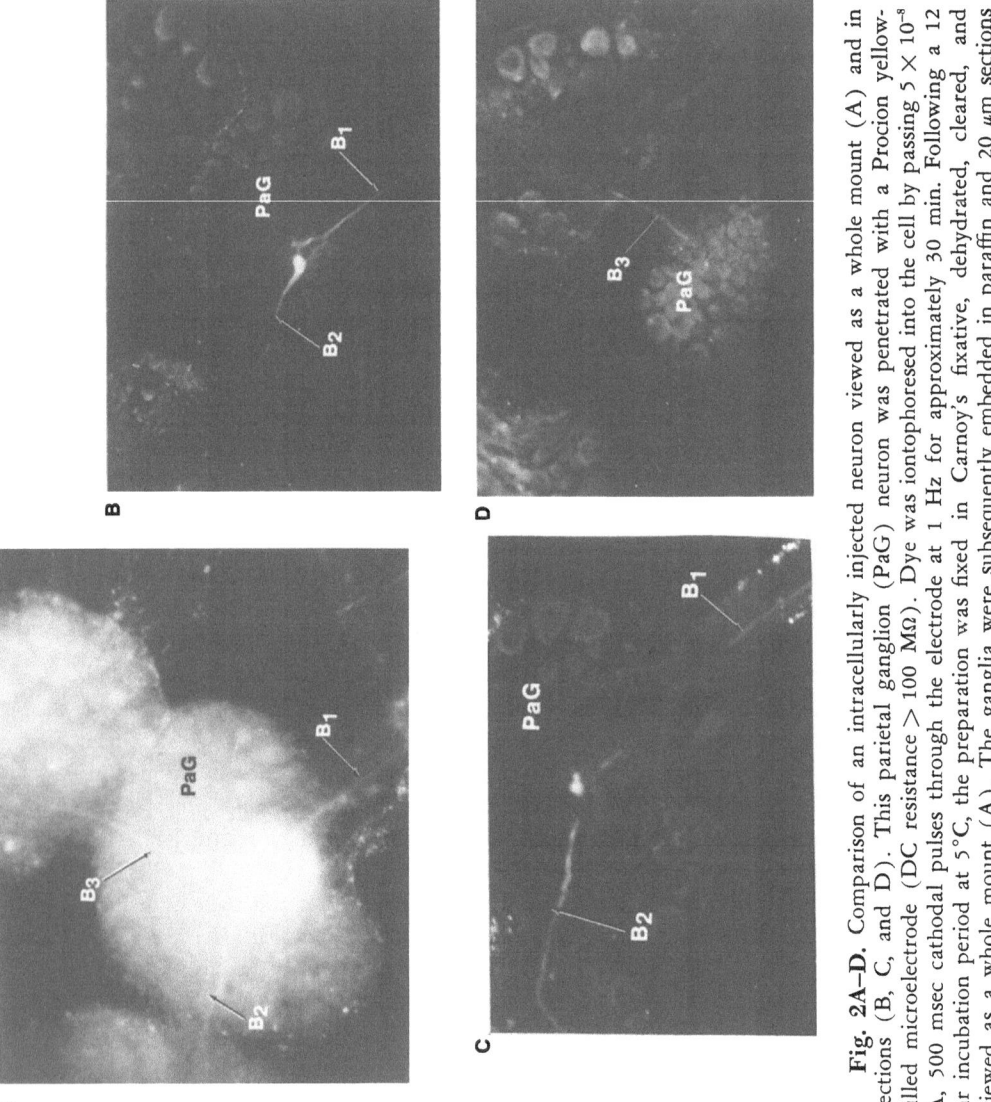

Fig. 2A–D. Comparison of an intracellularly injected neuron viewed as a whole mount (A) and in sections (B, C, and D). This parietal ganglion (PaG) neuron was penetrated with a Procion yellow-filled microelectrode (DC resistance > 100 MΩ). Dye was iontophoresed into the cell by passing 5×10^{-8} A, 500 msec cathodal pulses through the electrode at 1 Hz for approximately 30 min. Following a 12 hr incubation period at 5 °C, the preparation was fixed in Carnoy's fixative, dehydrated, cleared, and viewed as a whole mount (A). The ganglia were subsequently embedded in paraffin and 20 μm sections were taken. B, C, and D are examples of such sections. As can be seen by comparing A with B, C and D, the fine parts of the main axons branches (B1, B2, and B3) are clearly visible in the whole mount

volume and is quite distinct from the electrophoretic technique used by Kerkut and by us. Second, his preparations are cultured in a specific medium for 3 or 4 days, following injection, to allow time for diffusion or axonal transport, as the case may be.

It is our impression, based on a limited comparison of the two techniques, that pressure injection results in a darker, and possibly more complete, fill than that obtained by iontophoresis. This difference might be a function both of the larger quantity of dye which can be introduced into the soma with the pressure technique and of the geometry of the neuron. A rapid decrease in diameter of a neurite would greatly increase the axial resistance of that neurite and might result in blockage. Such blockage would be overcome more readily by an increase in pressure than an increase in current, since an increase in pressure acts to distend the neuron (up to twice the normal size in the soma), resulting in a decreased resistance to flow of the stain. On the other hand, pressure injection can rupture the neuron so that the dye no longer remains within that particular neural element.

Procion yellow has been used successfully in a number of molluscan preparations. In addition to those cited above, Willows (personal communication) has used both pressure and iontophoretic injection to study the morphology of neurons of *Tritonia diomedia.* It is generally agreed that success is often associated with the appearance of an orange hue following the beginning of injection and that, to obtain a complete fill, the soma must appear bright orange following dye injection. The fulfillment of these criteria depends upon the use of an electrode that does not block. Blockage of the electrode is less likely to occur if the tip is large (i.e., 1–2 μm in diameter) and has a resistance of about 10 MΩ when filled with a saturated solution of Procion yellow. However, large tip size and low electrode impedance are not essential for success; we have successfully used electrodes (ca. 0.5 μm tip diameter) with resistance of over 100 MΩ. The resistance of Procion-filled electrodes, as with other microelectrodes, may be significantly decreased if the technique of fiber filling is employed (Tasaki et al., 1968). Lower resistance electrodes show less distortion of waveforms and allow greater membrane polarization due to the increased ability to pass current. The addition of such fibers does not appear to effect one's ability to penetrate neural elements. Other, and possibly more important, procedures that are generally employed to prevent blocking are the filtering of the Procion solution through micropore filters of the order of 0.22 μm or less and the cleaning of the microelectrode glass. A good deal of frustration may be eliminated if each electrode is checked for dye-passing ability before impaling the neuron. This simply involves observation (low power) of the submerged tip of the electrode while applying pressure and/or current. Finally, plugging of the electrode seems to be less frequent if penetration is made while a low current or pressure is applied to force solution out of the microelectrode.

There is little information available on the parameters of intrasomatic injection. For Procion, we have used 0.5 sec pulses at 1 Hz of hyperpolarizing current at 5×10^{-8} A for iontophoretic injection. As stated above, currents of higher intensity and longer duration may be employed. As for pressure injection, the lowest pressures are usually associated with a lower probability of blockage. However, they also introduce smaller amounts of stain. We have used approximately 40 psi in most cases and employ the apparatus shown in Chapter 20.

Axonal Iontophoresis

We have used both cobalt and Procion yellow as stains in the axonal iontophoretic paradigm. Cobalt seems to offer an advantage because some spillage of the stain often occurs. Since the cobalt chloride must be reacted with sulfide to obtain the cobalt sulfide precipitate, excess cobalt chloride may be washed off before precipitation. In contrast, Procion yellow adheres to many tissues and is not easily washed off, often causing problems of excess fluorescence which obscures the paths of axons. This advantage of cobalt sulfide must be weighed against the distortion of the microanatomy which seems inherent in the reaction of tissues with the ammonium sulfide necessary to form the precipitate. Although we have not used it, sodium sulfide, which may cause less deleterious effects, could be substituted for the ammonium sulfide. Procion yellow, in our opinion, offers the advantage of greater resolution when compared to cobalt. Although we have employed a wide range of parameters for axonal iontophoresis, no significant advantages of a particular frequency, intensity, or duration of application of current have been apparent. However, a recent adaptation of the axonal iontophoretic technique in the absence of current (see Chapter 20) appears to give the best results.

In summary, it appears that all the present techniques of intracellular staining can be applied to molluscan preparations.

References

Bultitude, K. H.: A technique for marking the site of recording with capillary microelectrodes. Q. J. microsc. Sci. 99, 61–63 (1958).

Cottrell, G. A.: Direct postsynaptic responses to stimulation of serotonin-containing neurons. Nature 225, 1060–1062 (1970).

———: Synaptic connections made by two serotonin-containing neurons in the snail (Helix pomatia) brain. Experentia 27, 813–815 (1971).

Iles, J. F., and B. Mulloney: Procion yellow staining of cockroach motor neurones without the use of microelectrodes. Brain Res. 30, 397–400 (1971).

Kater, S. B., C. B. Heyer, and J. P. Hegmann: Neuromuscular transmission in the gastropod mollusc Helisoma trivolvis: identification of motoneurons. J. Comp. Physiol. 74, 127–139 (1971).

———, and C. R. S. Kaneko: An endogenously bursting neuron in the gastropod mollusc, Helisoma trivolvis: characterization in vivo. J. Comp. Physiol. 79, 1–14 (1972).

———, and C. H. F. Rowell: Integration of sensory and centrally programmed components in the generation of cyclical feeding activity of Helisoma trivolvis. J. Neurophysiol. 36, 142–155 (1972).

Kerkut, G. A., M. C. French, and R. J. Walker: The location of axonal pathways of identifiable neurones of Helix aspersa using the dye Procion yellow M-4R. Comp. Biochem. Physiol. 32, 681–690 (1970).

Mulloney, B., and A. Selverston: Antidromic action potentials fail to demonstrate known interactions between neurons. Science 177, 69 (1972).

Pitman, R. M., C. D. Tweedle, and M. J. Cohen: Branching of central neurons: intracellular cobalt injection for light and electron microscopy. Science 176, 412–414 (1972).

Sakharov, D. A., and J. Salánki: Study of neurosecretory cells of Helix pomatia by intracellular dye injection. Experientia 27, 655–656 (1971).

Stretton, A. O. W., and E. A. Kravitz: Neuronal geometry: determination with a technique of intracellular dye injection. Science 162, 132–134 (1968).

Tasaki, I., Y. Tsukahara, S. Ito, M. J. Wayner, and W. Y. Yu: A simple, direct and rapid method for filling microelectrodes. Physiol. Behav. 3, 1009–1010 (1968).

CHAPTER 11

Morphological Identification of Single Cells in the Fish Retina by Intracellular Dye Injection

Akimichi Kaneko

The vertebrate retina is a highly complicated nervous structure. Cells are arranged in three distinct layers: the layer of photoreceptors; the inner nuclear layer which consists of horizontal, bipolar and amacrine cells; and the layer of ganglion cells, the output end of the retina. Signals received at the receptor mosaic are transmitted from one layer to the next in sequence. Extensive processing of signals is performed within this complicated nervous network. As a result, the ganglion cells are highly differentiated for detecting various features of the external environment, such as contrast, color, movement and so on. The analysis of this nervous network and its physiological role has been the interest of many investigators. Of various methods applied, the electrophysiological recording techniques have been the most powerful for this type of study, since one can record from individual neurons and compare the behavior of one type of cell with the next.

In the electrophysiological study of the vertebrate retina, extracellular recording has been most often used and has produced many important studies, but its application is limited only to cells which generate all-or-none action potentials. Except for ganglion cells, it appears that retinal cells do not generate spike discharges but signal by graded potential changes which can only be detected with an intracellular recording method. Such studies have been attempted from two approaches. One is to select a favorable preparation such as the mudpuppy retina where the cells are very large and are relatively easily penetrated with microelectrodes, as first demonstrated by Bortoff (1964). Another approach for successful intracellular penetration is to use extremely fine glass pipettes together with a jolting device designed by Tomita (1965). The jolter, in principle, accelerates the retinal tissue against a micropipette. This is equivalent to advancing the microelectrode into the cell with an acceleration. An acceleration of about 100 g (gravitational acceleration) was found effective. Recently Baylor and Fuortes (1970) succeeded in penetrating turtle cones by passing an oscillating electric pulse current into the glass pipette placed very close to the cell. Presumably the pipette was accelerated by deformation caused by the current.

With these techniques, reports on intracellular recordings are accumulating for the retinas of various animals, such as the mudpuppy (Werblin and Dowling, 1969; Toyoda et al., 1969; Norton et al., 1970), the frog (Toyoda et al., 1970; Matsumoto and Naka, 1972), the turtle (Baylor and Fuortes, 1970; Baylor et al., 1971) and fish

(Tomita, 1965; Tomita et al., 1967; Kaneko and Hashimoto, 1967, 1969; Kaneko, 1970, 1971a, b; Kaneko and Yamada, 1972).

Stains Used for Identification of Single Retinal Cells

Reliable identification of penetrated cells is essential for a correlation between morphology and response characteristics of every cell. In a tissue like the retina, where an electrode cannot be manipulated under direct visual control, stain injection from the recording electrode and subsequent histological examination has been the most powerful method.

For faithful identification of the penetrated cells, the following conditions, at least, must be satisfied: (1) that the electrode stay within the same element during both recording and stain injection; (2) that the stain be retained in this element alone, in visible form, after preparation for histological examination; (3) that the structure of the marked cell and neighboring tissue be adequately preserved for morphological identification after passing current and preparing the tissue histologically; and (4) that no false localization occur by leakage of stain into the extracellular spaces and retention by other cellular and extracellular compartments.

Many kinds of stains have been used for identification of individual retinal cells. The author's experience has been limited mainly to Niagara Sky Blue and Procion yellow, and only these will be described in full detail. It is now clear that Procion yellow is superior to any other stain so far available, but the comparison between the two stains will contrast the characteristics of the two.

Niagara Sky Blue

Niagara Sky Blue 6B was first used by Potter et al. (1966) for their study of the connection between cells of the developing squid. The advantage of this stain is that it can stay in the injected cell during the course of later histological procedures, since it is highly soluble in water but only slightly in organic solvents. We used this stain to identify single carp cones (Kaneko and Hashimoto, 1967).

Pipettes were filled by boiling them in a 4% (w/v) aqueous solution of the stain for more than 10 min. After recording from a cell, the stain was injected by the arrangement shown in Fig. 1. The recording electrode was connected to the negative terminal of a 135 V battery (B) through a 40 MΩ current-limiting resistor (R_L). When the current was switched on, it set up forces at the electrode tip which caused a small portion, 1–2 μm in length, to break off, but the tip diameter was still about 0.5 μm, small enough to stay inside the cell. Although the electrode resistance decreased after its tip was broken, the amount of current was limited to less than 3×10^{-6} A by the resistor in series.

The stained cell appeared as a tiny blue dot which could be traced throughout the subsequent procedures. When the dye was ejected into the extracellular space, it was visible at first but diffused away within seconds. As a visual aid in handling the preparation, two large red dots were placed on the retina about 200 μm apart, one on each side of the injected cell. The retina was then fixed overnight at about 2°C in aldehyde mixture acidified to pH 4 (Potter et al., 1966). After dehydration in acetone, the retina was cleared in propylene oxide and embedded in Epon (Luft, 1961). Serial sections were cut at a thickness of about 10 μm, stained with 0.5% basic fuchsine

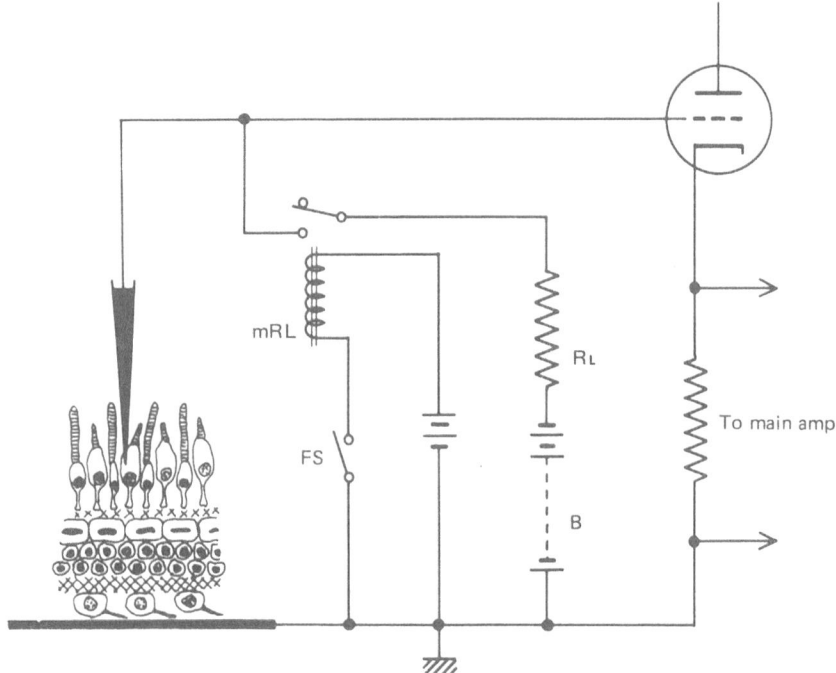

Fig. 1. Schematic drawing of the arrangement for intracellular injection of Niagara Sky Blue. After recording, a microrelay (mR$_L$) is activated and the electrode is connected to the negative terminal of a 135 V battery (B) through the current limiting resistor (R$_L$). FS, a foot switch. (Reproduced from Kaneko and Hashimoto, 1967, by permission of Pergamon Press.)

alcoholic solution and examined under the light microscope. Since it was possible to select sections containing the stained cell using a dissection microscope, only a few sections were examined under the high power microscope. An example of Niagara Sky Blue injection will be presented (see below).

Procion Yellow

Cells marked with Niagara Sky Blue can be accurately localized, but since it fails to diffuse into the dendritic and axonal trees, positive identification is not possible if cell bodies of more than one type are present in a given region, such as the inner nuclear layer. In 1968, Stretton and Kravitz introduced a new dye capable of revealing the entirety of the cell. When it was injected into the cell body, the dye diffused into the finest branches and thereby revealed the entire cell in much the same way as does the silver impregnation technique of Golgi. This dye, Procion yellow M4RS, has been applied to retinal cells with great success (Kaneko, 1970, 1971a; Kaneko and Yamada, 1972; Matsumoto and Naka, 1972; see also Chapter 12).

Electrodes

Micropipettes are made from a 9 cm Pyrex glass tubing of about 0.8 mm outer and 0.5 mm inner diameter. It is heated at the center by an iridium-platinum ribbon heater 50 μm thick and 4 mm wide. During heating a weak tension is applied to the glass tubing to pull its ends apart. After a few seconds, a strong pull is applied and a pair of pipettes is obtained. In order to make pipettes of the small tip diameter (less than 0.1 μm) which is essential for intracellular recording from small retinal cells, one has to adjust various factors of the puller; of many factors the width of the heater ribbon and the strength of the final pull are the most influential.

The pipettes are then filled with electrolytes or stain solution. The conventional way of filling pipettes was to boil them in the solution to be filled for a few minutes. However, filling pipettes by boiling is not applicable to Procion dyes since they are decomposed by heating.

Various ways of filling pipettes without heating have been suggested, but the method developed by Tasaki et al. (1968) was found most useful for pipettes of shallow taper and small tip diameter. In short, bundles of 10 or more glass fibers are inserted manually into the tubing before setting it on the puller. After a pipette is made, the glass fibers are removed from the body of the pipette to provide space to fill it, but parts remain in the tip. A hypodermic syringe is filled with dye solution, and the needle is inserted into the pipette until the needle comes to the neck of the pipette. The dye solution is now injected slowly, and it goes to the tip by the capillary action in a few seconds. Sometimes a small air bubble remains near the neck but it does not disturb the characteristics of the electrode.

Later M. Murakami (personal communication) developed a method of making glass tubing containing a linear filament inside. This makes it unnecessary to insert glass fibers into each separate length. The new glass tubing is made from a long Pyrex tubing of 6 mm outer and 4 mm inner diameter on a setup designed by Dr. A. L. Byzov (Fig. 2). A glass filament of 0.8 mm diameter is inserted into the thick tubing before setting. The tubing is suspended vertically and a part of the tubing near the bottom is heated in a cylindrical heater. The bottom end is now pulled out from the heater while the upper end is slowly advanced. The ratio of the speed of pulling out to that of insertion determines the diameter of the thin tubing. The inner glass filament fuses to the inner wall of the tubing and is pulled with it. When this tubing is made into micropipettes, this glass filament serves as a pathway for solution from the neck to the tip by its capillary action. This type of pipette is filled by the same method as applied to pipettes filled with fiber glass.

Recording

Microelectrodes filled with 4% (w/v) aqueous solution of Procion yellow have been used both to record intracellularly from individual cells and to inject the dye. These electrodes can be used in the conventional manner, but a minor difficulty with them for recording is that the resistance is usually two to three times higher than that found in pipettes filled with 3M KCl. The resistance may reach 300 to 500 MΩ.

Electrophoretic Dye Injection

Since in aqueous solution Procion yellow has negative charges, a current from the tissue to the electrode (a negative current from the electrode) drives the dye into the cell. Stretton and Kravitz (1968) used intermittent current but a steady current works as well. A large amount of current will damage the cell and result in the leakage of the dye from the impaled cell to the surrounding tissue. The optimum current varies

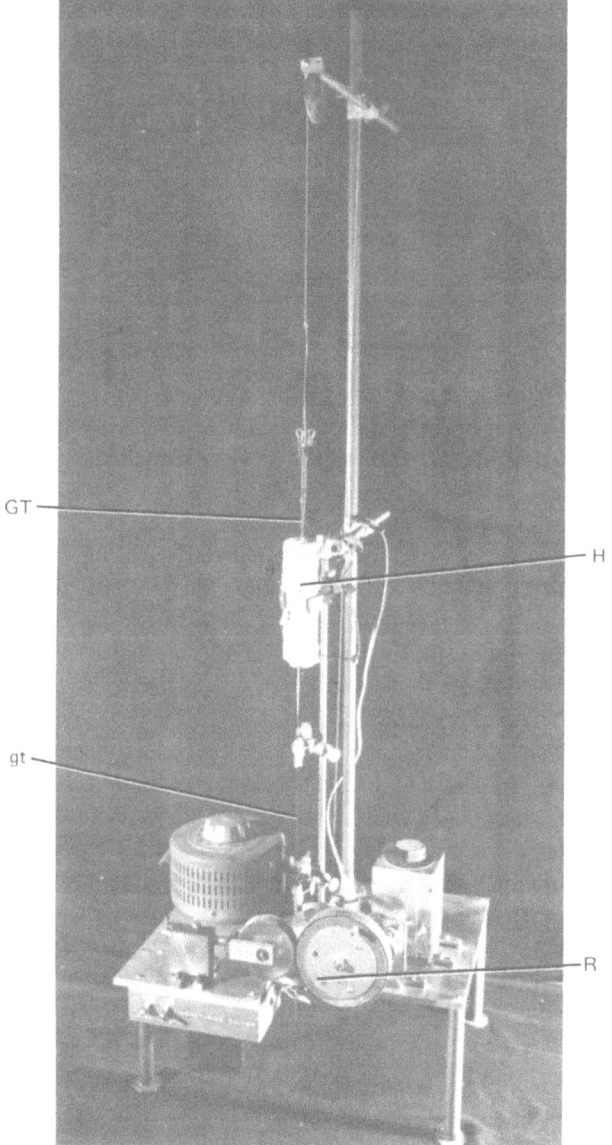

Fig. 2. Photograph showing the setup for making a thin glass tubing with a linear filament inside. A thick tubing (GT) is slowly fed into the cylindrical heater (H), while the bottom end (gt) is being pulled out by the roller (R). (Courtesy of Drs. A. L. Byzov and M. Murakami.)

from cell to cell, but in the author's experience 2 to 3×10^{-9} A was optimal for small cells like those in the carp and goldfish retina. This amount of current did not damage the cell: more than half of the cells continued to respond after the injection. The maintenance of responses after injection is strong evidence that the electrode stayed in the same cell during and after the dye injection.

· The more dye injected by a long application of current, the more brightly the cell fluoresces. Usually in the retina, the duration of current application is limited to a few minutes, since it is extremely difficult to hold a cell much longer. Fortunately, injection for such a short time was found to be enough to visualize the injected cells.

For a successful injection it is necessary to monitor the amount of current by a meter inserted in series with the circuit, since the electrode resistance increases when current is passed. Calculation from the supply voltage and limiting resistance often fails to give the true value of the current. For a steady current injection, a constant current source is recommended.

Fixation and Other Subsequent Histological Procedures

The injected retinal cells are usually invisible under the dissection microscope since the amount of injected dye is limited. As a visual aid, two large spots are made with india ink or similar substance on each side and at equal distances from the injected cell. These spots are visible and help us localize the injected cell when the tissue is trimmed before embedding.

In order to let the dye diffuse within the injected cell, the preparation is kept in a moist chamber for 10 to 15 min before fixation. The retina is then fixed according to the original method (Stretton and Kravitz, 1968) or with slight modification. In short, the retina is fixed overnight in aldehyde mixture acidified to pH 4 with acetate buffer. After the tissue has been dehydrated in graded acetone-water mixtures, it is cleared by propylene oxide and embedded in Maraglas epoxy resin. Paraffin can be used, too. Epon is not recommended since its autofluorescence masks weak fluorescence from fine processes of the injected cells. The retina is sectioned with a microtome equipped with a steel knife. Ten μm sections have been found most convenient. Serial sections are mounted with a nonfluorescing medium, such as Lustrex or Bioleit.

The sections are examined under a fluorescence microscope. The illuminator should be equipped with a filter which passes only the deep violet region of the spectrum, but quartz optics are not necessary. A dark field condenser is helpful in minimizing transmission of the excitation light into the viewing microscope. The microscope should be equipped with a yellow filter (e.g., Wratten 8) to selectively transmit the fluorescence wavelengths and to block the excitation spectrum. Care should be taken with the intensity of the light source: with very bright illumination the fluorescence fades away in several minutes.

The preparations can be stored for years if maintained in good conditions.

Application of Niagara Sky Blue and Procion Yellow to the Vertebrate Retina

Morphological Identification of Single Retinal Cells

Structure of the Fish Retina

The structure of the fish retina is well described by Ramón y Cajal (1893) and illustrated in Fig. 3. The left section shows the vertical retinal elements, the most

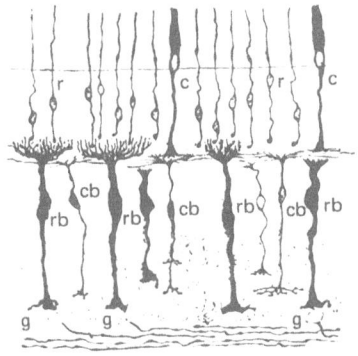
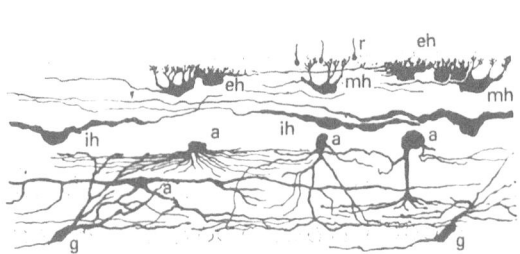

Fig. 3. Vertical section of the teleost fish retina stained by the Golgi technique. Vertical elements (left section) and horizontal elements (right section) shown separately; for a complete model both groups must be superimposed. c, cones; r, rods, cb, small bipolar cells; rb, large bipolar cells; eh, external horizontal cells; mh, intermediate horizontal cells: ih, internal horizontal cells; a, amacrine cells; g, ganglion cells. (Reproduced from Ramón y Cajal, 1893.)

direct pathway from receptors to ganglion cells. Two types of receptors are seen, rods and cones. Bipolar cells are also classified into two types, large bipolar and small bipolar cells. The right section shows the horizontal elements which are the routes of lateral connections within the retina. To this group belong horizontal cells of three different types and amacrine cells. Most vitread lie ganglion cells which send their axons to the higher central nervous system.

Receptors

Intracellular recordings from vertebrate photoreceptors were first made by Tomita (1965) in the carp retina. As shown in Fig. 4A, photoreceptors are hyperpolarized by illlumination. We, in Tomita's laboratory, demonstrated that these responses come from single cones by injecting Niagara Sky Blue (Kaneko and Hashimoto, 1967). Figure 4B is a photomicrograph of an injected cone. This picture shows that the cone myoid and nuclear region are deeply stained, while the outer segment and the ellipsoid remain colorless: Niagara Sky Blue stayed in the part of the cell where it was injected. Fortunately, since only the receptors have their cell bodies in this layer of the carp retina, this stain was sufficient to identify the cell as a receptor.

There are two types of receptors: rods and cones. In the carp both the outer and the inner segment of rods are very slender, and it is unlikely that they are ever successfully penetrated.

Carp and goldfish are known to have well developed color vision. Marks (1965) measured the absorption spectra of single cones in the goldfish with a microspectrophotometer. He detected three types of pigments having different sensitivity maxima (455, 530 and 625 nm), each contained in different cells. By intracellular recording Tomita et al. (1967) also showed three types of cones in the carp. As illustrated in Fig. 5, these responses were always hyperpolarizing regardless of wavelengths. Each cell had a response maximum at a different region of the spectrum; red (611 nm), green (529 nm) and blue (462 nm), in good agreement with spectrophotometry.

Intracellular recordings from receptors have been reported also in the mudpuppy

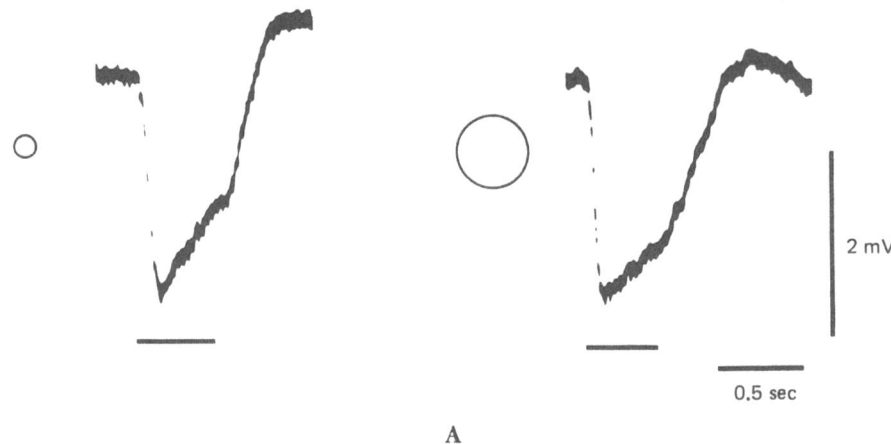

2 mV

0.5 sec

A

B

Fig. 4A and B. A shows responses from a single carp cone to a 100 μm white light spot (left) and to a broad illumination of about equal intensity. Response shapes were slightly distorted by recording through a CR-coupled amplifier. Horizontal line below the traces indicates the period of illumination. Negativity downwards. B is a photomicrograph showing a single cone injected with Niagara Sky Blue. (Reproduced from Kaneko and Hashimoto, 1967, by permission of Pergamon Press.)

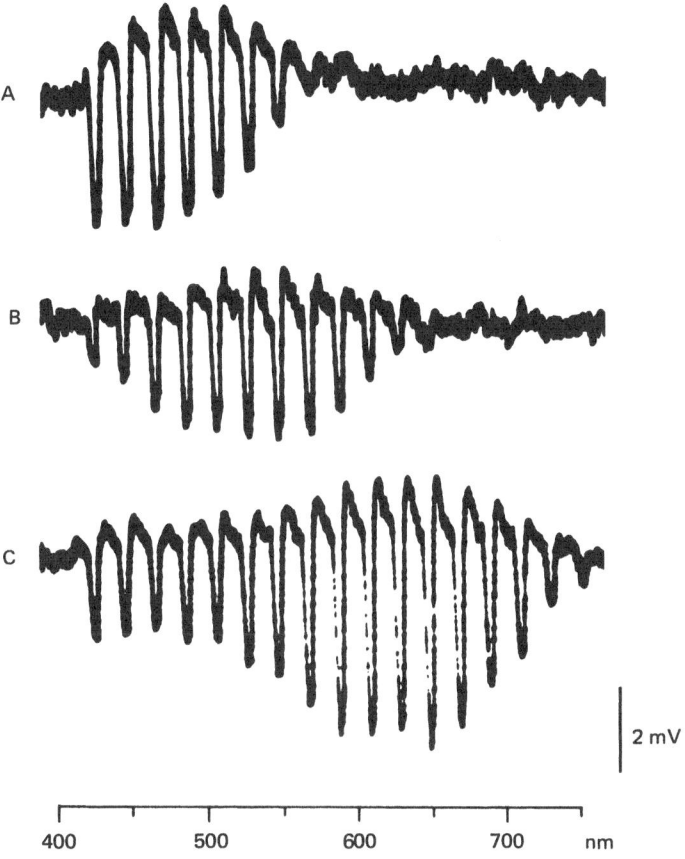

Fig. 5A–C. Responses from three types of cones to light spots of various wavelengths. Light flash of 0.3 sec in duration was presented about every one sec. Quantum flux of each monochromatic light was adjusted equal. Recording with a CR-coupled amplifier. A. Blue-sensitive cone; B. Green-sensitive cone; C. Red-sensitive cone. (Reproduced from Tomita et al., 1967, by permission of Pergamon Press.)

(Werblin and Dowling, 1969), the gecko (Toyoda et al., 1969), the turtle (Baylor and Fuortes, 1970; Baylor et al., 1971), the frog (Toyoda et al., 1970) and the axolotl (Murakami and Pak, 1971; Grabowski et al., 1972). These were identified by stain injection; some investigators used Niagara Sky Blue and others Procion yellow.

Receptors of these animals all showed hyperpolarizing responses to illumination; no difference was found in the response polarity between cones and rods. The receptor potential of most of the animals did not show obvious spatial summation (see Fig. 4A). Only in the turtle cones, Baylor et al. (1971) demonstrated an excitatory convergence from a near surrounding region of about 40 μm in radius and an inhibitory interaction from the far surround.

Horizontal Cells

Horizontal cells are found throughout the vertebrates. There are several morphological subtypes; for example, goldfish and carp have three types, namely external, intermediate and internal horizontal cells (Ramón y Cajal, 1893). Typically, horizontal cells are flattened cells lying parallel to the retinal surface. Thin dendrites come out of the sclerad surface and make contact with receptors as the lateral elements of the triad structure of the synapses (Stell, 1967).

S-potentials, graded sustained responses from the retina of many vertebrates, were found to originate in horizontal cells (MacNichol and Svaetichin, 1958; Mitarai, 1958; Oikawa et al., 1959; Werblin and Dowling, 1969; Steinberg and Schmidt, 1970). By injecting Procion yellow into cells giving S-potentials in the fish, the author confirmed these findings and demonstrated that all three types of horizontal cells generated S-potentials in the goldfish and carp (Kaneko, 1970; Kaneko and Yamada, 1972).

In the photopic retina there are two kinds of S-potentials; a luminosity (L-) type, which is hyperpolarizing to monochromatic light regardless of wavelength, and a chromaticity (C-) type which changes its polarity depending on wavelength. L- and C-type S-potentials have been recorded from both external and internal horizontal cells; in fact no consistent morphological differences have been found between cells generating the L- and C-type responses. Fig. 6B is a photomicrograph of an external horizontal cell. In addition to the cell body and dendrites extending to receptor terminals, a long axon is seen running horizontally. Fig. 7 shows an injected internal horizontal cell. It is an unusual looking cell without dendritic or axonal processes (e.g., Stell, 1967, 1972; Witkovsky and Dowling, 1969). It is not clear from where they receive their input or to which cells they project, and Stell proposes that they be called 'tubular cells' since they are not typical horizontal cells (Stell, 1972).

L-type S-potentials recorded in the photopic retina have a sensitivity peak at about 620 nm, the sensitivity maximum of red-sensitive cones (Tomita et al., 1967). From this observation and those by Watanabe and Hashimoto (1965) and by Witkovsky (1967), it is suggested that the external horizontal cells receive input from cones but not from rods.

In the scotopic carp retina, a different kind of L-type S-potential has been recorded (Kaneko and Yamada, 1972). These responses are characterized by lower thresholds and a slower time course than the L-type S-potentials recorded in the photopic retina. By injecting Procion yellow from the recording electrodes, these cells have been identified as intermediate horizontal cells (Fig. 8).

The spectral sensitivity of the intermediate horizontal cell, peaking at about 520 nm, agrees with the absorption spectrum of the rod pigment, porphyropsin (Munz and Schwanzara, 1967; Bridges, 1967). Neither selective adaptation nor strong background illumination have any effect on the spectral sensitivity peak. Furthermore much

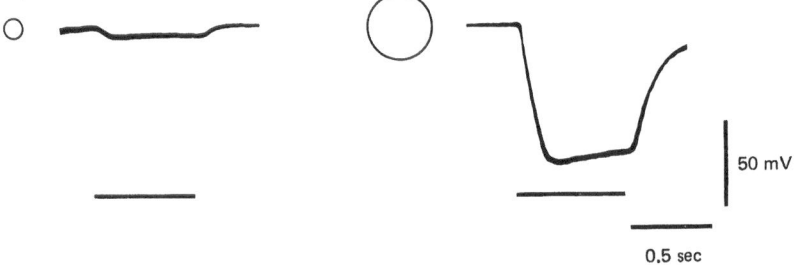

Fig. 6A.

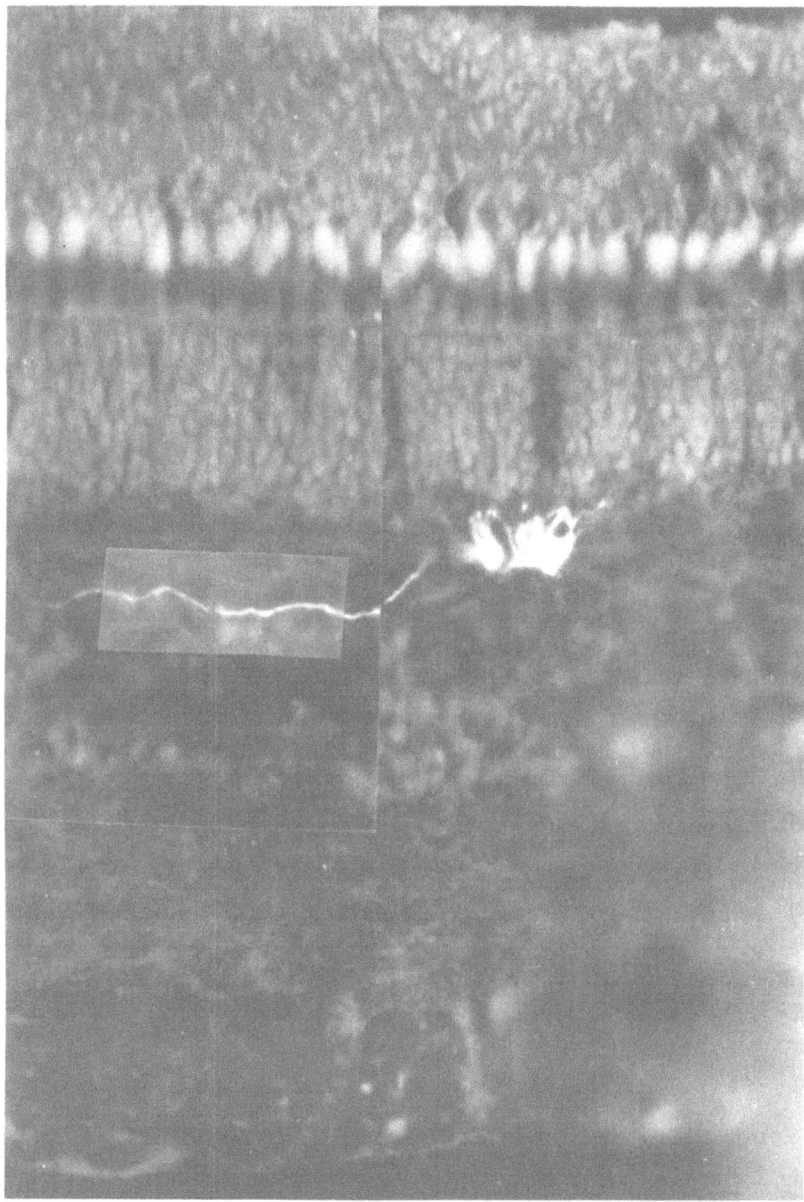

Fig. 6B.

Fig. 6A and B. A shows photopic S-potentials to a 100 μm white light spot (left) and to a broad illumination (right) of about equal intensity. D.C. recording. B is a montage photomicrograph showing an external horizontal cell injected with Procion yellow. Similar responses to those in A were recorded from this cell. (A reproduced from Kaneko and Hashimoto, 1967, by permission of Pergamon Press, and B from Kaneko, 1970, by permission of *J. Physiol.*)

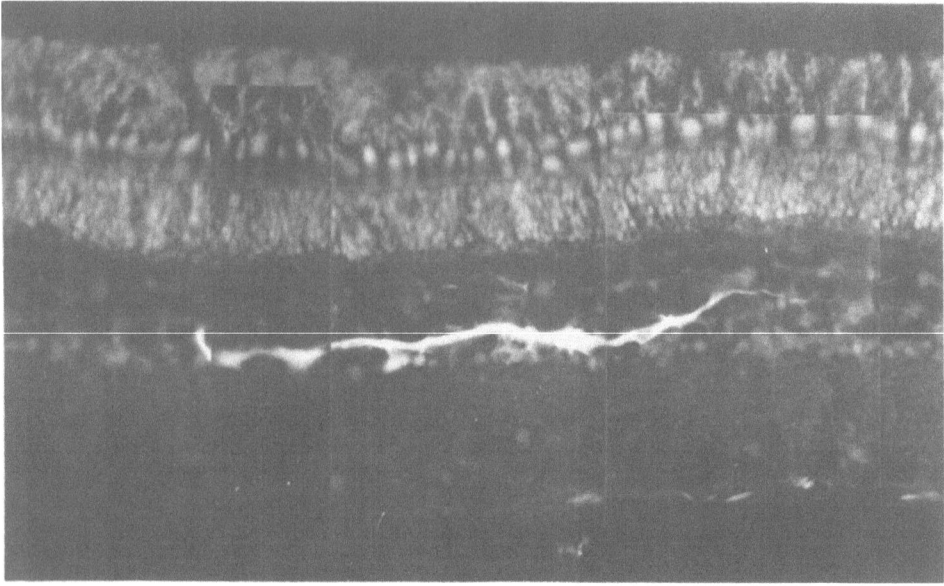

Fig. 7. A montage photomicrograph of an internal horizontal cell injected with Procion yellow. Photopic S-potentials were also recorded from this type of cell. (Reproduced from Kaneko, 1970, by permission of *J. Physiol.*)

stronger background light suppresses these cells entirely. From this observation we concluded that the intermediate horizontal cells of the carp receive rod inputs. This supports the conclusion drawn from histological evidence by Ramón y Cajal (1893) and by Stell (1967).

In contrast to responses from receptors, S-potentials are characterized by a large spatial summation within a uniform receptive field (Naka and Rushton, 1967; Norton et al., 1968). For example, as shown in Fig. 6A, there is an obvious difference in amplitude between the response to a 100 μm light spot and that to broad illumination. The receptive field of a horizontal cell apparently exceeds the dendritic field of a single horizontal cell. This property will be discussed below.

Bipolar Cells

Intracellular recordings from bipolar cells were first made in the mudpuppy (Werblin and Dowling, 1969). In the goldfish, the author recorded similar responses and confirmed by injecting Procion yellow that these responses were coming from bipolar cells (Kaneko, 1970). Figure 9 is a photomicrograph of a bipolar cell which showed similar responses to those in Fig. 10. The responses of bipolar cells are different from the responses of horizontal cells (S-potentials) in their receptive field organization. Cells of one group are depolarized by a small light spot and are hyperpolarized by an annulus. Cells of another group respond in the opposite way: hyperpolarization to a spot and depolarization to an annulus, such as shown in Fig. 10. By analogy to ganglion cells, a depolarizing response will be called an 'on' response and a hyperpolarizing response an off' response,

despite a lack of spike discharges. Morphologically there was no obvious difference
between on-center cells and off-center cells.

The receptive field center of bipolar cells was about 100 to 200 μm in diameter,
the same order of magnitude as the dendritic field (Stell, 1967; Kaneko, 1970, 1971b).
The edge of the surround was not well defined but was estimated to be about 1.5

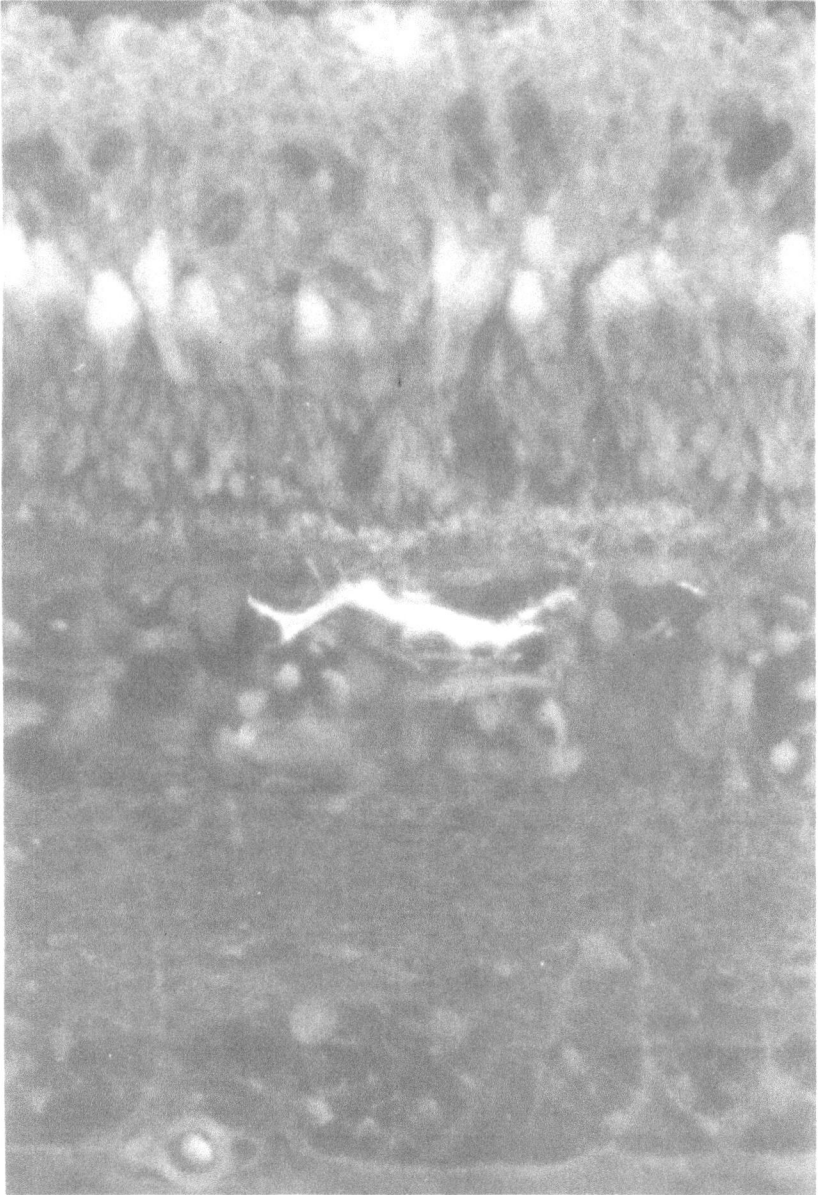

Fig. 8. Photomicrograph of an intermediate horizontal cell injected with Procion yellow.
S-potentials showing rod inputs were recorded from this type of cell. (Reproduced from
Kaneko and Yamada, 1972, by permission of *J. Physiol.*)

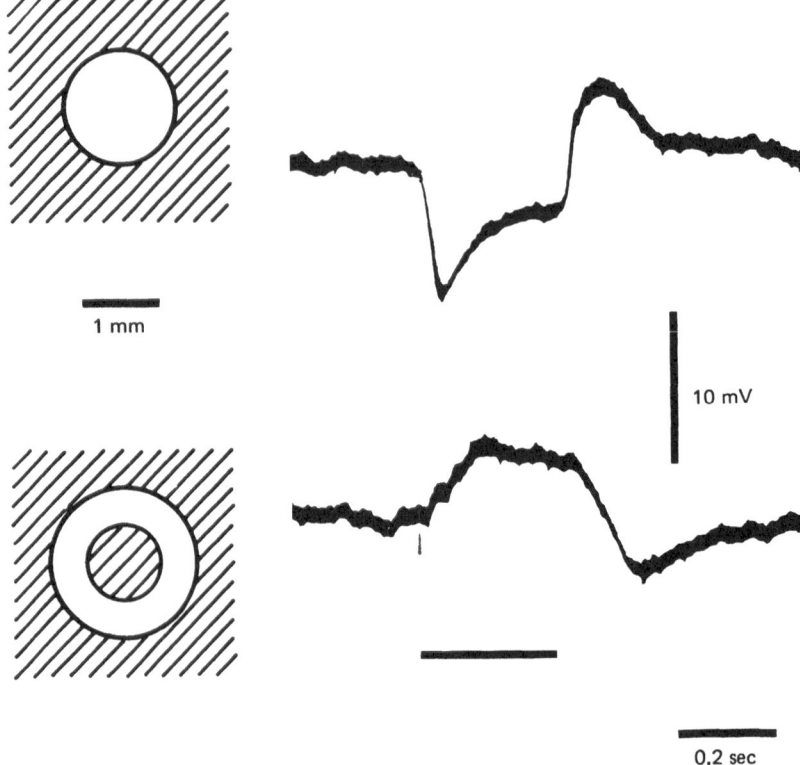

Fig. 9. Intracellular recordings from a bipolar cell which is hyperpolarized by a spot and depolarized by an annulus. Horizontal lines indicate the period of illumination. Both types of illumination are white light of about equal intensity. D.C. recording, positivity upwards. (Reproduced from Kaneko, 1970, by permission of *J. Physiol.*)

mm in diameter, far larger than the dendritic field of a single bipolar cell. The effect of the surround must therefore come through lateral convergence of surrounding cells, most likely horizontal cells, as suggested by Werblin and Dowling (1969) for the mudpuppy.

Amacrine Cells

From amacrine cells two types of responses were recorded, a transient type (Fig. 11) similar to those in the mudpuppy (Werblin and Dowling, 1969), and a sustained type (e.g., Fig. 1d of Kaneko and Hashimoto, 1969). Both kinds of cells have large receptive fields, more than 2.5 mm in diameter; no distinct center-surround organization was found. Fig. 12 shows a camera lucida drawing of an injected amacrine cell.

Identification of Two Neighboring Cells

To study the electrophysiological interactions between two cells it is necessary to penetrate both simultaneously; the cells can then be identified by the injection of two

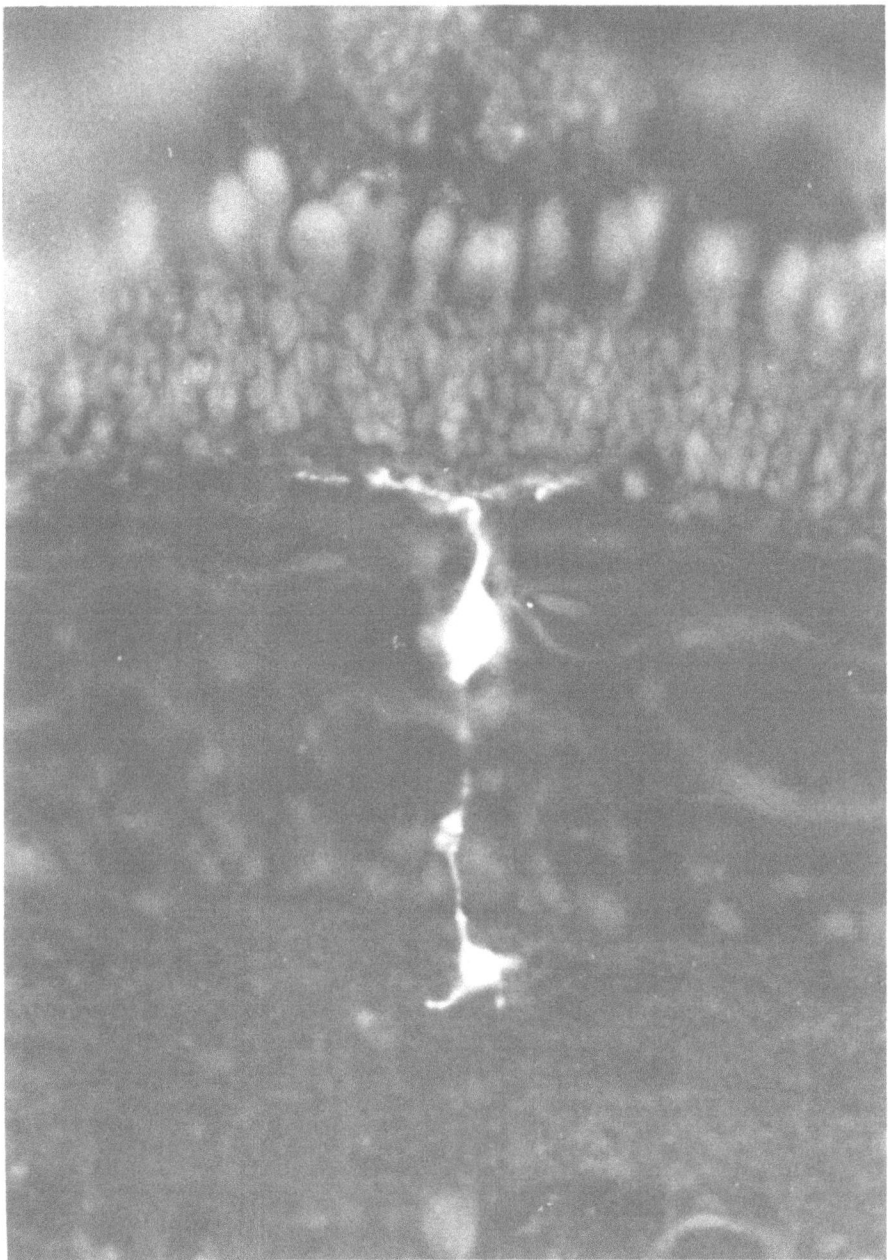

Fig. 10. A photomicrograph showing a bipolar cell from which the same type of response as shown in Fig. 9 was recorded. (Reproduced from Kaneko, 1970, by permission of *J. Physiol.*)

different dyes. This type of application of Procion dyes to the dogfish retina will be presented (Kaneko, 1971a).

The receptive field associated with a fish S-potential covers a large area of the retina (Tomita et al., 1958; Norton et al., 1968; see Fig. 6A), far exceeding the overall size of a single horizontal cell including its axon and dendrites. This suggests that signals from remote receptors converge through pathways other than direct receptor-horizontal cell contacts. The hypothesis of electrotonic lateral spread between S-potential units (Naka and Rushton, 1967) appears very likely, since tight junctions, commonly seen between cells that show electrical connections, are found between adjacent horizontal cells in fishes (Yamada and Ishikawa, 1965; Stell, 1967; O'Daly, 1967; Witkovsky and Dowling, 1969).

In the experiments on the dogfish, I used specially constructed electrodes. Two kinds of microelectrodes, a double-barrelled pipette and a single pipette, were joined side by side by applying cement about 500 μm behind the tips; their spacing was adjusted so that the two shafts were anywhere from 20 to 400 μm apart and the tips were roughly opposite each other (see the schematic drawing of Fig. 13). One of the double-barrelled pipettes was filled with 3M KCl and was used as a current electrode. The other barrel, Voltage electrode 1, was filled with Procion brilliant red H3BNS. The single pipette, Voltage electrode 2, was filled with another dye, Procion yellow M4RS. Voltage electrodes 1 and 2 were used both for recording potentials and for injecting dyes.

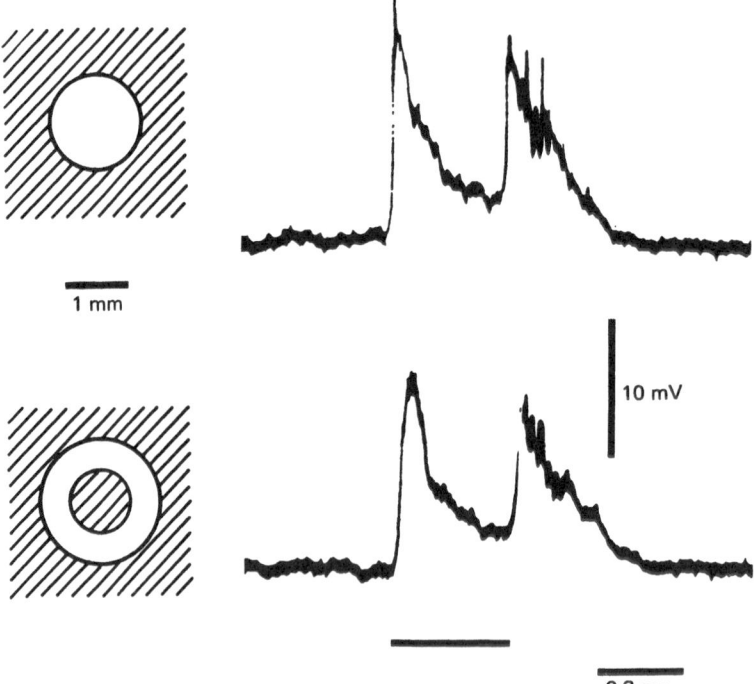

Fig. 11. Intracellular recordings from an amacrine cell. The same experimental conditions as in Fig. 9. (Reproduced from Kaneko, 1970, by permission of *J. Physiol.*)

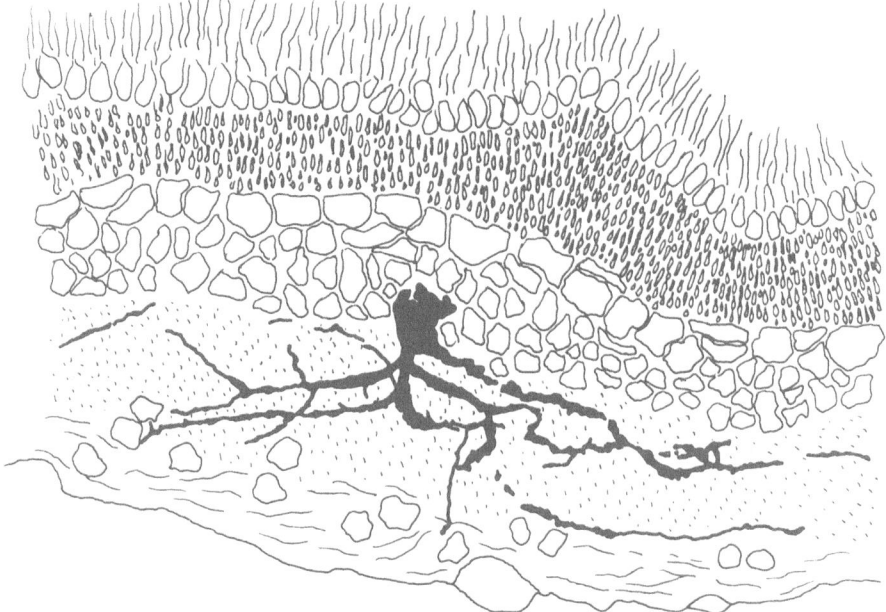

Fig. 12. A camera lucida drawing of an amacrine cell from which responses similar to those shown in Fig. 11 were recorded. (Reproduced from Kaneko, 1970, by permission of *J. Physiol.*)

Fig. 13 shows that the current injected into an external horizontal (cell 1) could be detected from a neighboring horizontal cell (cell 2). Here a hyperpolarizing square pulse current was applied through the current electrode. First the tips of both pipettes, separated by 150 μm, were positioned in the extracellular space near the surface of the isolated retina (Fig. 13A). The polarization in V_1 is the voltage drop across the coupling resistance of the double-barrelled pipettes. In V_2 no polarization was seen, there being only a brief transient related to capacitive coupling between the two electrodes. The set of electrodes was now advanced, and the pipettes penetrated two horizontal cells; polarization of cell 1 was now detected in cell 2 (Fig. 13B). The amount of polarization was 18 mV in cell 1 and 2.8 mV in cell 2, giving a coupling ratio of 0.15. These values remained the same both in the darkness and during illumination. No rectification was found between the two cells.

Types and locations of these two cells were then identified by injecting Procion brilliant red into cell 1 and Procion yellow into cell 2. The yellow-orange fluorescence of Procion yellow contrasted well with the reddish-brown fluorescence of Procion brilliant red. The micrograph (Fig. 14) shows two tangentially sectioned external horizontal cells side by side making two direct contacts with their processes. Similar electrical coupling was found in all the pairs of external horizontal cells. No coupling was seen between cells of different types.

In order to estimate the spatial distribution of potentials induced by the passing of current, many pairs of cells separated by various distances were studied. The location of the two cells was determined morphologically as in Fig. 14. This was necessary since the cell distance expected from the spacing between electrode tips was often different from that determined by morphology. Fig. 15 is a diagram showing the coupling ratio in various pairs of cells separated by different distances. Generally, larger coupling ratios were found between cells closer together, but the ratio did not vary regularly with intercellular distance.

Although one outlying cell by itself may have only a small effect on a central cell because the coupling ratio of two cells separated by several interposed cells is so small, the net effect of a spot covering many surrounding cells may be substantial, since the number of surrounding cells rapidly increases with distance. It is concluded from these observations that the large spatial summation observed in S-potentials is related to this electrical coupling between horizontal cells, as suggested by Naka and Rushton (1967).

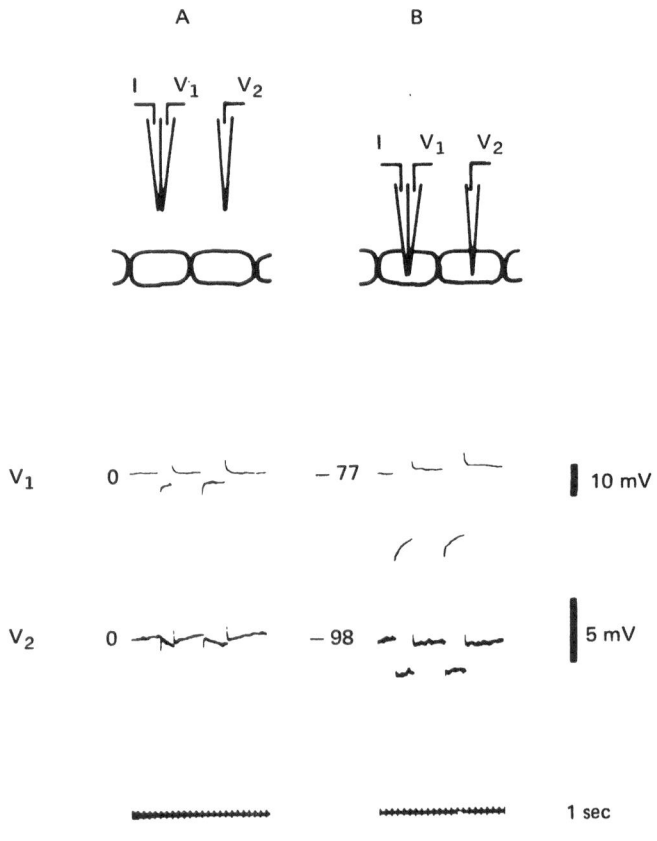

Fig. 13A and B. Simultaneous recordings from two external horizontal cells. In A, no polarization was detected when both electrodes were placed extracellularly. Small deflection in V_1 produced by a voltage drop across the coupling resistance of the double-barreled electrode produced by an extrinsic current of -55×10^{-9} A. In B, polarization of cell 1 spread to cell 2. (Reproduced from Kaneko, 1971a, by permission of *J. Physiol.*)

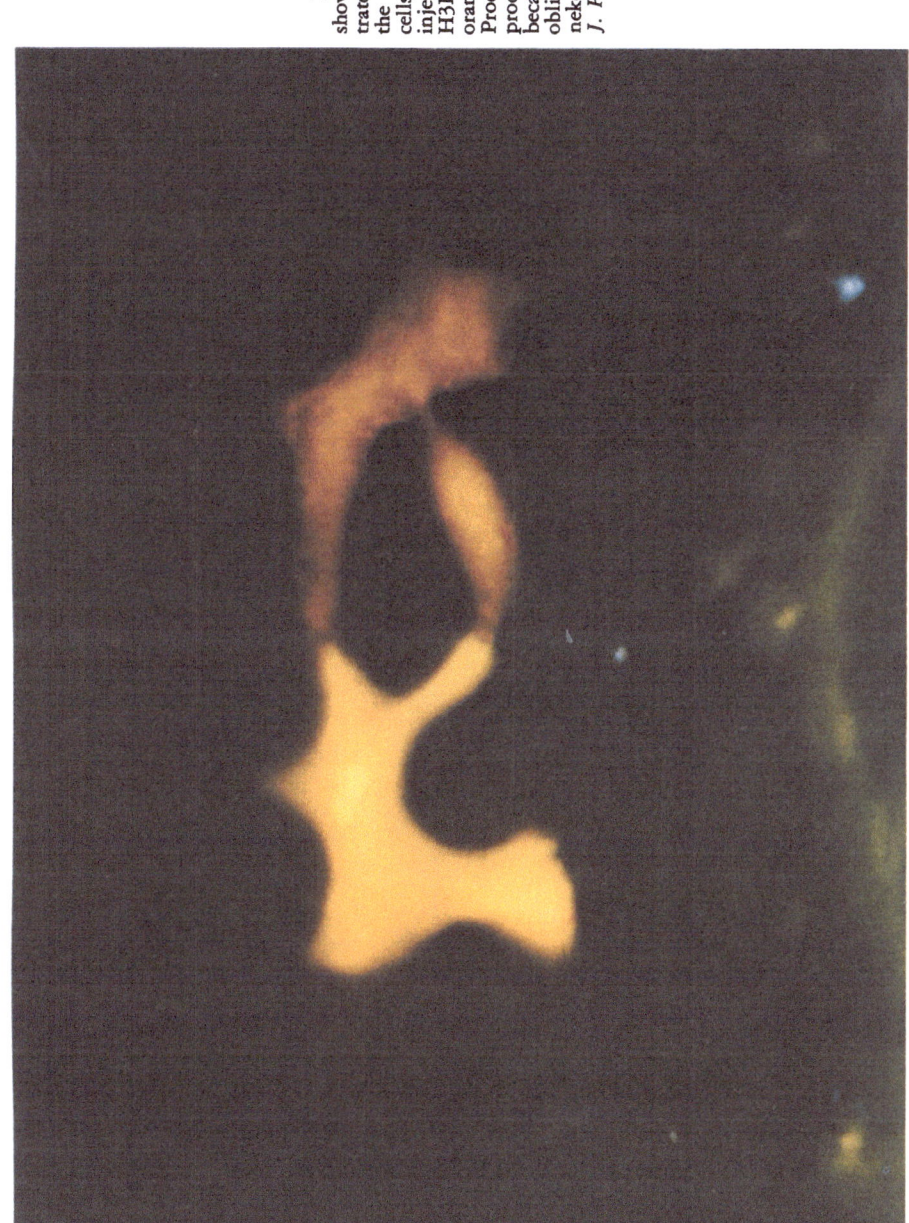

Fig. 14. Photomicrograph showing the location of two penetrated cells. Tangential section at the level of external horizontal cells. The cell appearing red was injected with Procion brilliant red H3BNS. Another cell appearing orange-yellow was injected with Procion yellow M4RS. Some processes of each cell are missing because the section was slightly oblique. (Reproduced from Kaneko, 1971a, by permission of *J. Physiol.*)

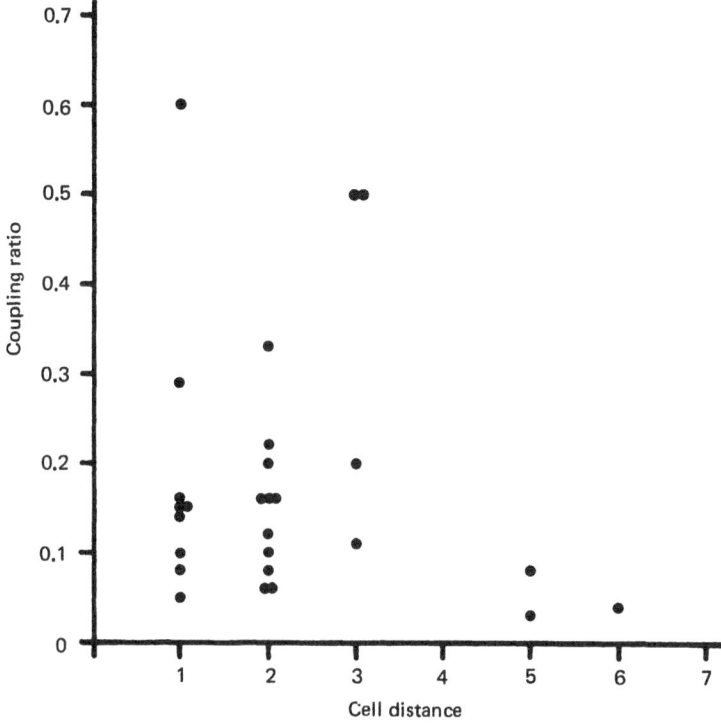

Fig. 15. Coupling ratios in pairs of cells separated by different number of cells. A cell distance of '1' signifies that a pair of cells were located next to each other; '2' that they were separated by one cell and so on. Each pair was obtained from a different piece of retina. (Reproduced from Kaneko, 1971a, by permission of *J. Physiol.*)

Diffusion of the Dye through Cell Contacts

Intercellular coupling can also be demonstrated by examining the passage of small molecules from one cell to another (see Chapter 8). Procion yellow has been found to diffuse across an electrical synapse (Payton et al., 1969) and between cultured cells in which electrical connections were demonstrated (E. F. Furshpan and D. D. Potter, personal communication). Also in the dogfish, the author found Procion yellow diffusing from an external horizontal cell to adjacent ones where electrical coupling was demonstrated (Kaneko, 1971a). The dye diffused only to other external horizontal cells, and not to different types of cells where no coupling was seen. Fig. 16 shows an example in which injected Procion yellow spread from an external horizontal cell (Y) to two of the surrounding cells (y). Usually the five or six surrounding external horizontal cells were not equally bright. In another example, the dye was observed to have diffused not only to a neighboring cell, but apparently from that cell to one beyond. If all the external horizontal cells were uniformly interconnected electrically, dye might have been seen in each of the surrounding cells. In fact, the amounts of the dye that diffused from the injected cell to each of the neighboring cells were far from equal, and the dye failed completely to enter some cells. This may suggest that junctions between

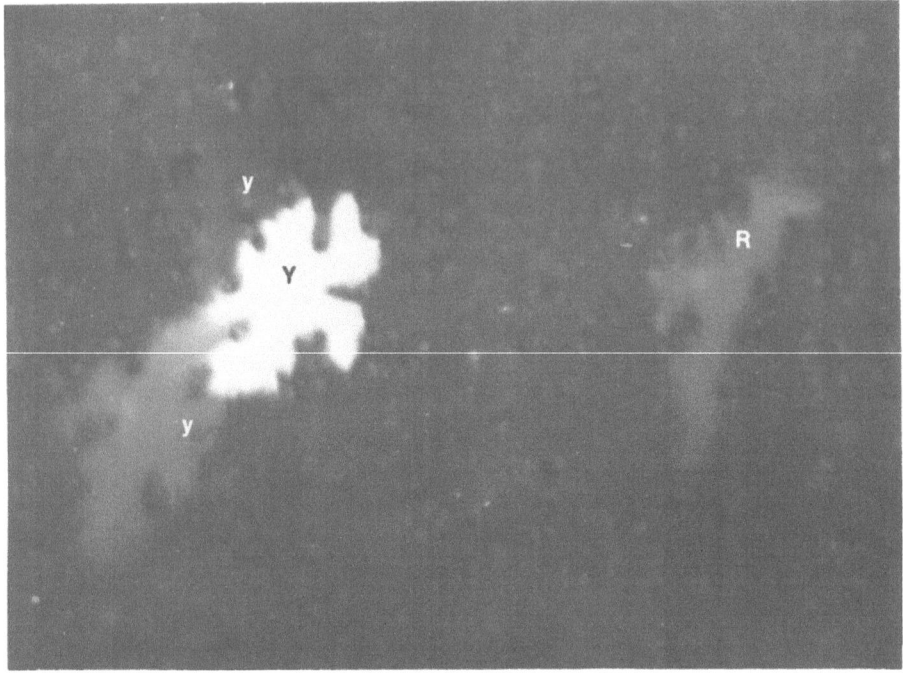

Fig. 16. Photomicrograph showing external horizontal cells. Note the diffusion of Procion yellow from an injected cell (Y) into two of the surrounding cells (y). R, the cell injected with Procion brilliant red. (Reproduced from Kaneko, 1971a, by permission of *J. Physiol.*)

different cells differ in permeability. This is consistent with the finding that pairs of cells having similar cell distances showed different coupling ratios.

Closing Remarks

With most of the earlier methods of stain injection, the localizations were not always positive because of difficulty in injecting the stain or recovering the marked cells, or because the stain stayed in the part of the cell where it was injected. They failed to meet the criteria for reliable identification discussed previously, especially the important criterion that the response be recorded after the injection of the stain.

Clearly Procion yellow is the best stain so far tested, and meets almost all the criteria for reliable identification. Furthermore, as described by Stretton and Kravitz (1968), Procion yellow "spreads into fine branches of cells, survives fixation and routine histological procedures, and permits reconstruction of cell shapes."

If similar excellent techniques are extended to the field of electron microscopy (see Chapter 5), one may determine the connections of specific cells of known functional organization at the microstructural level. Such an attempt has recently been made by injecting cobalt ions into an insect ganglion (Pitman et al., 1972 and see also Chapter 6), and its full application to other nervous systems is very much expected.

Acknowledgments. I thank Dr. M. Murakami for allowing me to use his unpublished photograph and Dr. J. A. Coles for carefully reading the manuscript. The original research reported in this paper was supported by USPHS grants 5 RO1 NB 06421, 5 RO1 NB 02260, 5 RO1 NB 05554 and EY 00017-06, by Matsunaga Science Foundation and by fellowship grants from the China Medical Board of New York, Inc. and from the Foundations' Fund for Research in Psychiatry, Inc. G69-444.

References

Baylor, D. A. and M. G. F. Fuortes: Electrical responses of single cones in the retina of the turtle. J. Physiol., Lond. *207*, 77–92 (1970).

———, ——— and P. M. O'Bryan: Receptive fields of cones in the retina of the turtle. J. Physiol., Lond. *214*, 265–294 (1971).

Bridges, C. D. B.: Spectroscopic properties of porphyropsins. Vision Res. 7, 349–369 (1967).

Bortoff, A.: Localization of slow potential responses in the Necturus retina. Vision Res. *4*, 627–636 (1964).

Grabowski, S. R., L. H. Pinto and W. L. Pak: Adaptation in retinal rods of axolotl: intracellular recordings. Science *176*, 1240–1243 (1972).

Kaneko, A: Physiological and morphological identification of horizontal, bipolar and amacrine cells in goldfish retina. J. Physiol., Lond. *207*, 623–633 (1970).

———: Electrical connexions between horizontal cells in the dogfish retina. J. Physiol., Lond. *213*, 95–105 (1971a).

———: Physiological studies of single retinal cells and their morphological identification. Vision Res., Suppl. *3*, 17–26 (1971b).

——— and H. Hashimoto: Recording site of the single cone response determined by an electrode marking technique. Vision Res. 7, 847–851 (1967).

———, ———: Electrophysiological study of single neurons in the inner layer of the carp retina. Vision Res. *9*, 37–55 (1969).

——— and M. Yamada: S-potentials in the dark-adapted retina of the carp. J. Physiol., Lond. *227*, 261–273 (1972).

Luft, J. H.: Improvements in epoxy resin embedding methods. J. biophys. biochem. Cytol. *9*, 409–414 (1961).

MacNichol, E. F. Jr. and G. Svaetichin: Electrical responses from the isolated retinas of fishes. Am. J. Opthal. *46*, 26–40 (1958).

Marks, W. B.: Visual pigments of single goldfish cones. J. Physiol., Lond. *178*, 14–32 (1965).

Mitarai, G.: The origin of the so-called cone potential. Proc. Japan Acad. *34*, 299–304 (1958).

Munz, F. W. and S. A. Schwanzara: A nomogram for retinene$_2$-based visual pigments. Vision Res. 7, 111–120 (1967).

Murakami, M. and W. L. Pak: Intracellularly recorded early receptor potential of the vertebrate photo-receptors. Vision Res. *10*, 965–975 (1970).

Matsumoto, N. and K.-I. Naka: Identification of intracellular responses in the frog retina. Brain Res. *42*, 59–72 (1973).

Naka, K.-I. and W. A. H. Rushton: The generation and spread of S-potentials in fish (Cyprinidae). J. Physiol., Lond. *192*, 437–461 (1967).

Norton, A. L., H. Spekreijse, H. G. Wagner, and M. L. Wolbarsht: Responses to directional stimuli in retinal preganglionic units. J. Physiol., Lond. *206*, 93–107 (1970).

———, ———, M. L. Wolbarsht, and H. G. Wagner: Receptive field organization of the S-potential. Science *160*, 1021–1022 (1968).

O'Daly, J. A.: ATPase activity at the functional contact between retinal cells which produce S-potential. Nature, Lond. *216*, 1329–1331 (1967).

Oikawa, T., T. Ogawa and K. Motokawa: Origin of the so-called cone action potential. J. Neurophysiol. *22*, 102–111, (1959).

Payton, B. W., M. V. L. Bennett and G. D. Pappas: Permeability and structure of junctional membranes at an electrotonic synapse. Science *166*, 1641–1643 (1969).

Pitman, R. M., C. D. Tweedle and M. J. Cohen: Branching of central neurons: intracellular cobalt injection for light and electron microscopy. Science *176*, 412–414 (1972).

Potter, D. D., E. J. Furshpan and E. S. Lennox: Connections between cells of the developing squid as revealed by electrophysiological methods. Proc. natn. Acad. Sci. U.S.A. *55*, 328–336 (1966).

Ramón y Cajal, S.: La rétine des vertébrés. La Cellule, *9*, 119–246 (1893).

Steinberg, R. H. and R. Schmidt: Identification of horizontal cells as S-potential generators in the cat retina by intracellular dye injection. Vision Res. *10*, 817–820 (1970).

Stell, W. K.: The structure and relationship of horizontal cells and photoreceptor-bipolar synaptic complexes in goldfish retina. Am. J. Anat. *121*, 401–424 (1967).

———: The morphological organization of the vertebrate retina. In: Handbook of sensory physiology, VII/2 Physiology of photoreceptor organs (Ed. M. G. F. Fuortes), Berlin: Springer-Verlag, 112–213 (1972).

Stretton, A. O. W. and E. A. Kravitz, Neuronal geometry: determination with a technique of intracellular dye injection. Science *162*, 132–134 (1968).

Tasaki, K., Y. Tsukahara, S. Ito, M. J. Wayner and W. Y. Yu: A simple, direct and rapid method for filling microelectrodes. Physiol. and Behav. *3*, 1009–1010 (1968).

Tomita, T.: Electrophysiological study of the mechanisms subserving color coding in the fish retina. Cold Spring Harb. Symp. quant. Biol. *30*, 559–566 (1965).

———, A. Kaneko, M. Murakami, and E. L. Pautler: Spectral response curves of single cones in the carp. Vision Res. 7, 519–531 (1967).

———, T. Tosaka, K. Watanabe, and Y. Sato: The fish EIRG in response to different types of illumination. Jap. J. Physiol. 8, 41–50 (1958).

Toyoda, J., H. Hashimoto, H. Anno, and T. Tomita: The rod response in the frog as studied by intracellular recording. Vision Res. *10*, 1093–1100 (1970).

———, H. Nosaki, and T. Tomita: Light-induced resistance change in single photoreceptors of *Necturus* and *Gekko*. Vision Res., 9, 453–463 (1969).

Watanabe, K. and Y. Hashimoto: S-potential in light and dark adaptation of the live carp. Proc. XXIII Intern. Congr. Physiol. Sci., 840 (1965).

Werblin, F. S. and J. E. Dowling: Organization of the retina of the mudpuppy, Necturus maculosus. II. Intracellular recording. J. Neurophysiol. *32*, 339–355 (1969).

Witkovsky, P: A comparison of ganglion cell and S-potential response properties in carp retina. J. Neurophysiol. *30*, 546–561 (1967).

——— and J. E. Dowling: Synaptic relationships in the plexiform layers of carp retina. Z. Zellforsch. mikrosk. Anat. *100*, 60–82 (1969).

Yamada, E. and T. Ishikawa: Some observations on the fine structure of vertebrate retina. Cold Spring Harb. Symp. quant. Biol. *30*, 383–392 (1965).

Discussion

DR. BAKER (Iowa): I have two questions. Have you been able to look at the effect of the Procion yellow injection on synaptic potentials?

Secondly, have you been able to measure the junctional resistance to ascertain if the gap junction is being injured by the Procion yellow injections?

DR. KANEKO: To the first question, probably what we are looking at is the synaptic potential, but it is not very typical of an EPSP or IPSP.

I think that the current injection damages the cell very much, although it is rather hard to say whether it is the current or dye which causes the damage. But after the injection, I can still record a response which is a good indication that the electrode remains in the same cell.

To the second point, I didn't try to calculate the transjunctional resistance.

DR. K. NAKA (Cal Tech): I think I can answer one of your questions. In the fish horizontal cells, it is possible to inject current into the S-space (through a Procion-filled electrode) to evoke discharges from the ganglion cells as well as to deposit the dye in the cell. It seems to me that Procion dye injection does not do any harm to the catfish horizontal cells.

DR. D. YORK (Queen's University): My problem, and that of several other people I have been talking with, concerns the success rate of these various injections. Would you care to comment on how many well stained cells you obtain out of the total attempted?

DR. KANEKO: I think that results depend on experience, and on what you want to find out. What I discovered was that sometimes the electrodes become plugged up and didn't pass current. Then you have to change the electrode. It is also very necessary to watch if the current is really passing or not. If you carefully check these points, I think that you can achieve more than 50% success.

CHAPTER 12

Morphological and Physiological Identification of Retinal Cells in the Turtle

Yoko Hashimoto, Takehiko Saito, William H. Miller, and Tsuneo Tomita

Our understanding of retinal function has benefited from the introduction of intracellular marking techniques that allow a particular cell to be fully characterized both as to its physiological function and morphology. The Procion yellow intracellular injection technique originally introduced by Stretton and Kravitz (1968) and perfected for use in the retina by Kaneko (1970, 1971a, b, and Chapter 11 of this volume) has proved most useful. We have used this technique to identify several retinal cell types and to investigate the origin of S-potentials in the turtle retina.

The luminosity (L) and chromaticity (C) types of S-potential were first described in the fish retina by Svaetichin (1956). Svaetichin found that the L-type response gave a slow, maintained hyperpolarization for all wavelengths of light stimulation while the C-type response gave hyperpolarization at some wavelengths and depolarization at others.

Subsequently there were many investigations of the origin of the S-potentials, the most recent of which were those of Kaneko (1970), Steinberg and Schmidt (1970), Werblin and Dowling (1969), and Matsumoto and Naka (1972). As a result of these studies, the S-potential has been definitely localized to the horizontal cells. These are tangentially oriented cells which synapse with both the photoreceptor cells and bipolar cells [Dowling and Werblin (1969, 1971), Witkovsky and Dowling (1969)], and which are located just vitread to the receptor cells. Our main finding reported in Miller, et al. (1973) and in this chapter is that the L- and C-type S-potentials originate in morphologically distinct cell types in the turtle retina.

Materials and Methods

All of the histological and physiological experiments were performed on the slider turtle, *Pseudemys scripta*. We enucleated the eye in dim light, removed the anterior pole, and quartered the eye cup. A small section of the eye cup was placed on a jolter (Tomita, 1965). We filled our microelectrodes with 6–8% (w/v) Procion yellow M4RS. Electrode resistances ranged from about 80 to over 300 MΩ. We found the higher resistance microelectrodes more satisfactory for recording from and injecting the cells that gave the C-type response. We recorded L- and C- type responses when the pipette was advanced about 120 μm after touching the internal limiting membrane. The characterization of the response as L- or C-type was made by testing the response with red light transmitted by a 690 nm interference filter. The L-type response is a hyperpolarization to red light while the C-type gives a depolarization of the cell when stimulated with this wavelength of light. We then injected the cell using from 1 to 3×10^{-9} A of negative current for a period of about 1 to 3 min. For example, we consistently had good results with pipettes of about 200 MΩ when we injected 1×10^{-9} A for about 3 min. We endeavored to inject only cells with a good amplitude of response and usually only analyzed the results of injections of cells that had a healthy response after the injection. Only one cell was injected for each histological preparation, and we followed the histological procedure of Kaneko (1970) closely. We modified the procedure slightly using Spurr's medium instead of Maraglas and glycerin instead of Lustrex for coverslip mounting. We incubated the preparation in Ringer's from 5 to 15 min after injection, and in our experience 5 min was sufficient for penetration of the entire cell by the dye. However, it should be noted that we did not find an axon in any of our horizontal cells as has been reported by Ramón y Cajal (1894) for certain horizontal cells.

In addition to the histological preparation of Procion yellow injected cells we fixed turtle retinas in 3% glutaraldehyde phosphate buffered in pH 7.4, postfixed in 2% OsO_4 in collidine buffer pH 7.4 and embedded in Spurr's medium and sectioned at 2 μm. These preparations were stained with Toluidine Blue for examination of the cells in the light microscope in order to make comparisons of the horizontal cells in conventional and fluorescence microscopy.

Results and Discussion

Conventional Light Microscopy

Cross sections of osmium-fixed material show two morphologically different cells whose cell bodies are in the inner nuclear layer and whose dendrites are in the outer plexiform layer. Our results show that these two cell types are correlated with the L- and C-type physiological responses. Fig. 1 illustrates the appearance of the L-type cell in conventional light microscopy. This cell (black arrow) has thick, relatively empty, lightly stained dendrites. Under oil immersion it is often possible to observe fine dendrites branching from the thick ones of the L-type cell and penetrating the outer plexiform layer to the receptor pedicles.

The C-type cell is shown in Fig. 2 (black arrow). Because the dendrites are thin and densely stained, they are difficult to distinguish from the substance of the outer plexiform layer in photographs. However, the lefthand C-type cell of Fig. 2 has a discernible border just above its nucleus. The dendrites are relatively easy to trace using oil immersion but are difficult to reproduce in black and white photographs. Note the relatively lightly stained dendrites of the L-type cells in the outer plexiform layer in Fig. 2.

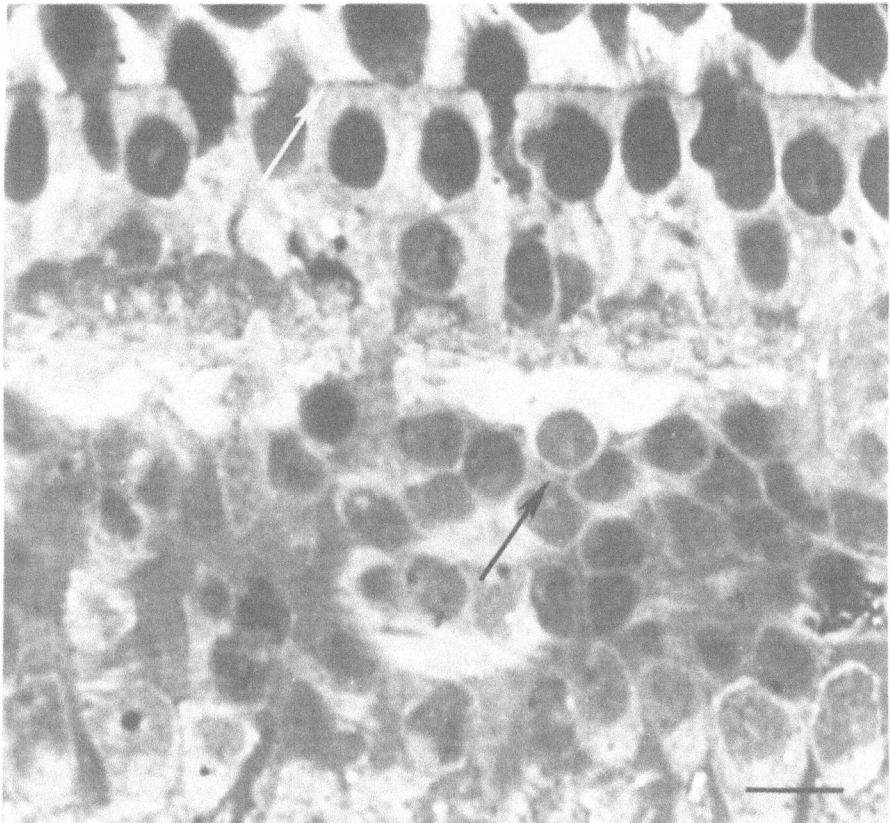

Fig. 1. Glutaraldehyde-osmium fixed turtle retina, 2 μm thick cross section. Black arrow points to L-type horizontal cell. White arrow indicates external limiting membrane immediately below which are the nuclei of the receptor cells. Cell bodies at bottom of figure are amacrine cells at vitread border of inner nuclear layer. Magnification marker, 10 μm. (From Miller et al., 1972, reproduced with permission from *Vision Research,* Pergamon Press.)

Procion Yellow Injected L- and C-type Cells

We have injected with Procion yellow about 100 L-type and 20 C-type cells as identified by their responses to red light stimulation. We observed these injected cells using fluorescence microscopy and in every case the cell type resembled the morphologically distinct cell types shown in Figs. 1 and 2. Examples of L- and C-type injected cells from tangentially sectioned retinas are shown in Fig. 3. The cells we examined within the L-type or C-type class showed a considerable range of structure. Nevertheless, the L-type cells identified by physiological recordings were always characterized by predominantly thick dendrites (Fig. 3A), although almost all L-type cells had some thin dendrites. The C-type cells, on the other hand, were all characterized by predominantly thin dendrites as shown by the example in Fig. 3B. The tangential section is the only one which allows a comprehensive view of the entire dendritic pattern of the cell.

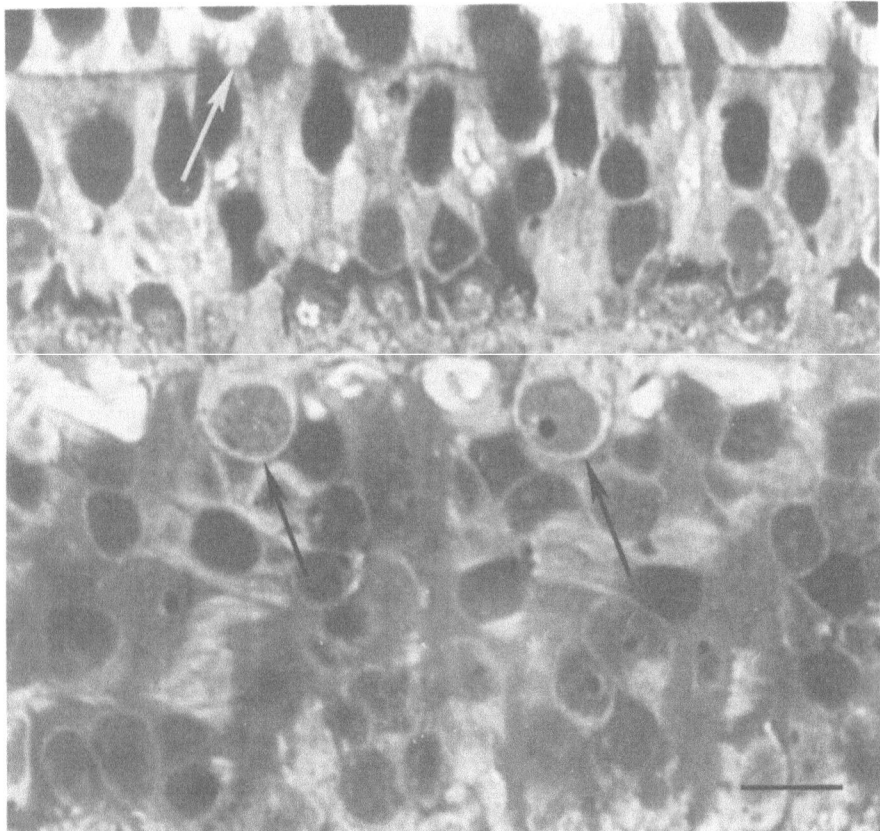

Fig. 2. Same as Fig. 1 except that it illustrates C-type horizontal cells (black arrows). Magnification marker, 10 μm. (From Miller et al., 1972, reproduced with permission from *Vision Research,* Pergamon Press.)

For instance, looking at the L-type cell of Fig. 3A, one can see how cross sectioning in certain planes would give the impression of a cell that has only relatively fine dendrites.

Fig. 4 shows two further examples of tangentially sectioned L- and C-type cells (Fig. 4A and B), and below them examples of cross sectioned L- and C-type cells (Fig. 4C and D). Observation of the cells in tangential and cross sectioned Procion yellow injected material compared with that of conventional osmium and glutaraldehyde microscopy strengthens our conclusion about the morphological nature of the L- and C-type cells in the turtle retina. Comparison of the tangential sections showing L- and C-type indicates that the dimensions of the dendritic fields are about the same in the two cell types.

Relation to Previous Findings

Horizontal cells in various species have been classified on the basis of their retinal location and dendritic patterns. Thus in the carp there are three types of horizontal cells classified as internal, intermediate, and external. The internal horizontal cell is located most sclerad, the external has its cell body most vitread, and the intermediate is located between the other two (Ramón y Cajal, 1894). Although previous studies in the goldfish did not show any differences in horizontal cell type responses (Kaneko, 1970), a more recent investigation reported in Chapter 11 of this volume has yielded the interesting information that the intermediate horizontal cell is connected exclusively with the rod photoreceptors. This is in agreement with Stell's (1967) morphological observation that the intermediate horizontal cells synapse with the rods in the goldfish.

The turtle retina, however, is almost entirely lacking in rods. In this connection it is interesting that according to Ramón y Cajal (1894) there are only two types of horizontal cells in another reptile, the lizard. While there may be differences between the lizard and the turtle, we can determine just two types of horizontal cells in the turtle. From a morphological standpoint, our L-type cell most closely resembles Ramón y Cajal's internal horizontal cell of the carp and our C-type cell most closely resembles Ramón y Cajal's external horizontal cell of the carp.

Procion Yellow Injection of Other Turtle Retinal Cells

The technique of tangentially sectioning retinas for analysis of Procion yellow-injected horizontal cells is successful primarily because of the tangential orientation of the horizontal cell dendrites. Being able to observe the entire cell in one section has obvious

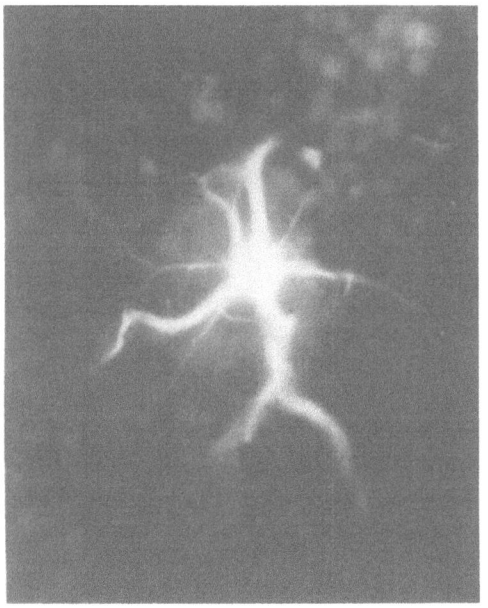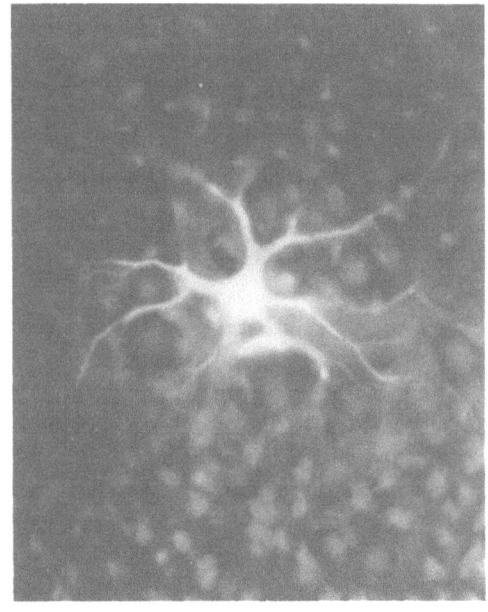

A   B   ——

Fig. 3A and B. A shows Procion yellow-injected L-type horizontal cell in tangential section. Note predominantly thick dendrites. B shows Procion yellow-injected C-type horizontal cell in tangential section. Note predominantly thin dendrites. Magnification marker, 10 μm, both pictures. (From Miller et al., 1973, reproduced with permission from *Vision Research*, Pergamon Press.)

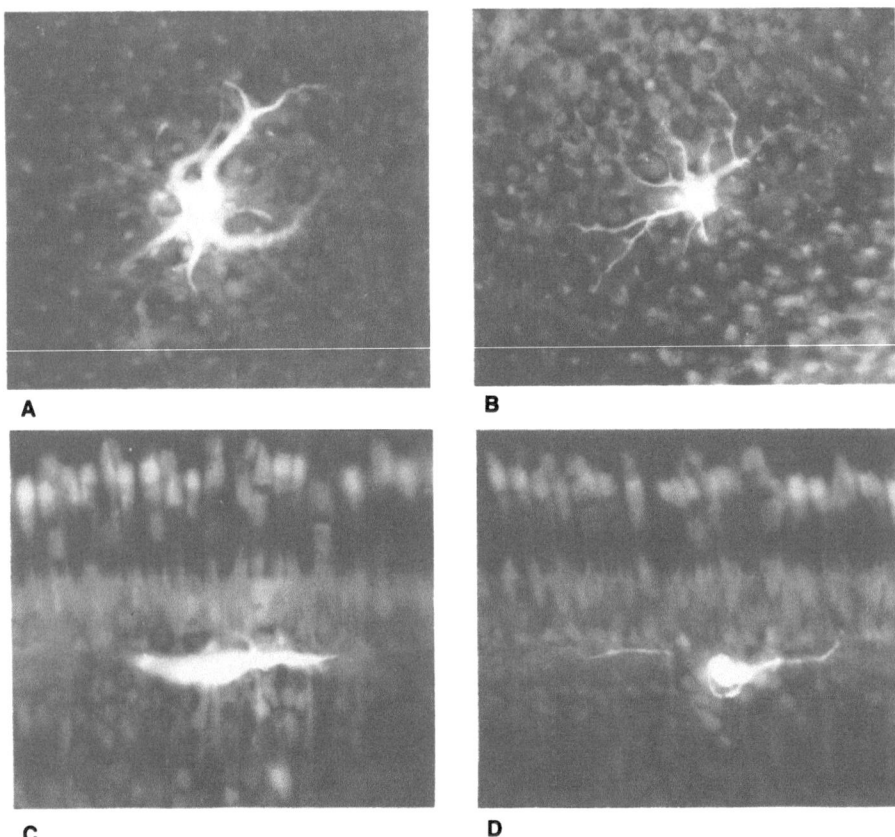

Fig. 4A–D. Procion yellow-injected horizontal cells. A. L-type, tangential section. B. C-type, tangential section. C. L-type, cross section. D. C-type, cross section. Magnification marker, 10 μm, all pictures.

advantages, and there are few retinal cell types that can be so easily observed. Nevertheless, we have injected isolated examples of other turtle retinal cells for comparison. Next to the horizontal cells, the amacrines are most suitable for study in tangential section because their dendritic tree has a large lateral component not too far vertically distant from its cell body. Fig. 5A shows an example of an amacrine cell. The physiological recording showed an on-off response with spikes of small amplitude. The cell was marked by passing 2×10^{-9} A of negative current for 1 min. We found the marked cell body at the vitread border of the inner nuclear layer and the dendritic tree in the inner plexiform layer. This cell closely resembles one of the amacrine cells described by Ramón y Cajal (1894, his Fig. 5, c; plate III) as an amacrine cell with twisted tuft.

We were unable to obtain good preparations of ganglion cells in tangential section because their dendrites permeate the entire inner plexiform layer. Fig. 5B shows an example of a large ganglion cell in cross section. This cell gave an on-off response with spikes of more than 60 mV amplitude. We injected it with 1×10^{-9} A for

1 min. We have several examples of such ganglion cells which show this characteristic "outstretched arms" dendritic pattern in cross section and in one case a short segment of the axon filled with Procion yellow.

Lastly we would like to report an observation concerning the Müller cells. We found that if we used rather low resistance electrodes (less than 80 MΩ) or if we passed too much current such that the injected cell was badly damaged, Müller cells would tend to take up the Procion yellow. For example, Fig. 5C shows such a Müller cell and all its processes from the external to the internal limiting membrane. The appearance of this cell is similar to that of the Müller cell of the lizard as depicted by Ramón y Cajal (1894).

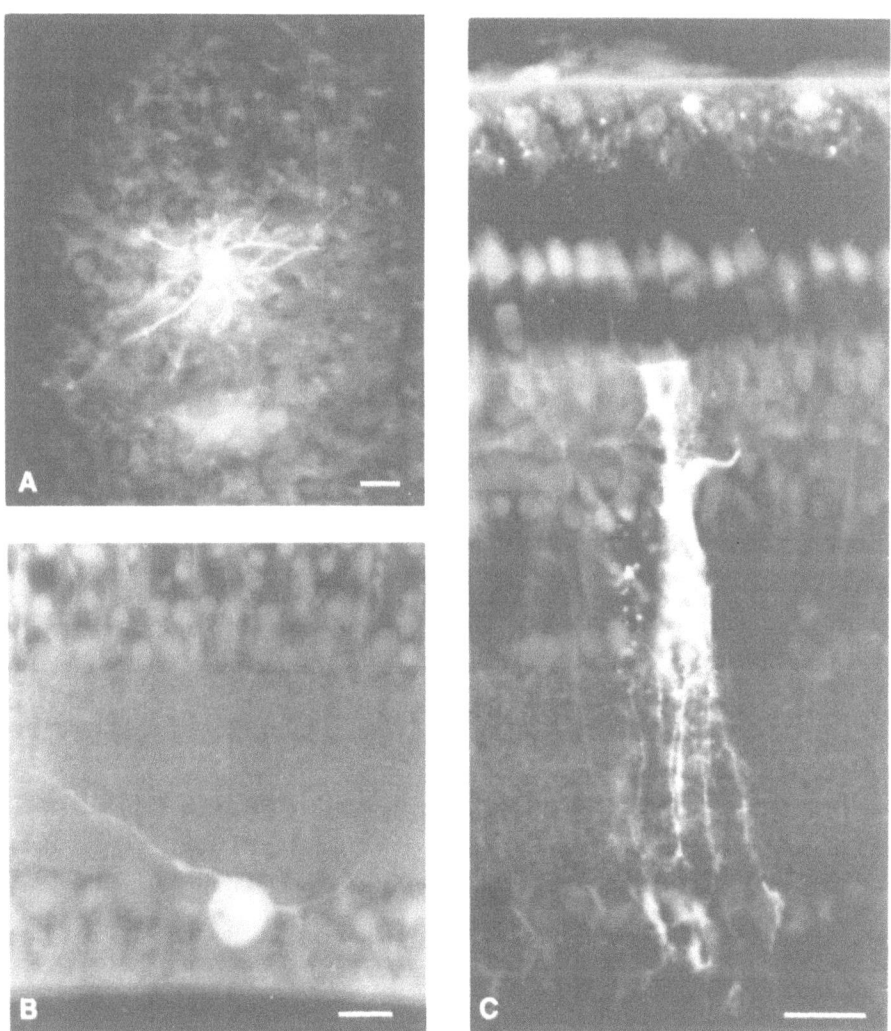

Fig. 5A–C. Procion yellow-injected cells of the turtle retina. A. Amacrine cell in tangential section. B. Large ganglion cell in cross section. C. Müller cell in cross section. Magnification marker, 10 μm for all pictures.

Acknowledgments. This work was supported by Research Grant EY00764 from the National Eye Institute, U.S. Public Health Service, and by a Fight for Sight, Inc. Fellowship to T. Saito.

References

Dowling, J. E., and F. S. Werblin: Organization of the retina of the mudpuppy, *Necturus maculosus.* I. Synaptic structure. J. Neurophysiol. *32,* 315–338 (1969).

—— ——: Synaptic organization of the vertebrate retina. Vision Res. Suppl. *3,* 1–15 (1971).

Kaneko, A.: Physiological and morphological identification of horizontal, bipolar and amacrine cells in goldfish retina. J. Physiol., Lond. *207,* 623–633 (1970).

——: Electrical connexions between horizontal cells in the dogfish retina. J. Physiol., Lond. *213,* 95–105 (1971a).

——: Physiological studies of single retinal cells and their morphological identification. Vision Res. *3,* 17–26 (1971b).

Matsumoto, N., and K.-I Naka: Identification of intracellular responses in the frog retina. Brain Res. *42,* 59–71 (1972).

Miller, W. H., Y. Hashimoto, T. Saito, and T. Tomita: Physiological and morphological identification of L- and C-type S-potentials in the turtle retina. Vision Res. *3,* 443–447 (1973).

Ramón y Cajal: Die retina der wirbelthiere. III. Die retina der reptilien. pp. 85–97. Wiesbaden: Verlag von J. F. Bergmann, 1894.

Steinberg, R. H., and R. Schmidt: Identification of horizontal cells as S-potential generators in the cat retina by intracellular dye injection. Vision Res. *10,* 817–820 (1970).

Stell, W. K.: The structure and relationship of horizontal cells and photoreceptor-bipolar synaptic complexes in goldfish retina. Am. J. Anat. *121,* 401–424 (1967).

Stretton, A. O. W., and E. A. Kravitz: Neuronal geometry: determination with a technique of intracellular dye injection. Science *162,* 132–134 (1968).

Svaetichin, G.: Spectral response curves from single cones. Acta physiol. scand. *39*(134), 17–46 (1956).

Tomita, T.: Electrophysiological study of the mechanisms subserving color coding in the fish retina. Cold Spring Harb. Symp. quant. Biol. *30,* 559–566 (1965).

Witkovsky, P., and J. E. Dowling: Synaptic relationships in the plexiform layers of carp retina. Z. Zellforsch. mikrosk. Anat. *100,* 60–82 (1969).

Discussion

DR. OGDEN (Utah): I would like to know if Drs. Kaneko and Hashimoto have tried whole mount preparations, and if so can you see the Procion cells in these preparations?

DR. KANEKO (Keio): We would like to do that, because it is so simple, but the autofluorescence from the receptors is so strong that it is not possible to get useful results.

DR. HASHIMOTO (Yale): We have found the same thing.

DR. BAKER (Iowa): Did I understand you correctly, Dr. Hashimoto, to say that it is not necessary to penetrate a Müller cell to have it filled? Therefore, if you have a lot of dye in the extracellular region these "glial-type cells" will pick it up?

DR. HASHIMOTO: Yes, we sometimes find this but it does not occur in every case.

CHAPTER 13

Morphological Identification of Simple, Complex and Hypercomplex Cells in the Visual Cortex of the Cat

David Van Essen and James Kelly

The visual cortex of the brain has been intensively studied for many years, and as a consequence a large body of data concerning its anatomy and physiology is now available for a variety of species. The neurons in area 17 of the cat respond to bars or edges of light in such a way that Hubel and Wiesel (1962, 1965) were able to classify the receptive field of each cell as simple, complex, or hypercomplex. The study of Golgi impregnated material from this region indicates that, within a given layer, most cells are either stellate or pyramidal, as judged principally by the shapes of the cell somata and the patterns of dendritic arborization (Ramón y Cajal, 1922). Given this information, it is natural to ask whether the structural and functional properties of cells in the visual cortex are related in a consistent manner. The technique of Procion yellow dye injection (Stretton and Kravitz, 1968) has enabled questions of this type to be tackled in a number of preparations, including the retinas of the goldfish (Kaneko, 1970, and Chapter 11) and turtle (Baylor, Fuortes and O'Bryan, 1971; Hashimoto et al., Chapter 12). In the present report, we describe the application of this technique to the analysis of cells in the visual cortex.

Methods

Adult cats were initially anesthetized with ketamine HCl and maintained in a state of light anesthesia by intravenous administration of thiopental sodium. The animal was artificially respirated after being paralyzed with succinyl choline, pneumothorax was induced, and the appropriate level of respiration was determined by continually monitoring the CO_2 level in the expired air. The animal was mounted in a Horsley-Clarke apparatus, the pupils were dilated with atropine, and contact lenses were used to focus the eyes upon a tangent screen. A 2×5 mm region of the primary visual cortex was then exposed and coated with a layer of 2.5% agar in saline solution. A transparent plexiglass chamber, filled with saline, was sealed around the opening in the skull. The sliding top plate of this closed chamber contained a small aperture filled with vaseline through which a microelectrode could be advanced into any region of the exposed cortex. The electrodes were mounted onto a stepping micromanipulator (AB Transvertex, Sweden) which could be advanced in finely controlled steps as small as 4 μm.

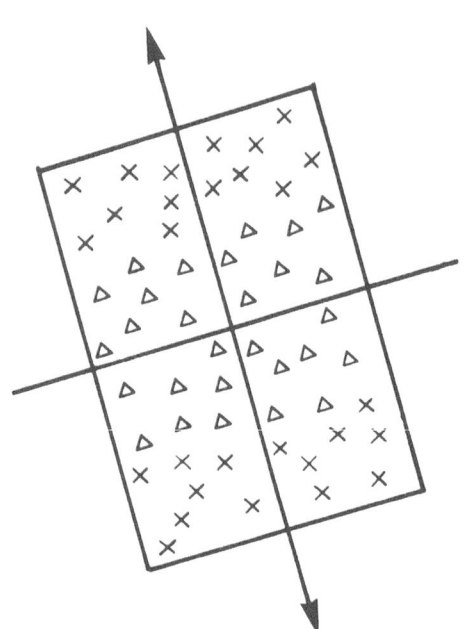

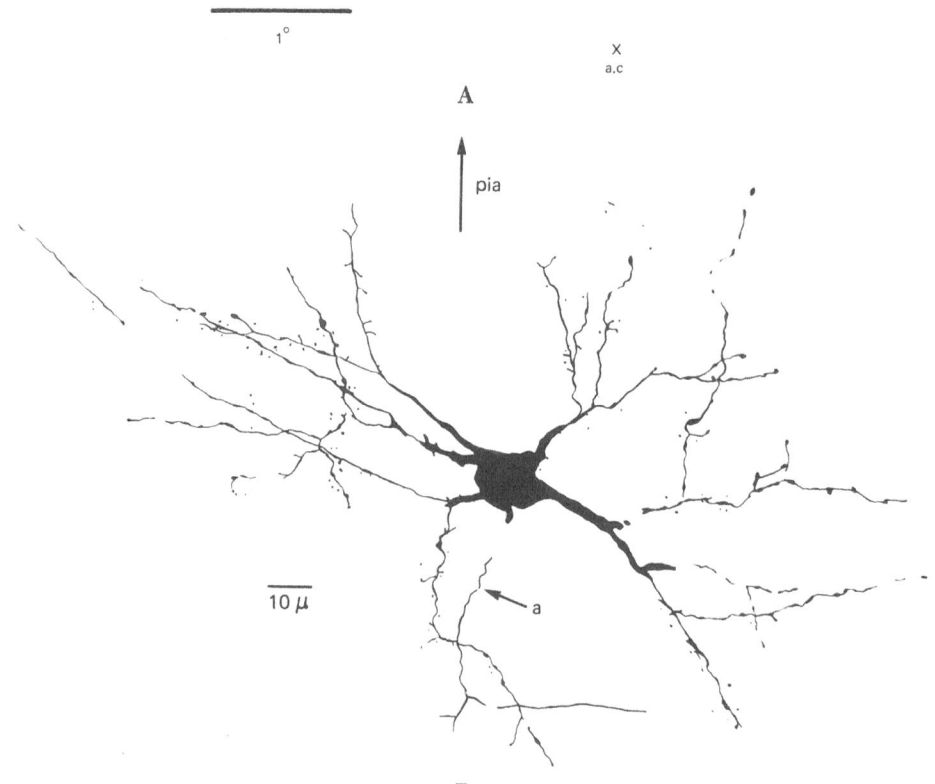

1°

A

×
a,c

↑ pia

$\overline{10\,\mu}$

a

B

In our early experiments we tried to inject cells using fine-tipped electrodes having resistances of 200–500 MΩ when filled with Procion yellow. These electrodes often provided stable intracellular penetrations but they did not pass enough dye to stain cells. Electrodes whose tips were broken against a glass slide until their resistances were 80–150 MΩ occasionally stained cells, but during intracellular penetrations there were frequently signs of cell injury and often there was too little time for adequate mapping of the receptive fields. Our most successful results have come from using low resistance electrodes with bevelled tips (Barrett and Graubard, 1970, and Chapter 18). Glass micropipettes were filled with a 6% (w/v) solution of Procion yellow M4RAN using a fiberglass filling technique. The tips were bevelled at an angle of 20–40° by lightly pressing the electrode for a few seconds against a rotating Arkansas grinding wheel (L. A. Clark Co.) mounted in a small lathe. The elliptical openings at the tips were usually about 1 to 3 μm in size and the electrode resistances were 20–80 MΩ. The advantage of this type of electrode is that the resistance is low enough to permit a stable extracellular recording from an uninjured cortical cell, yet the tip remains sharp enough to penetrate it.

In a typical experiment many single units were recorded from extracellularly as the electrode was lowered into the cortex. The receptive field of each cell was mapped using spots and slits of light projected onto the tangent screen facing the animal. After the characteristics of a cell had been determined from extracellular recording, an attempt was made to penetrate the cell by advancing the electrode in small steps. The penetration of a cell was often facilitated by passing large (10–30 $\times$ 10^{-9} A) current pulses through the electrode after each mechanical advance. If a cell was impaled its receptive field was quickly remapped to be absolutely sure that the extracellular and intracellular recordings were from the same unit. The cell was then injected with Procion yellow by passing a steady hyperpolarizing current of 1–20 $\times$ 10^{-9} A for 1–30 min. The resting potentials recorded from neurons impaled with bevelled electrodes ranged from 5 to 50 mV. We were surprised to find that several cells were quite well stained even though their resting potentials were only 5–10 mV. Moreover, by passing a steady hyperpolarizing current it was often possible to reduce the injury discharges from an impaled cell to a level low enough that the responses to visual stimuli could be seen easily.

After each attempted injection of a cell the electrode was withdrawn and a new penetration was made at least 0.5 mm from any of the previous electrode tracks. The

---

Fig. 1A and B. A. Receptive field map of a simple cell that was subsequently injected with Procion yellow. This cell was driven by stimuli to the left eye; it responded best to a narrow slit of light held in a 2:00–8:00 o'clock orientation and moved in either direction orthogonal to the slit (arrows). Stationary spots and slits of light also drove this unit. X, areas giving "on" responses; triangles, areas giving "off" responses. The cell showed only 5 mV resting potential after penetration, but it was successfully stained by passing a steady 1 $\times$ 10^{-8} A current for 20 min. The large X indicates the projection onto the visual field of the area centralis (a.c.) of the left eye. B. Camera-lucida drawing of the same cell, which lay in two adjacent coronal sections of the cortex. Drawings were made using brightfield fluorescence illumination and a 100X oil immersion objective which allowed processes less than 1 μm in diameter to be visualized. Most of the branches emerging from the cell body show occasional faint appendages that are likely to be dendritic spines. One descending process (a) is more uniform and is probably the axon.

point of entry of each electrode was marked on an enlarged photograph of the surface of the cortex, particularly with reference to the pattern of the blood vessels, so that in each experiment up to a dozen separate penetrations could be reliably identified. At the end of the experiment an electrode filled with a 4% solution of Chicago Blue was used to make several superficial dye marks that served as reference points during histological processing. The animal was perfused with 10% formalin in saline and serial frozen sections (80 μm thick) were cut from area 17 and examined with fluorescence microscopy. In most experiments only one or a few cells were actually stained, and the reference dye marks allowed each of these to be assigned to a particular electrode track. For each injected cell an adjacent Nissl stained section was examined to be sure that the cell was in area 17 and to determine which cortical layer it occupied.

Results of Procion Dye Injection

To date we have successfully injected 16 physiologically identified neurons and one glial cell. In order to illustrate our results we will describe in detail two cells whose

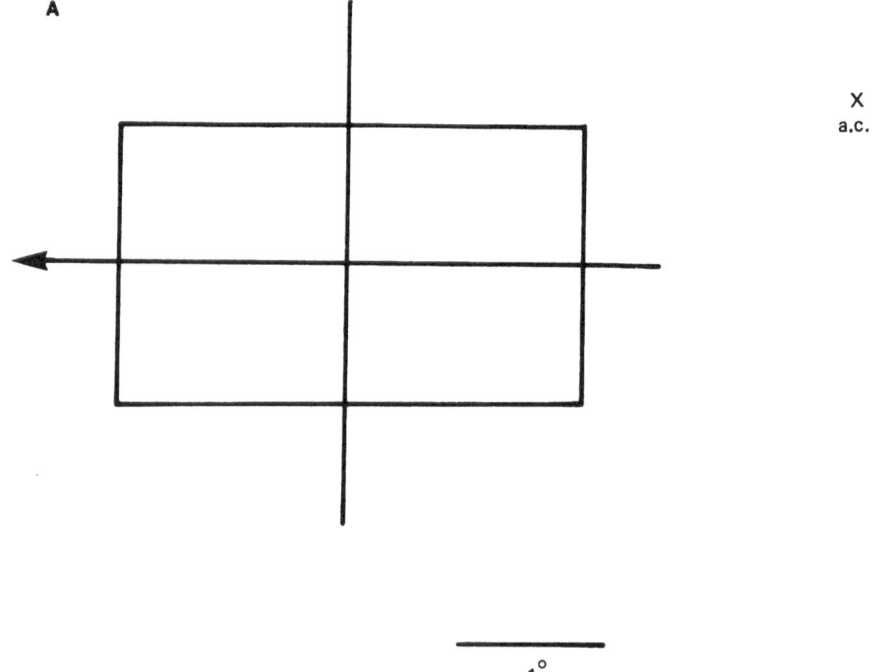

Fig. 2A–C. A. Receptive field of a complex cell that responded optimally to a vertical slit. The cell was driven equally well by both eyes and had a moderate directional preference (right to left). Stationary vertical slits produced discharges at "on" and "off," but stimuli more than 30° from vertical gave little or no response. The cell was penetrated without previous extracellular identification; the resting potential was 40 mV. To inject the cell, a 1–2 ×10⁻⁹ A hyperpolarizing current was passed for 15 min. B. Camera lucida drawing of the same cell. The cell is similar in appearance to the large pyramidal cells seen in Golgi preparations. C. Partial photographic montage of this cell, which was a large pyramid situated in layer V.

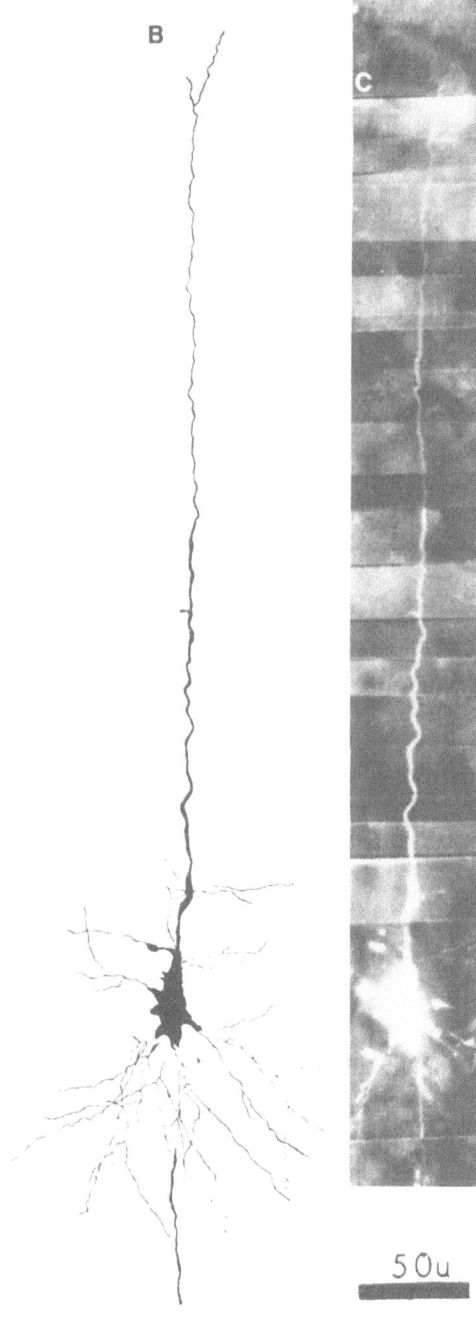

Fig. 2B and C.

properties were particularly well defined. The first of these cells responded optimally to a narrow slit of light moved slowly across the receptive field in a 2:00 to 8:00 o'clock orientation (Fig. 1A). This unit was clearly a simple cell, as defined by Hubel and Wiesel (1962), since on the basis of the responses to stationary spots and slits, its receptive field could be subdivided into a central inhibitory (off) zone flanked by antagonistic excitatory (on) regions.

Histological examination of the cortex revealed a single fluorescent neuron in the region where this unit was injected. The cell was located 350 μm below the pial surface, in layer IV a + b of Otsuka and Hassler (1962) as determined from the cytoarchitecture of an adjacent Nissl-stained section. A camera lucida drawing of this neuron (Fig. 1B) shows a cell body about 15 μm in diameter and somewhat irregular in shape. Most of the processes radiating from the soma appear to be dendrites, since they divide into finer branches that are beaded and show occasional dendritic spines. One descending process, labeled a, is more uniform in diameter, bears no spines, and is probably the axon. This neuron has the characteristic features of stellate cells seen in Golgi preparations of the visual cortex (Ramón y Cajal, 1922; O'Leary, 1941).

Fig. 2 illustrates a pyramidal cell injected in a separate experiment. The optimal stimulus for this unit was a vertical slit moving from right to left across the receptive field (Fig. 2A). Stationary vertical slits produced a brisk discharge at on and off when positioned anywhere within the receptive field, but a rectangle filling the field yielded only a weak response, indicating that this was not a simple cell. Nor was the cell hypercomplex, since the response to a slit was not diminished when the stimulus was lengthened so as to extend above and below the receptive field. It thus fulfilled the criteria for a complex unit. A camera lucida reconstruction and a photographic montage of this cell are shown in Fig. 2B and C. The pyramidal-shaped cell body is situated in layer V, 700 μm below the pial surface. A long apical dendrite radiates upward at least as far as layer III, and several fine basal dendrites and horizontal processes extend from the cell body.

In some of the injected cells either the functional or the morphological properties of the unit were not as clear-cut as in the two examples just described. For example, it was often difficult to map the receptive field of a unit because its responses either were weak or were partially obscured by injury discharges produced by the electrode penetration. In addition, some cells were incompletely stained and could not easily be classified as either stellate or pyramidal. About 20 injected cells have been discounted altogether for such reasons. We have divided the 16 remaining neurons into those whose classification was unequivocal and those which were identified functionally and anatomically, but with less certainty.

Table 1. Summary of Identified Cells

		Definite	Probable
Simple cells:	stellate	4	1
	pyramidal	0	1
Complex cells:	stellate	1	0
	pyramidal	4	3
Hypercomplex cells:	stellate	1	0
	pyramidal	1	0

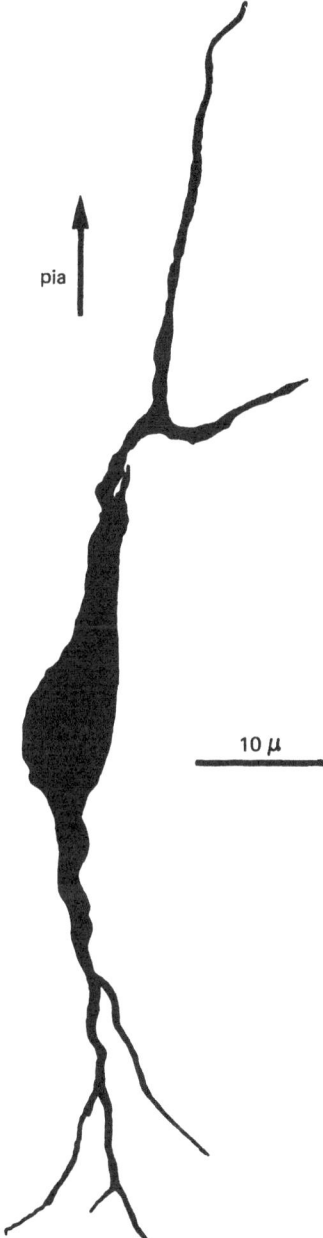

Fig. 3. Camera-lucida drawing of an injected silent cell which showed no evidence of impulse actively or synaptic potentials. The small cell body is located 1180 μm below the pial surface in layer V, and its morphology indicates that it is an oligodendrocyte.

Most of the simple and complex cells that were injected conformed to a general pattern, as shown in Table I. Of the five clearly identified simple units, four were definitely stellate cells and the fifth was probably stellate also; one probable simple cell was definitely a pyramid. Five injected cells had well-defined complex receptive fields; four of these cells were pyramidal and one was stellate. The three probable complex units were all pyramidal cells. We have in addition stained two hypercomplex units, one of which was pyramidal and the other stellate. Both of these hypercomplex cells, and also the two exceptions to the general pattern seen for simple and complex cells, were all found in cortical layers II and III, suggesting that the correlation between cell structure and function may be different from one cortical layer to the next.

In addition to recording from cortical neurons, we have often encountered silent cells that have resting potentials of 60–90 mV, and in these cells we have not observed impulse activity or synaptic potentials. Such cells are frequently penetrated by fine-tipped high resistance (200 MΩ) electrodes, but only rarely by bevelled electrodes. It has been difficult to stain these cells with Procion yellow, but in one experiment a silent cell was successfully stained and recovered. Fig. 3 is a camera lucida drawing of this cell, which lies 1180 μm below the pial surface, in layer V of Otsuka and Hassler. Two main processes arise at opposite poles of the small soma and run in an orientation parallel to the bundles of axons which course through layer V. Even though the staining of the finest branches of this cell may not have been achieved, the small number of stout processes which emerge from the cell soma and their orientation, coupled with the absence of a distinct axon, indicate that this cell is an oligodendrocyte.

There was a wide variation in the way the silent cells in the cortex responded to visual stimuli. Many of the cells thought to be glia gave no response, but some responded to visual simuli with graded depolarizations 1 to 6 mV in amplitude. These slow depolarizations are presumably mediated by an increase in extracellular K$^+$ produced by activity in surrounding neurons (Kuffler, Nicholls and Orkand, 1966; Orkand, Nicholls and Kuffler, 1966; Karahashi and Goldring, 1965; Baylor and Nicholls, 1969; Dennis and Gerschenfeld, 1969; Miller and Dowling, 1970). Although we do not have direct evidence bearing upon the mechanism of this response, it is noteworthy that a few cells with small receptive fields (1° to 3°) showed a definite preference for stimuli of a particular orientation. In one experiment the responses of a neuron and a silent cell were recorded sequentially in the same penetration. The receptive fields of the two cells overlapped, and it was clear that the maximal response of the silent cell was evoked by a bar of light swept across the field in the orientation preferred by the previously recorded neuron. From results of this kind, it seems likely that the response patterns of glial cells are determined in part by the orientation column (Hubel and Wiesel, 1962 and 1963) in which they reside.

Discussion

Low-resistance bevelled microelectrodes make it possible to inject dye into neurons after the functional properties of the uninjured cell have been mapped with the electrode lying extracellularly. Although our electrodes may be biased towards selectively impaling relatively large cortical cells, we have been able to inject and recover cells whose somata are less than 10 μm in diameter. In a well-injected cell, dye spreads into even the finest processes, so that in the vicinity of the soma nearly as much detail can be seen as in a Golgi-impregnated neuron. The results obtained up to this point support the idea that the morphological and physiological properties of cortical neurons are closely related. Most injected simple cells were stellate, and most complex cells were pyramidal. A larger sample of cells obviously is needed before one can say whether such a simple pattern is true for all layers of the cortex. This type of approach, however, does promise to increase our understanding of the different stages of information processing in the visual pathway, particularly when used in conjunction with the intracellular recording of synaptic events.

Acknowledgments. We are grateful to Drs. David Hubel and Torsten Wiesel for their aid and encouragement in this project. This project was supported by NIH Fellowships NINDS 5F02N551936-01 and NEI F02EY51537-01 and NIH Grants Nos. SR01 EY00605-13 and SR01 EY00606-08.

References

Barrett, J. N., and K. Graubard: Fluorescent staining of cat motoneurons *in vivo* with bevelled micropipettes. Brain Res. *18*, 565–568 (1970).

Baylor, D. A., M. G. F. Fuortes, and P. M. O'Bryan: Receptive fields of cones in the retina of the turtle. J. Physiol., Lond. *214*, 265–294 (1971).

——, and J. Nicholls: Changes in extracellular potassium concentration produced by neuronal activity in the central nervous system of the leech. J. Physiol., Lond. *203*, 565–569 (1969).

Dennis, M. J., and H. Gerschenfeld: Some physiological properties of identified mammalian glial cells. J. Physiol., Lond. *203*, 211–222 (1969).

Hubel, D. H., and T. N. Wiesel: Receptive fields, binocular interaction and functional architecture in the cat's visual cortex. J. Physiol., Lond. *160*, 106–154 (1962).

—— ——: Shape and arrangement of columns in the cat's striate cortex. J. Physiol., Lond. *165*, 559–568 (1963).

Kaneko, A.: Physiological and morphological identification of horizontal, bipolar and amacrine cells in goldfish retina. J. Physiol., Lond. *207*, 623–633 (1970).

——: Electrical connexions between horizontal cells in the dogfish retina. J. Physiol., Lond. *213*, 95–105 (1971).

Karahashi, Y., and S. Goldring: Intracellular potentials from "idle" cells in cerebral cortex of cat. Electroenceph. clin. Neurophysiol. *20*, 600–607 (1965).

Kuffler, S. W., J. Nicholls, and R. Orkand: Physiological properties of glial cells in the central nervous system of amphibia. J. Neurophysiol. *29*, 768–787 (1966).

Miller, R. I., and J. E. Dowling: Intracellular responses of the Müller (glial) cells of mudpuppy retina: their relation to b-wave of the electroretinogram. J. Neurophysiol. *33*, 323–341 (1970).

O'Leary, J. L.: Structure of the area striata of the cat. J. comp. Neurol. *75*, 131–164 (1941).

Orkand, R., J. Nicholls, and S. W. Kuffler: Effect of nerve impulses on the membrane potential of glial cells in the central nervous system of amphibia. J. Neurophysiol. *29*, 788–806 (1966).

Otsuka, R., and R. Hassler: Über Aufbau und Gliederung der corticalen Sehsphare bei der Katze. Arch. Psychiatr. Nervenkr. *203*, 212–234 (1962).

Ramón y Cajal, S.: Studien uber die sehrinde der katze. J. Psychol. Neurol. *29*, 161–181 (1922).

Stretton, A. O. W., and E. A. Kravitz: Neuronal geometry: determination with a technique of intracellular dye injection. Science *162*, 132–134 (1968).

Discussion

DR. KRIPKE (Utah): Dr. Van Essen how far have you succeeded in tracing the axons of your cells?

DR. VAN ESSEN (Harvard): Not very far at all. Usually we can trace the axon only for a few hundred microns, and in many injected cells we cannot even tell which process is the axon. In only one cell have we actually seen the axon entering the white matter.

DR. BENNETT (Albert Einstein): Many people have asked, "How much current did you pass to get dye into the cells?"

However, one often observes the passage of current without movement of dye.

The question really should be "How much current are you passing, and what is the transfer number of the dye?"

DR. VAN ESSEN: I think from our experience, if the resistance of the electrode goes up drastically, then we can feel confident that it will not inject a cell. But, unfortunately, the converse is not true. That is, even some low-resistance electrodes don't seem to pass current very well.

CHAPTER 14

Procion Yellow Staining of Functionally Identified Interneurons in the Spinal Cord of the Cat

E. Jankowska and S. Lindström

Interneurons probably constitute the majority of the neurons in the spinal cord. Until recently, however, none of the interneurons described in anatomical studies could be assigned to any functionally defined neuronal pathway and none of those analyzed in physiological studies had been identified morphologically. Such identifications have now become possible with the technique of intracellular staining with Procion yellow developed by Stretton and Kravitz (1968 and Chapter 2). We have used this staining technique for different types of interneurons located in the dorsal horn, intermediate region, and ventral horn of the cat spinal cord, but in this presentation we will restrict ourselves to a description of results obtained on two physiologically well defined groups of cells: (1) the interneurons mediating reciprocal inhibition of motoneurons from group Ia muscle spindle afferents (Jankowska and Lindström, 1972) and (2) Renshaw cells (Jankowska and Lindström, 1971; van Keulen, 1971). The morphological characteristics of these cells will be summarized and the utility of Procion yellow for staining small cells and axons in the mammalian spinal cord will be discussed.

The Ia inhibitory interneurons were found to have functional properties which distinguish them from all other types of interneurons (Hultborn et al., 1971a, b; see Fig. 2A–C) and it was shown that the individual cells with such properties do indeed terminate on motoneurons and evoke unitary IPSPs in them (Jankowska and Roberts, 1971b). For Renshaw cells, which are characterized by a typical high frequency discharge following stimulation of motor axons (Renshaw, 1946; Eccles et al., 1954; see Figs. 2D and H), the evidence that they project to motoneurons and mediate their recurrent inhibition is similarly strong but still indirect (Eccles et al., 1954; see also Willis, 1971).

Comments on the Technique of Dye Injection

The general techniques of recording and staining as well as criteria used for identification of the interneurons are described elsewhere (Jankowska and Lindström, 1971, 1972).

The micropipettes used for recording and dye injection were pulled in a conventional manner and their tips were broken to 1.5–2.5 μm. The pipettes were filled with freshly made 5–6% (w/v) aqueous solution of Procion yellow M4R and had resistances of 15–25 MΩ. The same electrode could be used for staining of several cells. While adapting the technique of injection of Procion yellow to small cells and fibers, usually difficult to impale, we found that the following precautions are of importance: (1) To limit the current used for injection of the dye to $10–25 \times 10^{-9}$ A, depending on the cell size; when stronger currents are used, the cells are more frequently lost and/or damaged, with the result that the dye might leak from them. (2) To use continuous constant current (slowly rising) rather than pulses, since the use of pulses increases the risk of losing the cells. The current was passed for periods of up to 20 min. (3) To record synaptically or antidromically evoked responses of the cell during the entire dye injection to ascertain that the dye was injected intra- and not extracellularly, and to have the proof that the right cell was being injected. (4) To monitor continuously the current-passing capability of the electrode.

The period between dye injection and sacrifice of the animal was 1–11 hr. To allow a longer time for dye migration along the axons, the spinal cords were perfused *in situ* with cold Ringer solution and kept for 16–32 hr at 4°C before fixation (Jankowska and Lindström, 1970a).

The spinal cords were fixed in 10% neutralized formalin, dehydrated in ethyl alcohol, and embedded in paraffin. Sections were cut at 15 μm, deparaffinized, and mounted in Entellan (Merck). The cells were viewed through a Leitz microscope with a Ploem incident light fluorescent illuminator equipped with a K495 dichroic beam splitting mirror, a GB12 exciting filter (3 mm) and a K510 absorbing filter.

Location of Various Types of Interneurons

The approximate location of several types of interneurons recorded in the spinal cord had earlier been determined from the position of the electrode tips. Some of these interneurons appeared to be grouped in fairly well defined zones of the gray matter, e.g., Renshaw cells (Eccles et al., 1954; Thomas and Wilson, 1965; Willis, 1971), cells responding with long latency and a long discharge to stimulation of high threshold flexor reflex afferents (FRA) after application of DOPA (Jankowska et al., 1967), cells with small or large cutaneous receptive fields in the dorsal horn (Wall, 1967), dorsal horn neurons specifically excited by noxious and termal stimuli (Christensen and Perl, 1970), and Ia inhibitory interneurons (Hultborn et al., 1971b).

The tendency to be located in relatively narrow and more or less separated areas appeared even more striking in the samples of cells stained with Procion yellow. The Ia inhibitory interneurons were found in an area just dorsal and dorsomedial to motor nuclei (Fig. 1A, filled circles) in lamina VII of Rexed (1954), which corresponds to the middle zone of terminal projections of group Ia afferents (Fig. 1B-II) described by Szentágothai (1967). The area of location of Ia inhibitory interneurons is clearly separated from the area in the intermediate region of the spinal gray matter in which another group of interneurons, previously supposed to mediate the Ia reciprocal inhibition of motoneurons (Eccles et al., 1956) are located (Fig. 1A, open circles). The latter have somewhat different functional properties (cf., Hultborn et al., 1971b) and were concluded to project to cells other than motoneurons (Hultborn et al., 1971b; see also Hongo et al., 1972). This conclusion is supported by their separate location, although functional subgroups of the same population of cells also may have different locations, as found in the case of cells of origin of the dorsal spinocerebellar tract (Lindström and Takata, unpublished).

The stained Renshaw cells were located more ventrally and medially than the Ia inhibitory interneurons, in the area where the motor axons gather before they leave the gray matter. Fig. 1A shows the location of 7 Renshaw cells stained in L7 (also see van Keulen, 1971). This location corresponds very well to that indicated by the original physiological evidence (Renshaw 1946; Eccles et al., 1954), and to the area of terminal branching of motor axon collaterals according to Szentágothai (1967, cf. Fig. 1B).

Thus, it may be that some types of interneurons in the spinal cord form nuclei in which they are only to a limited extent intermixed with other cells. If so, this fact would be of great importance for anatomical studies in which physiological identification is not possible. In the case of Ia inhibitory interneurons and of Renshaw cells, there seem to be hardly any other types of cells in their immediate vicinity as judged from tracking and recording within the areas of their location.

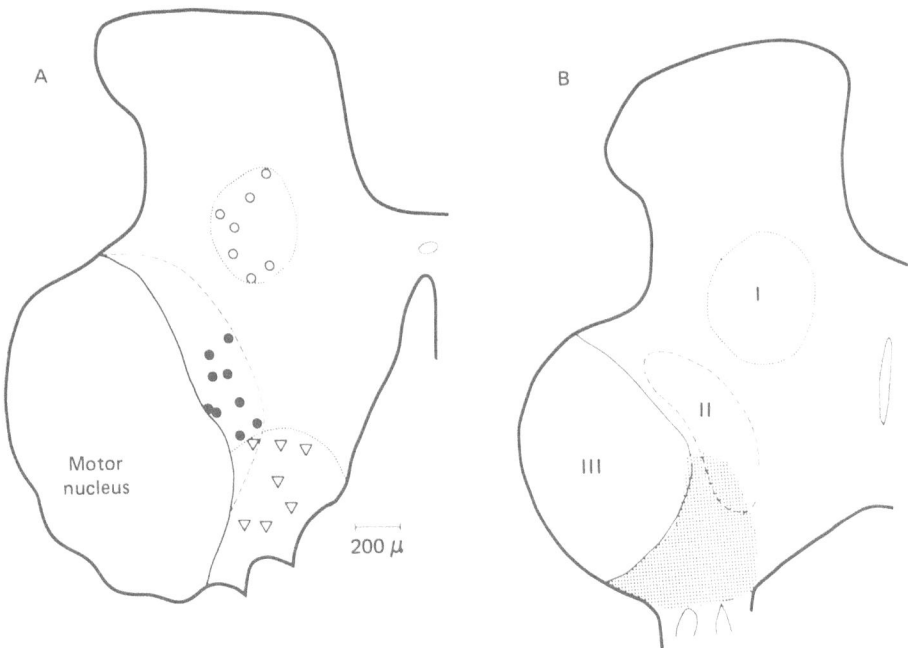

Fig. 1A. Semi-schematic diagram of location of three groups of cells stained with Procion yellow in relation to the borders of the motor nuclei and the gray matter. ●, Ia inhibitory interneurons, ○, Ia interneurons not inhibited via motor axon recurrent collaterals; △, Renshaw cells. Only cells stained in L7 are indicated. The dashed lines enclose the areas of location of the three groups of cells, considering also the location of cells stained in L6. B. Three areas of massive termination of group Ia afferents: in the intermediate region (I), a semilunar region outside motor nuclei (II), and in the motor nuclei (III), according to Szentágothai (1967). Pointiled area corresponds to the field of arborization of initial collaterals of motor axons according to Szentágothai (1967).

General Characteristics of Renshaw Cells and of Ia Inhibitory Interneurons

Soma Size

In both groups of cells, the somata were found to have a fairly uniform size. The Ia inhibitory interneurons are of medium size (20 $\times$ 30 μm, Fig. 4A and C), while the Renshaw cells probably are among the smallest cells in the spinal cord (10–20 μm, Fig. 4 B and D). The given values are without correction for shrinkage (about 15%) due to histological procedures. In respect to their size, the Ia inhibitory interneurons and Renshaw cells can thus be differentiated from larger cells like α-motoneurons, cells of origin of the ventral spinocerebellar tract (Jankowska and Lindström, 1970a) and some commissural cells (Jankowska and Lindström, 1970b) but not from other medium-sized and small interneurons in the ventral horn. γ-motoneurons were reported (Bryan et al., 1972) to be the same size as Ia interneurons.

Dendritic Tree

In well stained cells the dendrites were traced until they tapered to about 1 μm diameter and terminated in a drop-like manner (Fig. 4 F). The Ia inhibitory interneurons had long, usually straight, dendrites with only few bifurcations (Figs. 2, 3 and 4A, C). They projected within a radius of 200–500 μm, with the maximal extension dorsoventrally and the least rostrocaudally (compare cells in Fig. 3B with cell 1 in Fig. 2 and cell 1 in Fig. 3A). The ratio of dendritic extension radii in these two directions was approximately 2 : 1. In this respect the Ia inhibitory interneurons would differ from most of the interneurons in lamina V through IX of Rexed, reported by Scheibel and Scheibel (1966, 1969) to have little or no extension along the rostrocaudal axis. The Ia inhibitory interneurons described here would also differ from cells with a predominantly longitudinal or a symmetrical orientation of dendrites (Scheibel and Scheibel, 1966, see their Fig. 4).

The Renshaw cells had thinner and more tortuous dendrites which extended for about 150–200 μm from the soma without any preferred direction (Fig. 2, cell 2, and Fig. 4B). This also means that Renshaw cells, similarly to Ia inhibitory interneurons, have a longer rostrocaudal extension of dendrites than postulated by Scheibel and Scheibel (1966, 1969) for most of the interneurons.

Apart from the shape of the dendritic tree, the number of main dendrites and the type of branching seem to be common for both groups of cells analyzed by us and for most spinal interneurons described morphologically (Ramón-Moliner, 1962; Gelfan et al., 1970; Matsushita, 1970).

Axonal Projections

The axons were differentiated from dendrites on the basis of the following properties (see Fig. 4C, E, and G): (1) a uniform diameter in contrast to the tapering of dendrites, (2) an initial narrow segment, (3) an envelope of nonfluorescent space corresponding

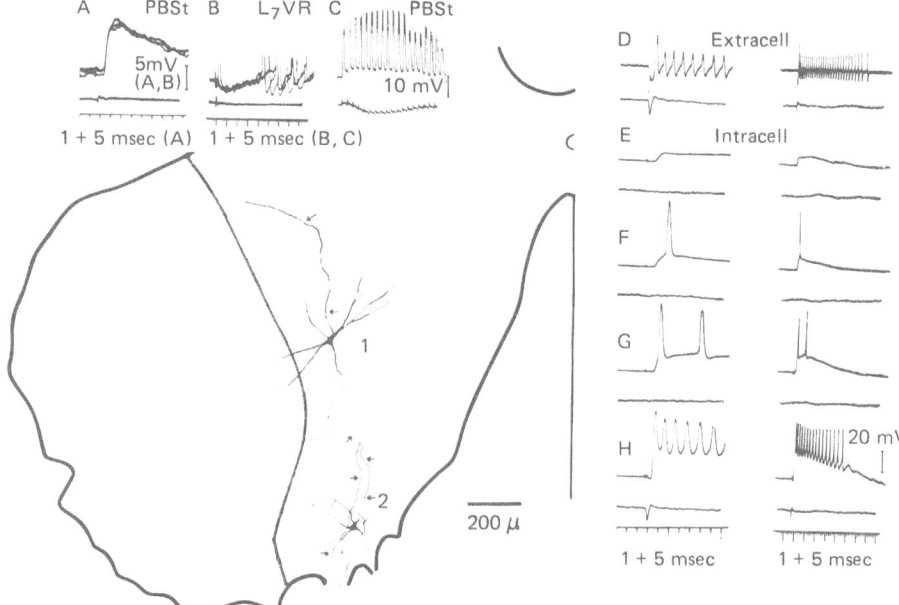

Fig. 2. Reconstruction of dendrites and axons (indicated by arrows) of a Ia inhibitory interneuron (cell 1) and of a Renshaw cell (cell 2). The dashed line indicates the extent of the motor nuclei. The records from the Ia inhibitory interneuron (A–C) illustrate the criteria used for identification of Ia inhibitory interneurons (Hultborn et al., 1971b). A. Monosynaptic excitation from low threshold afferents in posterior biceps-semitendinosus nerve (PBst). B. Inhibition following stimulation of the L7 ventral root. C. Ability to follow a high frequency orthodromic stimulation. In D–H are records from the Renshaw cell, taken simultaneously with two different sweep speeds. Those in D were taken just before the penetration. Records in E–H show responses of the cell to increasing strength of stimulation of the L7 ventral root (cf. sizes of incoming volleys in the records from the surface of the spinal cord, lower trace in each pair). At threshold only an EPSP was evoked. Note increase in its amplitude with increasing strength of stimulation and the decrease in latency of the first spike.

to the myelin sheath (see below), (4) a gradual weakening of the intensity of the fluorescence with the distance from the soma (Jankowska and Lindström, 1970, 1971), (5) a projection over considerably longer distances than the dendrites, and (6) a smooth contour and a winding trajectory of the axis cylinder (cf. Jankowska and Lindström, 1970). The latter feature was very characteristic, though probably an artifact due to shrinkage during histological procedures.

Both the Ia inhibitory interneurons and the Renshaw cells had myelinated axons. The external diameters (as judged from the nonfluorescent space around the stained axis cylinders) were 6–14 μm for the Ia inhibitory interneurons and 6–8 μm for the Renshaw cells. The axons of Ia inhibitory interneurons could be traced over a distance of up to 1.0–1.5 mm and in one exceptional case (cell 2, Fig. 3A) over 2.0 mm. For Renshaw cells the longest distances were less than 1.0 mm. The axons of several Ia interneurons could be traced to the ventral or lateral funiculus (Fig. 3) and even for some distance within the white matter. The axons of the Renshaw cells could be

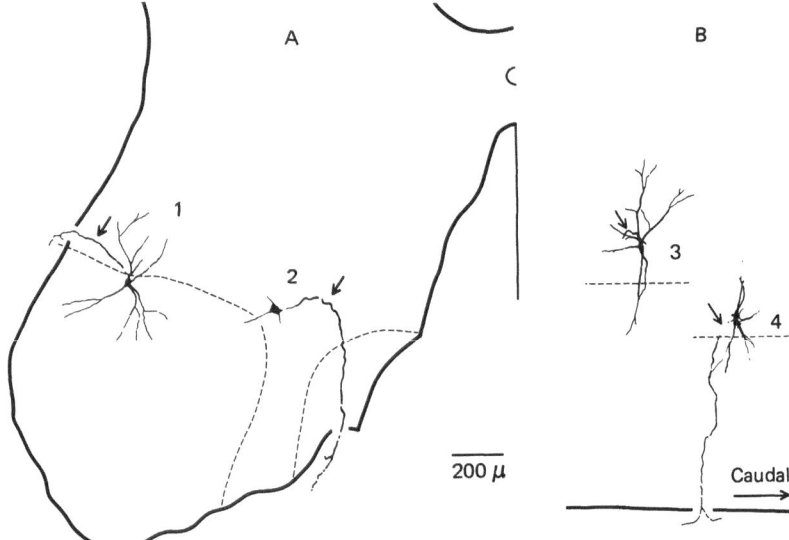

Fig. 3. Reconstruction of dendrites and axons of four Ia inhibitory interneurons in a transverse (A) and in a sagittal (B) plane. In cell 2 the dye was injected into the axon and migrated back to the soma. The continuous line indicates the contour of the ventral horn in A and its ventral border in B. The dashed lines indicate the extent of motor nuclei in A and their dorsal border in B. Arrows point to the axons. That of cell 3 projects transversely to the plane of reconstruction (Modified from Jankowska and Lindström, 1972).

followed to the border between the ventral horn and the ventral funiculus. This is in agreement with physiological data which show that many if not all of these interneurons belong to funicular cells (Jankowska and Roberts, 1972a; Jankowska and Smith, 1973). Axon collaterals (Fig. 4H) and axonal branching (Fig. 3) were observed only infrequently. In Renshaw cells, the paucity of early collaterals probably was due to the deficiency of Procion staining since the axons of the Renshaw cells would be expected to branch profusely at the same level (Eccles et al., 1954) and they were generally less satisfactorily stained than those of other cells. However, in one cell (Fig. 2, cell 2) an axon collateral was found to be given off before the main axon turned ventrally towards the white matter. This collateral headed toward the motor nuclei. In the case of Ia interneurons a small number of early collaterals might be genuine since many of them should terminate on motoneurons located several millimeters from their somata (Hultborn et al., 1971b). The paucity of early collaterals, if confirmed, might be a characteristic feature of Ia inhibitory interneurons since most funicular cells in lamina VII were claimed to have profusely branching initial collaterals (Szentágothai, 1967; Scheibel and Scheibel, 1969).

The axonal projections may characterize the Ia inhibitory interneurons better than other features. Physiological studies have indicated that in the extreme case the interneurons should send an axon along one of the funiculi over a distance of 15–20 mm, before it would reenter the ventral horn and start to branch within the motor nuclei, terminating on a considerable proportion of motoneurons (Jankowska and Roberts,

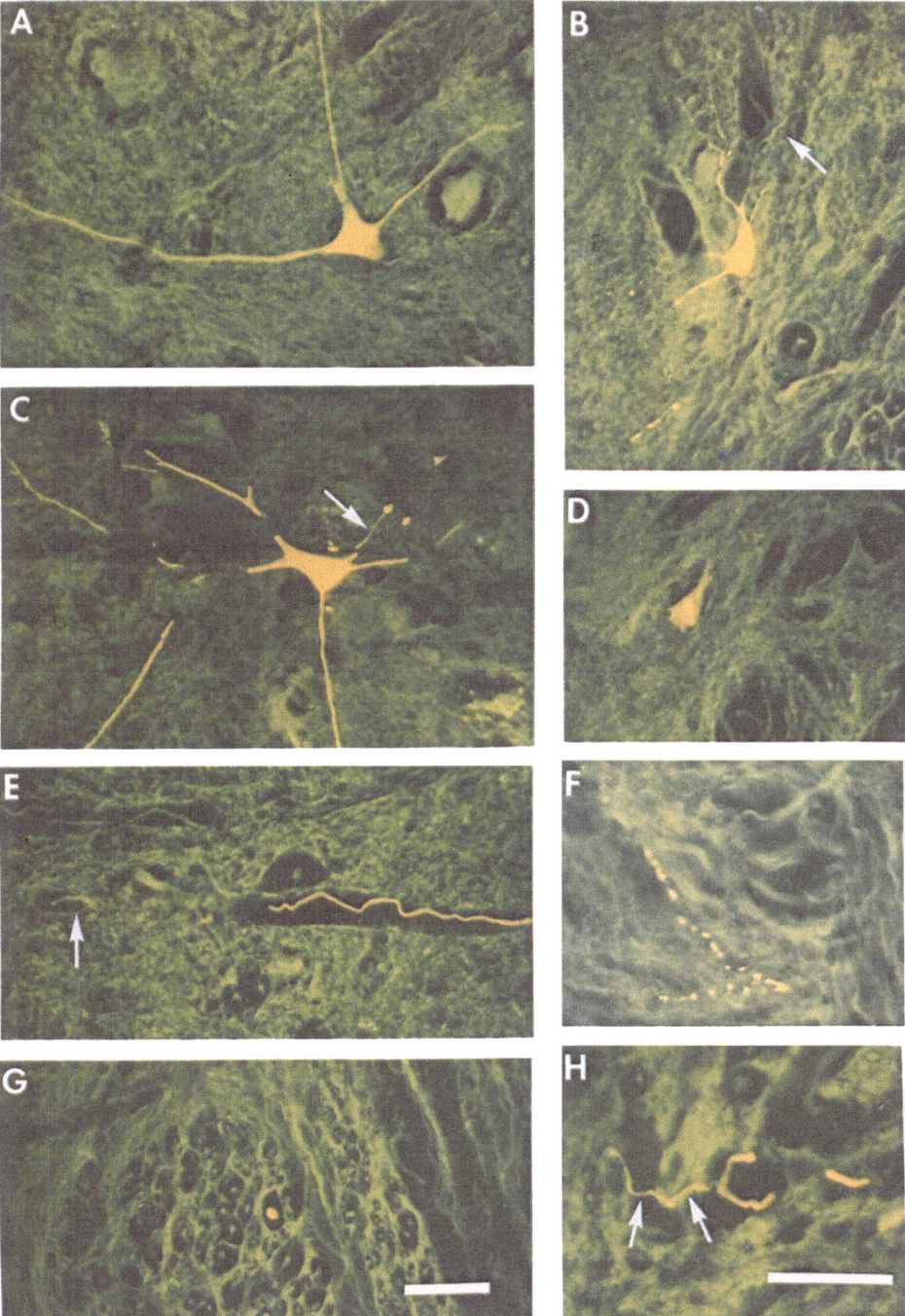

Fig. 4. Microphotographs of Procion yellow-stained neurons (15 μm sections). A and C. Ia inhibitory interneurons. B and D. Renshaw cells. E. Two parts of an axon of a ventral spinocerebellar tract cell in the gray matter. F. Distal parts of a thin dendrite of a Renshaw cell. G. Transversely sectioned axon of a Ia inhibitory interneuron in the ventral funiculus. H. A part of an axon of a spinocervical tract neuron with an axon collateral. The arrows point to the axon in B, the thin initial segment of the axon in C, a distal fragment of the axon in E, and the axon collateral in G. Note the nonfluorescent space around the axons in E and G, indicating the myelin sheath, and the decrease in intensity of the fluorescence between more proximal and more distal pieces of the axon in E. The calibration bar in G is 50 μm for A–E and G; the one in H is 50 μm for F and H. Kodak Ektachrome high-speed color film (daylight), exciting filter BG 12 (3 mm) and absorbing filter K 510.

1971a, b), some ventral spinocerebellar tract (VSCT) cells (Gustafsson and Lindström, 1970), and other Ia interneurons (Hultborn, 1972). Depending on their input, some Ia inhibitory interneurons should terminate on motoneurons located fairly close and some on those located far away from their somata. The expected variability of the projections of Ia interneurons has been seen in the sample of cells stained with Procion yellow. For instance, it was observed that they project to either the ventral or the lateral funiculus (Fig. 2). This was related more to the location of the cell body (the projection was usually radial to the nearest funiculus, Ramón y Cajal, 1909) than to the input and final destination.

Using the axonal projections and their termination on motoneurons, together with other features revealed by Procion yellow staining, it would probably be possible to recognize tentatively the Ia inhibitory interneurons and Renshaw cells among the cells stained with conventional histological techniques. Until more is known about them, however, the physiological criteria still would be indispensable for their proper identification.

The Value of Procion Yellow for Intracellular Staining of Individual Cells

Procion yellow proved to be most useful in identifying the cells which had been recorded from, as practically all injected cells could subsequently be found. The quality of staining did, however, vary. The amount of injected dye and the site of penetration of the cells seem to be decisive; the injection of the dye into the soma gave better staining of soma-dendrites, while its migration along the axon was more effective when it was introduced intra-axonally (cf. cells 1 and 2 in Fig. 3). The degree of the damage to the cells (in some cases there was evidence that the dye diffused through the hole made in the membrane) was also of importance, but sometimes even cells with very poor resting membrane potential were well stained. In the worst cases, only the location and the size of the somata of injected cells could be ascertained. In the well stained cells (about half of all those injected) the dendritic trees and the initial trajectory of axons could also be reconstructed. However, the analysis of fine details of their structure appeared difficult and would require other techniques. For instance, spine-like elements were observed only rarely, although for the ventral horn interneurons, these have been clearly revealed with other techniques (Scheibel and Scheibel, 1969).

The main drawback of Procion yellow staining was undoubtedly its insufficient migration along the main axons as well as their branches. The trajectory of the axons could be reconstructed only for their initial part, about 2–3 mm in the largest cells (motoneurons and VSCT cells) and about 0.5–1.5 mm in the smallest (Renshaw cells and Ia inhibitory interneurons) because the intensity of the dye decreased very quickly (see Fig. 4E). The very thin initial segment of most axons (Fig. 4C) might have been a limiting factor for the diffusion of the dye. After injection into the axons themselves, they could be traced over somewhat longer distances, up to 4–5 mm. However, in both cases the thin axon collaterals were weakly stained and could be followed only for very short distances. Cells with long axons probably constitute a majority of the interneurons in the spinal cord since most of them seem to be funicular cells (cf. Scheibel and Scheibel, 1966, 1969). Even those with fairly local effects (like Renshaw cells) may have axons extending over several millimeters (Ryall et al., 1971; Hultborn et al., 1971b, Jankowska and Smith, 1973). Thus, to stain them and to reveal their terminals and synaptic contacts, other techniques must be used.

In spite of the above mentioned limitations of Procion yellow staining, we find it most useful both for expanding and supporting the results of physiological studies. In addition to staining interneurons we used Procion yellow for staining a number of other cells [cells of origin of the ventral (Jankowska and Lindström, 1970) and dorsal (Lindström and Takata, unpublished) spinocerebellar tracts and of the spinocervical tract, and some large afferent fibers. Procion yellow proved valuable for morphological identification of all these types of cells and fibers, provided that they were successfully impaled. One reason for its great usefulness lies in the fact that the micropipettes filled with Procion yellow can be used not only for staining but also for routine recording in many types of experiments. They do not damage the cells unless the dye is injected (cf. Jankowska and Lindström, 1970) and do not block much more than the electrodes filled with KCl or K-citrate. For instance, the records obtained from cell 2 in Fig. 2 before its staining are among the best records from Renshaw cells. The only drawback of electrodes filled with Procion. yellow is a relatively high impedance (15–30 MΩ if compared with 3–6 MΩ, for electrodes of the same size filled with K-citrate). Further it is of great advantage that, once injected into the cells, Procion yellow usually does not leak from them, either *in situ* or after fixation and cutting the spinal cord, and that it is very resistant to different histological procedures. The fluorescence of the injected structures not exposed to the excessive light was found to endure for more than two years, with only moderate fading. The resistance of Procion yellow to histological procedures allows the combination of the intracellular staining of individual cells with conventional staining techniques. The possibility of counterstaining will be very useful, e.g., for defining the external diameters of stained myelinated fibers or revealing terminals of lesioned degenerating fibers in relation to Procion yellow-stained cells.

References

Barker, D., P. Bessou, E. Jankowska, B. Pagès, and M. Stacey: Distribution des axones fusimoteurs statiques et dynamiques aux fibres musculaires intrafusales chez le chat. C. R. Acad. Sc. Paris *275*, 2527–2530 (1972).

Bryan, R. N., D. L. Trevino, and W. D. Willis: Evidence for a common location of alpha and gamma motoneurons. Brain Res. *38*, 193–196 (1972).

Christensen, B. N., and E. R. Perl: Spinal neurons specifically excited by noxious and thermal stimuli: marginal zone of the dorsal horn. J. Neurophysiol. *33*, 293–307 (1970).

Eccles, J. C., P. Fatt, and K. Koketsu: Cholinergic and inhibitory synapses in a pathway from motor-axon collaterals to motoneurons. J. Physiol., Lond. *126*, 524–562 (1954).

——— ———, and S. Landgren: The central pathway for the direct inhibitory action of impulses in the largest afferent nerve fibres to muscle. J. Neurophysiol. *19*, 75–98 (1956).

Gelfan, S., G. Kao, and D. S. Ruchkin: The dendritic tree of spinal neurons. J. comp. Neurol. *139*, 385–412 (1970).

Gustafsson, B., and S. Lindström: Depression of Ia IPSPs in spinal border cells by impulses in recurrent motor axon collaterals. Acta physiol. scand. *80*, 13A–14A (1970).

Holmqvist, B.: Crossed spinal reflex actions evoked by volleys in somatic afferents. Acta physiol. scand. *52*, Suppl. 181, 1–67 (1961).

Hongo, T., E. Jankowska, and A. Lundberg: The rubrospinal tract. IV. Effects on interneurones. Exp. Brain Res. *15*, 54–78 (1972).

Hultborn, H.: On the control of transmission in the reciprocal Ia inhibitory pathway to motoneurones. In: Mechanisms of Neuronal Integration in Nerve Centers. Ed. P. G. Kostyuk. (Kiev, in press).

———, E. Jankowska, and S. Lindström: Recurrent inhibition from motor axon collaterals of transmission in the Ia inhibitory pathway to motoneurones. J. Physiol., Lond. *215*, 591–612 (1971a).

——— ———: Recurrent inhibition of interneurones monosynaptically activated from group Ia afferents. J. Physiol., Lond. *215*, 613–636 (1971b).

Jankowska, E., M. G. M. Jukes, S. Lund, and A. Lundberg: The effect of DOPA on the spinal cord. 6. Half-centre organization of interneurones transmitting effects from the flexor reflex afferents. Acta physiol. scand. *70*, 389–402 (1967).

———, and S. Lindström: Morphological identification of physiologically defined neurones in the cat spinal cord. Brain Res. *20*, 323–326 (1970a).

——— ———: Intracellular staining of physiologically identified interneurones in the cat spinal cord. Acta physiol. scand. *79*, 4A–5A (1970b).

——— ———: Morphological identification of Renshaw cells. Acta physiol. scand. *81*, 428–430 (1971).

——— ———: Morphology of interneurones mediating Ia reciprocal inhibition of motoneurones in the spinal cord of the cat. J. Physiol., Lond. *226*, 805–823 (1972).

———, W. Roberts: An electrophysiological demonstration of the axonal projections of single spinal interneurones in the cat. J. Physiol., Lond. *222*, 597–622 (1972a).

——— ———: Synaptic actions of single interneurones mediating reciprocal Ia inhibition of motoneurones. J. Physiol., Lond. *222*, 623–642 (1972b).

———, D. O. Smith: Antidromic activation of Renshaw cells and their axonal projections. Acta physiol. scand. *88* (1973).

Matsushita, M.: Dendritic organization of the ventral spinal gray matter in the cat. Acta anat. *76*, 263–288 (1970).

Ramón-Moliner, E.: An attempt at classifying nerve cells on the basis of their dendritic patterns. J. comp. Neurol. *119*, 211–227 (1962).

Ramón y Cajal, S.: Histologie du Systeme Nerveux de L'homme et des Vertebrés. 2 vol. Paris: Maloine, 1909.

Renshaw, B.: Central effects of centripetal impulses in axons of spinal ventral roots. J. Neurophysiol. *9*, 191–204 (1946).

Rexed, B.: A cytoarchitectonic atlas of the spinal cord in the cat. J. comp. Neurol. *100*, 297–379 (1954).

Ryall, R. W., M. F. Piercey, and C. Polosa: Intersegmental and intrasegmental distribution of mutual inhibition of Renshaw cells. J. Neurophysiol. *34*, 700–707 (1971).

Scheibel, M. E., and A. B. Scheibel: Spinal motoneurons, interneurons and Renshaw cells. A Golgi study. Arch. ital. Biol. *104*, 328–353 (1966).

——— ———: A structural analysis of spinal interneurons and Renshaw cells. In: The Interneuron. Ed. M. A. B. Brazier, UCLA Forum Med. Sci. No. 11. pp. 159–208. Los Angeles: University of California Press, 1969.

Stretton, A. O. W., and E. A. Kravitz: Neuronal geometry: determination with a technique of intracellular dye injection. Science *162*, 132–134 (1968).

Szentágothai, J.: Synaptic architecture of the spinal motoneuron pool. Electroenceph. clin. Neurophysiol. Suppl. *25*, 4–23 (1967).

vanKeulen, L. C. M.: Identification des cellules de Renshaw par l'injection intracellulaire de la substance fluorescente "Procion Yellow." J. Physiol., Paris *63*, 131A (1971).

Wall, P. O.: The laminar organization of dorsal horn and effects of descending impulses. J. Physiol., Lond. *188*, 403–423 (1967).

Willis, W. D.: The case for the Renshaw cell. Brain Behav. Evol. *4*, 5–52 (1971).

Discussion

DR. JANKOWSKA (Göteborg): I would like to give you another example of application of Procion yellow: the staining of muscle fibers in cat muscle spindles. Together with Dr. P. Bessou and Dr. B. Pagès from Toulouse, we injected Procion yellow into single intrafusal fibers. These are of two types, large nuclear bag and smaller nuclear chain, and are innervated by two types of γ-motor fibers, which increase either dynamic or static responses of spindle afferents. The problem of whether one type of γ-fibers terminates on only one type of intrafusal muscle fibers or on both is still controversial. The identification of injected fibers by Dr. D. Barker and Dr. M. Stacey from Durkham revealed that both nuclear bag and nuclear chain fibers are innervated by static fibers. Terminations of dynamic γ-fibers, only two so far, have been found on nuclear bag fibers.

Procion yellow was injected into those intrafusal muscle fibers in which junction or action potentials were evoked from an isolated γ-fiber, previously identified as static or dynamic. The technique of injection was similar to that used by Dr. Lindström and myself for spinal interneurons, i.e., by application of low intensity $(10–20 \times 10^{-9}$ A) continuous constant current. Further details are given by Barker et al. (1972).

DR. BAKER (Iowa): The amount of stain migration along axons you have shown is the best we have seen in vertebrate systems and is, as you pointed out, still unsatisfactory. We have had similar difficulties attempting to fill axon collaterals (which are known to exist) in the Purkinje cells of birds and cats. We have employed both microelectrode injection and axonal iontophoresis. Is there, then, some unknown problem with vertebrate neurons?

DR. KRAVITZ (Harvard): This is a more general question. I wonder if our problems with vertebrate neurons involve difficulties in fixation or embedding procedures. Technical considerations such as differences between perfusion versus immersion in fixative, pH of fixative, and even the embedding material can have a possible effect on our ability to visualize processes.

DR. BARRETT (Colorado): Reducing the background fluorescence might make it possible to follow axons and axon collaterals for greater distances. We obtained fairly low levels of background fluorescence in frozen sections of cat spinal cord using a 4% paraformaldehyde fixative at pH 7 (see Chapter 18). Visualization of small axonal branches might also be improved by using the interference barrier filter, mentioned by Dr. Hashimoto, to block the yellow-green background fluorescence.

DR. KRAVITZ: Just to comment on that, in fact, if you look at whole mounts of lobster ganglia, you see cells with some of their branches filled, but when you cut serial thin sections through the ganglion, you see dye going much farther, possibly twice as far.

I think the background fluorescence is very important, and I think cutting that background fluorescence down by either working with thinner sections or by using a fixative that depresses the background fluorescence is important in determining what you are going to see.

DR. DE BAULT (Iowa). I would like to comment on the problem of autofluorescence.

If you have ever observed or actually measured the excitation and emission spectra of Procion yellow and formalin induced autofluorescence, they are somewhat similar with considerable overlapping; yet their excitation and emission maxima, as well as their fluorescence efficiencies, are different enough so that one can eliminate most, if not all, of the autofluorescence by using appropriate barrier filters.

If one uses red barrier filters with Procion yellow, one can mask the green autofluorescence very well, and the Procion yellow should stand out red against darkened background.

The cut-off wavelength for the barrier filter which is most probably the best is around 610 nm. However, the tail of the emission spectra for Procion yellow goes all the way up to 680–700 nm, in my experience, and barrier filters of even higher value can be used. In general, the higher the cut-off wavelength of the barrier filter, the higher the contrast between Procion yellow fluorescence and the background autofluorescence. Another factor that could help to reduce the autofluorescence is to try to block the mercury lines below 425 nm as they are more excitatory to the autofluorescence than to the Procion yellow fluorescence. The use of an Xe lamp, which produces substantially less short wavelength irradiation than does a comparable Hg lamp and thus is easier to block, would also be another way to increase the contrast between Procion yellow fluorescence and background autofluorescence.

DR. SELVERSTON (University of California, San Diego): It has been our experience that the amount of axonal filling we get in crustacean materials is a function of how much is in the soma, since this will flow into the processes during the filling, and the amount of time that is allowed for migration. Thus the amount of axonal filing is related to the size of the cell body since that will act as the principal reservoir for the dye that is going to diffuse out.

I would suspect in your case where the cell body looks very small, you simply don't get enough dye in to migrate very far into the processes.

DR. BARRETT: Perhaps one reason that axons tend to be poorly stained after somatic dye injections is that most of the applied current flows out into the dendritic tree, carrying the dye with it. Dr. Murphey's experiments suggest that single axons might be better stained by injecting dye directly into the axon. We injected Procion yellow iontophoretically into Ia afferent fibers at their entrance into the spinal cord, and were able to trace them for about 1 cm.

DR. LINDSTRÖM: May I ask you a question? Were you able to trace, for instance with Ia afferents, the thin collaterals going down in the ventral horn amongst the neurons?

DR. BARRETT: We could see some of the thin collaterals of the injected Ia fibers descending toward the ventral horn, but they usually became indistinguishable before reaching it. We could never ascertain whether the collateral ended in a synapse or just faded into the background fluorescence. Fine collaterals might be easier to trace in the reduced background fluorescence of very thin sections. The less reactive Procion H dyes might migrate farther than Procion yellow, but some method would have to be devised to increase the reactivity of the H dyes during fixation, such as that discussed after Dr. Stead's paper.

CHAPTER 15

Procion Yellow and Cobalt as Tools for the Study of Structure-Function Relationships in Vertebrate Central Nervous Systems

Rodolfo Llinás

The relationship between neuronal morphology and neuronal function may be approached from two different points of view. It has been stated in the past that neurons which have similar inputs and whose axons terminate on similar target cells have similar functions ("l'identité ou la dissemblance physiologique des neurones se jugera exclusivement par la similitude ou la différence de leurs relations"—Ramón y Cajal, 1909, p. 149). This statement summarizes the view held by many circuit analysts in neurobiology today, namely that the shape of neurons has little to do with their integrative properties. In a holistic sense, therefore, this approach postulates that connectivity is the dominant morphological parameter determining the functional properties of the nervous system.

Another viewpoint has been evolving in neurobiology during the past two decades: the first step in this direction was the recognition of the functional significance of synaptic distribution along the soma-dendritic components, especially in mammalian motoneurons (Fadiga and Brookhart, 1962; Rall, 1959, 1962, 1964, 1967; Terzuolo and Llinas, 1966; Burke, 1967; Lux, 1967; Rall et al., 1967; Smith et al., 1967; Kuno and Miyahara, 1969; Jack et al., 1970, 1971; Kuno and Llinás, 1970b; Lux et al., 1970; Nelson and Lux, 1970; Burke and Bruggencate, 1971; Jack and Redman, 1971). More recently, the concept of neuronal integration has been further complicated by (1) the demonstration of tonic and phasic properties of neurons (Kernell, 1966; Burke, 1968; cf. Granit, 1972), (2) the relating of cellular size to the electrical properties of neurons (Henneman et al., 1965a, b; Burke, 1968; Burke and Bruggencate, 1971; cf. Granit, 1970), (3) the presence of dendritic spikes (Eccles et al., 1958; Lorenté de Nó and Condouris, 1959; Terzuolo and Araki, 1961; Spencer and Kandel, 1961; Hild and Tasaki, 1962; Purpura, 1967; Llinás and Nicholson, 1969, 1971; Kuno and Llinás, 1970a; Martinez et al., 1970; Korn and Bennett, 1971, 1972) and (4) dendritic inhibition (Llinás and Terzuolo, 1965; Diamond, 1968; Kidokoro et al., 1968; Burke et al., 1971; Llinás and Nicholson, 1969, 1971; Kuno and Llinás, 1970a). Such findings have added complexity to the early view of neuronal integration as a simple algebraic summation of the synaptic input and indicate that, although neurons may have similar inputs and similar targets for their axons, they may, in fact, behave quite differently. These extensions of the concept of neuronal integration have resulted in the rebirth of interest in detailed neuronal morphology among neurophysiologists.

From a functional standpoint the form of a neuron represents, for some of us, a unique algorithm allowing neurons to "evaluate," in accordance with their functional properties, their synaptic input. Rather than a nervous system based strictly on connectivity, we appear to be gravitating towards the idea of a system where complex integration occurs at all levels, and where small differences between electrical properties of neurons may conceivably play a more decisive role than the similarities or differences which such neurons display in their connectivity (Llinás et al., 1971). More important, one may speculate that many of the fundamental questions facing neurobiologists (e.g. the nature of learning) may be related to small modifications in neuronal form and consequent function, rather than to major changes in neuronal connectivity.

Thus the Procion yellow staining techniques pioneered by Stretton and Kravitz (1968) and the more recent cobalt method (Pitman et al., 1972) should be considered as invaluable in the further analysis of structure-function relationships at the single cell level. The remainder of this paper will detail some methods which have been developed in our laboratory and discuss results obtained with these and other techniques and their potential for future analysis.

Intracellular Staining Techniques in the Vertebrate CNS

Microelectrode Injection

The experience of this laboratory has been largely in the field of cerebellar electrophysiology. It is difficult to penetrate the neurons of the cerebellum and rather unusual to record from an impaled cell for more than a few minutes (Eccles et al., 1966). This fact has led to the development of a special technique for the injection of Procion yellow from a micropipette (Llinás and Nicholson, 1971). Micropipettes are pulled with long thin shanks to avoid tissue damage. They are filled with 5–10% (w/v) Procion yellow dye by a cold filling technique such as the fiber glass method (Tasaki et al., 1968). The tips of the electrodes are lightly touched against the edge of a microscope slide under microscopic inspection. Such electrodes, if properly fractured, have a resistance of 30–80 MΩ.

In order to inject current, a relay is used between the recording amplifier and micropipette. Activation of the relay disconnects the amplifier and connects a dry battery capable of delivering up to 300 V to the microelectrode (negative polarity). When a cell has been penetrated and its electrical activity recorded and analyzed, the relay is activated for a period of 0.5 to 2.0 sec. This results in a large current (as high as 10^{-6} A) being passed through the electrode. Apparently the electrode leaves the cell during the pulse and usually becomes temporarily blocked electrically; thus the current pulse is extremely brief (well below 0.5 sec). Nevertheless, such a brief pulse is able to inject enough dye to fill a part, or on occasion most, of the dendritic tree of a Purkinje cell. This technique has been successfully used in the cerebellum of the alligator (Fig. 1G, H), ray, and sturgeon, in the tectum of the mudpuppy, in the electric lobe of the electric ray (Fig. 1A), and in the motoneurons of the cat (Fig. 1B). In many of the neurons studied, it was noticed that the nucleus stained more conspicuously than the rest of the soma. This may be due to the pH effect mentioned in Chapter

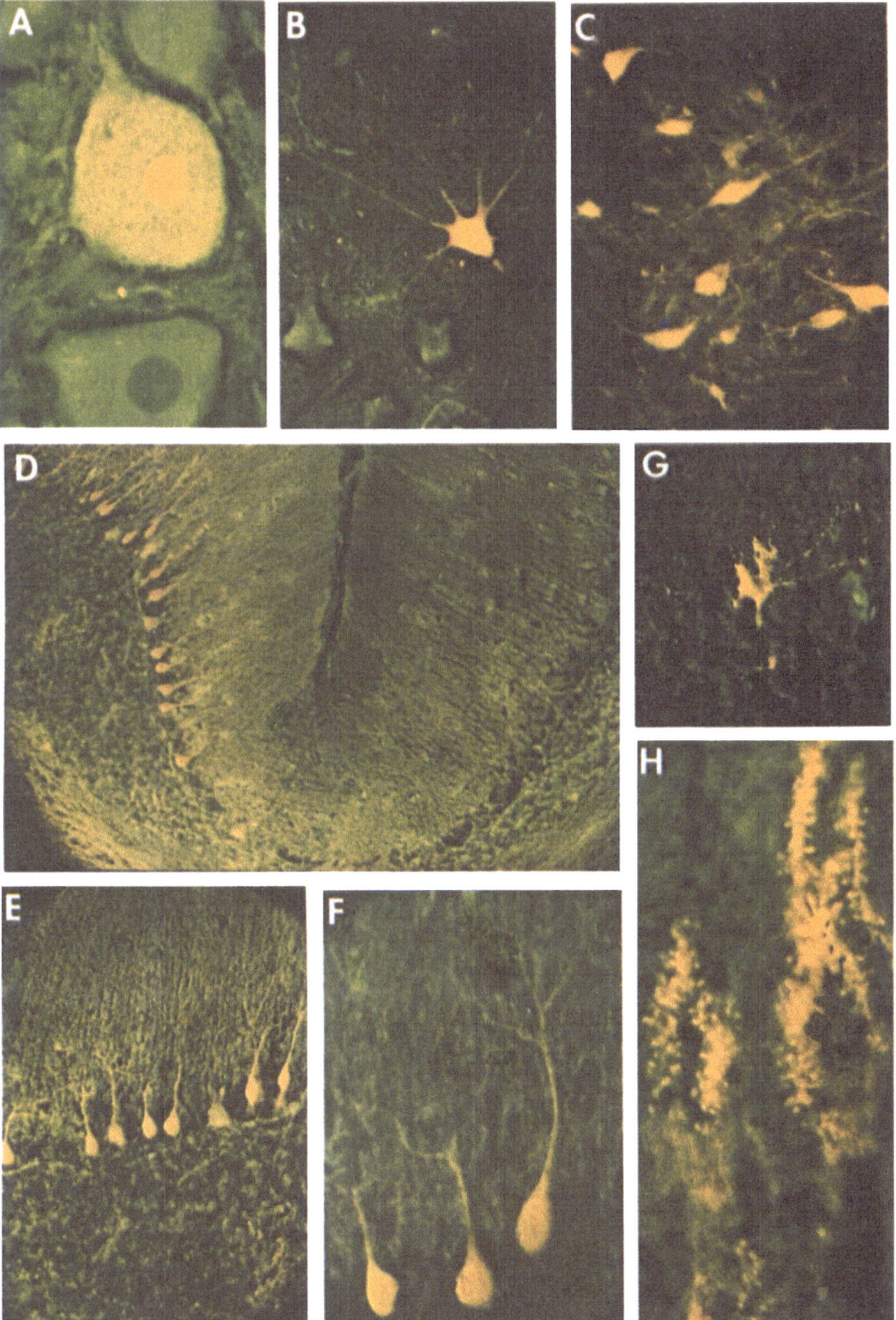

Fig. 1. Different types of vertebrate neurons stained with Procion yellow A. Motoneuron of the electric lobe of *Narcine braciliensis*. This cell was injected with only a small amount of dye. Note the high concentration of dye in the nucleus. B. Cat motoneuron from ventral horn in the lumbar spinal cord (L5). The cell was stained with rapid injection technique. C. Cells of origin of the vestibular efferent system in the frog vestibular nucleus by axonal filled iontophoresis from vestibular nerve (courtesy of Dr. W. Precht). D. Axonal iontophoresis of pigeon cerebellar Purkinje cells from underlying white matter. Note that the staining is restricted to axons projecting to site of gross Procion yellow injection. In E, details of Purkinje cells and of some of the mossy fiber afferents terminating in the granular layer. F. Further detail of the Purkinje cells shown in D. G. Stellate cell filled by intracellular microelectrode in the molecular layer of alligator cerebellar cortex using rapid injection. Cell was probably damaged by the injection. H. Intracellular Procion yellow stain of Purkinje cell branchlet in alligator cerebellum. The dye was introduced by the rapid injection method. Note that the very thin branchlets as well as the dendritic spines are filled with dye. For further details see text.

2, since the nucleus is known to be more basic than the cytoplasm. Even when the tissue was fixed as little as 1–2 hr after dye injection the dendritic tree was filled. With good histological fixation, the dendritic spines of Purkinje cells (diameter 0.2 μm, length 1 μm) are clearly visible (Fig. 1H). Axons are rarely seen although the large axons of the electromotoneurons of the electric ray have been filled. On occasion the animals have been fixed within 30 min of injection, and this has prevented excessive migration of the dye and thus revealed the site of penetration of the electrode (Llinás and Nicholson, 1971). Most frequently a large dendrite (diameter 5 μm) is the site of penetration in a Purkinje cell. Thus, entire dendritic segments may be filled by penetrating a branch of 5 μm diameter and passing dye for 0.5 sec or less.

For routine examination, material is fixed in an aldehyde fixative, embedded in paraffin and cut at 15 μm. Sections are de-paraffinized, mounted in Entellan (Merck, Darmstadt, Germany) and viewed in dark field fluorescence with a BG-3 exciter filter and a Zeiss 500 nm barrier filter. Photomicrographs at high magnification are made with a 40× Zeiss Planapo oil immersion objective. The diaphragm in the objective significantly reduces glare which would otherwise obscure features at this magnification. GAF Anscochrome 500 or Kodak Ektachrome 160 ASA color film is used.

Axonal Iontophoresis

Following the Iles and Mulloney (1971; see also Chapter 7) discovery that cells may be filled by iontophoretically passing the stain into the cut end of a nerve trunk, W. Precht (personal communication) used this method to demonstrate the cells of origin of the vestibular efferent system in the frog (Fig. 1C). (See discussion following this paper). A variation of the axonal iontophoretic technique has been utilized for the central nervous system (CNS) by R. Baker in this laboratory on the cerebella of pigeons and cats. Current of $0.5–1.0 \times 10^{-6}$ A is passed between a Procion-filled (5–10%, w/v) macroelectrode (100 μm tip diameter) and an indifferent electrode for 2–3 hr. In this way, collections of Purkinje cells whose axons project to the site of Procion yellow injection are filled in a retrograde manner (Fig. 1D–F).

Axonal iontophoresis with cobalt in the vertebrate CNS was achieved by passing current from a suction electrode containing cobalt chloride (100 mM) with a cathodic current of $1–3 \times 10^{-5}$ A for a period of 2–4 hr. The tissue was excised from the animal and bathed in a 0.5% solution of ammonium sulfide in Ringer's solution for 30 min to produce a black precipitate. The tissue was fixed in 70% alcohol, dehydrated with ascending alcohols and cleared in methyl salicylate. The resulting whole mount was either viewed directly with a conventional, incident or bright field, light microscope, or hand sectioned (at about 500 μm) before viewing.

Intracellular Staining with Procion Yellow Dye

Procion yellow dye can be used for several types of investigations in the CNS: (1) identification of elements of the CNS, (2) detailed morphology and its relation to neuronal integration, (3) connectivity of neuronal nets and tridimensional study of axonal convergence, and (4) as a pharmacological agent.

Identification of Elements of the CNS

The most common use of Procion yellow has been to identify a penetrated cell. Among these experiments, mammalian glial cells were identified by Dennis and Gerschenfeld (1969); this directly demonstrated the glial origin of the type of intracellular records reported previously as generated by glial elements (Phillips, 1956; Nicholls and Kuffler, 1964; Kuffler et al., 1966; Wardell, 1966; Hild and Tasaki, 1962; Kelly et al., 1967; Krnjevic and Swartz, 1967; Miller and Dowling, 1970). Glial elements have also been filled in the retina (Chapter 11). This emphasizes the importance of further characterization of the non-neuronal elements in the CNS.

In a different category, that related to identification of neurons, has been the study of motoneurons following intracellular dye injection (Barrett and Graubard, 1970; Chapter 18). This type of study has contributed, among other things, a comparison of the length and distribution of dendritic processes in neurons with those obtained by Golgi methods. More importantly, it may serve, given a large sample of cells stained, to characterize the types of neurons which can be routinely investigated in the CNS as opposed to others which, by virtue of their size or fragility, may seldom be successfully impaled.

Another study where identification of the cell has been important concerned the cells of origin of large ascending pathways, such as the ventral spino-cerebellar tract (Jankowska and Lindström, 1970) as well as spinal cord inhibitory interneurons (Chapter 14). Among the most interesting results are those obtained by Janksowska and Lindström (1971; Chapter 14) and van Keulen (1971) for Renshaw cells. This particular work deserves attention inasmuch as it has solved the involved issue of whether Renshaw cell inhibition is mediated through an interneuron or through a direct connection between motoneuron axon collaterals (see discussion by Werman, 1972). Jankowska and Lindström have successfully demonstrated that the Renshaw cells, which have a direct inhibitory action on motoneurons (Eccles et al., 1954), other Renshaw cells (Ryall et al., 1971a, b) and interneurons (Hultborn et al., 1970) are lodged in the ventral horn (among the motoneurons) and that their axons project through commissural and ascending and descending tracts to other segmental levels. The chief importance of this study has been the positive demonstration that Renshaw cells are not Golgi Type II but rather tract cells as hypothesized by Scheibel and Scheibel (1971). Controversy about the Renshaw cell was triggered by the suggestion that the cells should be Golgi type II cells (i.e., a neuron whose axons do not project far from the cell of origin) as contrasted with the anatomical picture which showed an almost complete absence of Golgi type II cells in the spinal cord. Inhibitory interneurons have also been recorded and stained in the cerebellar cortex. Fig. 1G shows an example of one such case: a stellate cell in the molecular layer of alligator cerebellar cortex.

Morphology and Neuronal Integration

Following the correct identification of the penetrated neuron, a most important parameter is the relationship of electrophysiological properties to neuronal form. The number of studies of this type in vertebrates is so far small. Such studies involve the measurement of specific resistivity of central mammalian neurons in vertebrates (Barrett and Crill, 1971; Chapter 18). Given a precise value of the actual membrane area and an accurate measurement of input impedance of a neuron, a calculation of the specific membrane resistivity may be obtained. This parameter is essential for the detailed description of functional properties of neurons and has yielded not only a description of passive resistivity but also a reasonably accurate measurement of the ratio between dendritic and somatic conductivity, ρ, as defined by Rall (1959).

Other studies with Procion yellow have aided the understanding of integrative properties of neuronal dendrites. Llinás and Nicholson (1971) have recently demonstrated dendritic all-or-none depolarizations generated by the electroresponsive properties in Purkinje cells. The origin of these responses was pin-pointed by means of intradendritic Procion yellow injection. Furthermore, such dendritic depolarizations were shown to be conducted in a noncontinuous manner, due to the fact that some dendritic sites appear to be excitable while others behave in a passive manner (Fig. 2C). It was also shown that conduction in a dendrite can be centrifugal and does not necessarily involve activation of nearby dendrites (independency properties of dendritic spikes) and that inhibition may occur directly on dendrites and is, therefore, not limited to the somatic region. This latter finding is in contrast to older views which assumed that inhibition is restricted to the soma.

Since, for the most part, Procion yellow only crosses the synaptic cleft in a chemically mediated synapse following cleft rupture (Llinás and Nicholson, unpublished observations in squid giant synapse), the use of Procion as a tool for demonstrating synaptic contact between neurons is of little use at light microscopical level. However, Payton et al. (1969) and Bennett (Chapter 8) have demonstrated that the presence of electrotonic coupling is accompanied by the movement of Procion from the presynaptic into the postsynaptic element, probably through a special site of close membrane apposition: the so-called "gap junction." In this manner a study of neurons that may be electrotonically coupled, as well as of the extent of the coupling in a three-dimensional manner, may be undertaken following Procion injection intracellularly.

Connectivity in Neuronal Nets, and Axonal Convergence

Intracellular Injection

Besides the intracellular injection of soma and dendrites, Procion yellow has been used to reveal axons. This use of the dye has been extremely important in examining the relationship between the distribution of a telodendrion and the functional property of the synapse it generates on a given target cell (cf. Chapter 11).

Still more elegant has been the recent demonstration of the spatial distribution of the presynaptic terminal onto a receiving neuron by means of Procion injection of one color and the actual morphology of the receiving cell by injection of a second Procion color (Chapter 11). This type of technique is essential in the further understanding of the functional meaning of somatic-dendritic distribution of synaptic input with regard to neuronal integrative properties.

Axonal Iontophoresis

The use of Procion yellow dye is not restricted to intracellular injection. Indeed, intracellular filling of neuronal elements has been obtained by simple electrophoretic

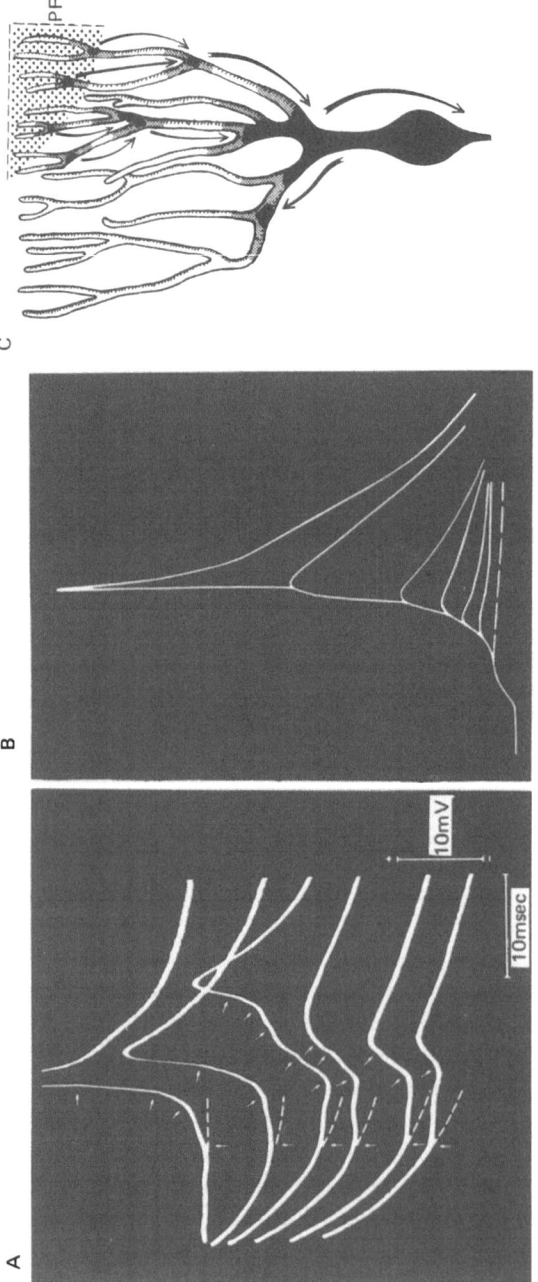

Fig. 2. Intracellular recording in alligator Purkinje cells. A. Action potentials recorded in Purkinje cell dendrite 200 μm from the surface. Successive hyperpolarizing current injected through the recording electrode revealed that the large dendritic spike shown on the first trace is actually produced by the addition of all-or-none components (arrows). This dendritic spike was generated by a dendritic EPSP produced by parallel fiber stimulation (upward arrow). As the hyperpolarization is increased, the different all-or-none depolarizing potentials are blocked in a sequential manner. B. Reconstruction of the intradendritic action potential showing the six all-or-none components shown in A. C. Diagram of mechanism of dendritic spike generation. Each all-or-none component is taken to be generated by a different hot spot (shown as a dark area) at or near dendritic bifurcation. The action potential is produced by the summation of all-or-none local responses which finally reach the soma and generate a full outgoing action potential. (Modified from Llinás and Nicholson, 1971).

movement of the dye from severed axonal elements into the cells of origin (Iles and Mulloney, 1971; Chapter 7). This form of Procion staining has been extremely successful in demonstrating the cells of origin of ventral root neurons in the frog, especially when describing the location of the cell of origin for particular ventral root filaments (Precht, personal communication). Precht and his associates (personal communication) have demonstrated the presence of efferent systems arising from the general area of the vestibular nuclei (Fig. 1C) and the auricular lobe of the cerebellum, in agreement with the anatomy and electrophysiology of the efferent system (Hillman, 1969; Llinás and Precht, 1969; Precht et al., 1971).

R. Baker of this laboratory has used axonal iontophoresis to inject white matter of the pigeon cerebellum (see Techniques section). This technique allows the visualization of axons such as those of Purkinje cells in the cerebellar cortex and will also stain the soma and may reach as high as two-thirds of the dendrite. This technique raises an interesting possibility regarding the convergence of axon terminals onto a particular point in the CNS. By applying Procion at one point in space in the CNS, it should be possible in principle to stain the cells which project to that particular point and thus to describe the spatial distribution of the neurons projecting to a given locus. This is of considerable interest, especially in the understanding of the organization of the cerebral and cerebellar cortices, and the projections of the organized planar ensemble of neurons in these cortices onto central nuclei. This convergence is still a little understood aspect of cortical neuromorphology. For instance, it has not been possible to determine to what extent neurons lying close together in the cortex project to neurons lying close together in the nuclei, or to ascertain the number and distribution of possible axons arising from neurons projecting to a common target. This three-dimensional description of axonal convergence appears to be central in the description of spatial distribution of neuronal activity and its relation to holistic function of the CNS.

Procion Yellow as a Pharmacological Agent

Among the most surprising properties reported for Procion yellow dye has been its supposed anesthetic actions following extracellular exposure to the dye (Payton et al., 1969; Payton, 1970), as well as after intracellular injection (Jankowska and Lindström, 1970; Chapter 18). Preliminary voltage clamp experiments in internally perfused axons for as long as 20 min showed, however, no appreciable change in resting potential, resting resistance, conduction velocity of the action potential, or in peak sodium and potassium currents (Frazier and Narahashi, personal communication). The Procion yellow for this internal perfusion was prepared in a concentration of 10^{-4} gm/cc of a solution containing 400 mM potassium, 337 mM glutamate, 15 mM H_2PO_4 and 333 mM sucrose at a pH of 7.3. Moreover, in a chemically mediated synapse, Llinás and Nicholson (unpublished observations) showed that Procion yellow injected electrophoretically in the postsynaptic fiber of a squid giant synapse may block the EPSP in a time span of 30–60 min; in several cases, however, synaptic transmission was unaffected after a lengthy injection of Procion yellow (Fig. 3). Since a clear block of transmission cannot be differentiated from the "spontaneous failures" which are sometimes observed in this preparation, the examples where transmission is maintained for hours after Procion injection strongly suggest that Procion dye does not block action potentials or synaptic

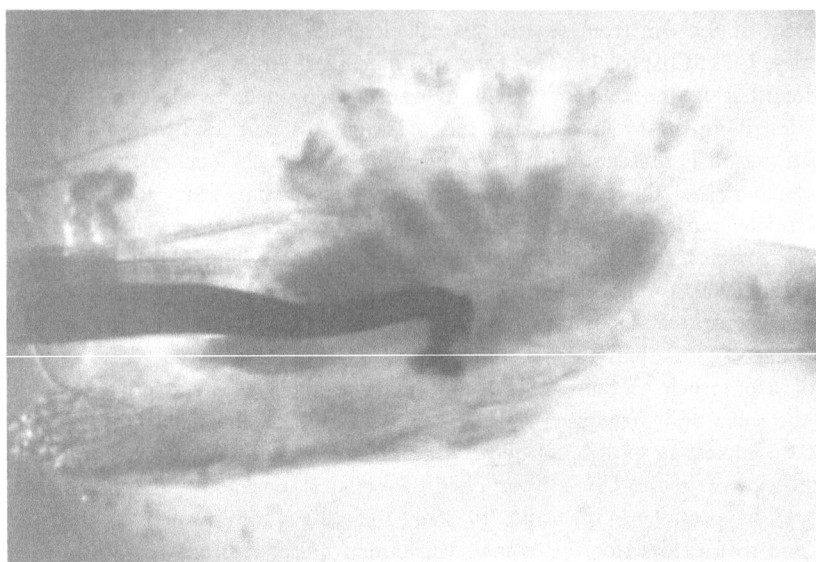

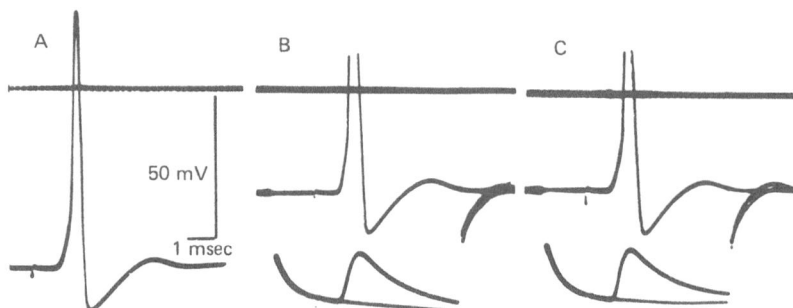

Fig. 3. Synaptic transmission at the squid giant synapse after Procion yellow injection postsynaptically; photomicrograph of the squid stellate ganglion showing the postsynaptic axon filled with Procion yellow. In A, intracellular recording from postsynaptic axon showing orthodromically evoked action potential. Dotted line indicates zero resting potential. In B and C, orthodromic action potential. The all-or-none EPSP shown in each record was disclosed by hyperpolarizing the membrane with an inward current pulse from a second intracellular microelectrode. Records B and C were taken at the beginning, and at 60 min after Procion injection and illustrate the lack of deleterious effect of Procion on the synaptic and/or action potentials in this preparation.

transmission in this system. Furthermore, experiments by J. T. Hackett in this laboratory indicate that Procion yellow injected postsynaptically in the myoneural junction of the frog does not reduce the size of miniature potentials or endplate potentials following iontophoretic dye injection for over 1 hr. This evidence indicates, therefore, that Procion yellow is not an anesthetic substance when applied intracellularly, at least within the time limits tested. However, the question as to whether this dye acts as an anesthetic when applied extracellularly must be further investigated.

Regardless of whether or not Procion dye is an internal anesthetic per se, the possibility of synthesizing an internal anesthetic compound permanently bound to Procion dye is of fundamental interest. One such compound may be an analogue of quaternary ammonium ions (tetraethyl or tetramethyl ammonium) which are known to block potassium activation in the frog and squid (cf. Hille, 1970), as well as some potassium-mediated synaptic potentials (Kehoe, 1972).

It would also appear possible to use analogues of anesthetics which block sodium and potassium permeabilities when injected intracellularly. N. Feder (personal communication) has suggested the possibility of binding procaine to Procion yellow. Other candidates are quaternary derivatives of lidocaine, QX-314, and QX-572, which block the action potentials in squid giant axon more effectively internally than from the outside surface, without affecting resting potential appreciably (Frazier et al., 1970). Their structures must be modified, however, since they will not bind to Procion yellow. Another anesthetic substance employed by these authors has been hemicholinium-3, whose internal anesthetic action, in this situation, is not related to inhibition of the synthesis of acetylcholine (MacIntosh, 1956). Of the three compounds, the most promising may be an analogue of QX-314, which was found by Frazier et al. to be virtually ineffective from the outside in blocking action potentials. This is of importance since one does not want to anesthetize the cells extracellularly due to possible microelectrode leakage. Although little is known about the action of these anesthetics on the synaptic ionic conductances, it may be possible that, as in the case of procaine (Maeno, 1966; Kordas, 1970), procaine and lidocaine (Maeno et al., 1971) and lidocaine (Steinbach, 1968a, b), these anesthetic agents may also affect synaptic ionic permeabilities. The important questions here are (1) whether it is possible to produce strong binding between these compounds with Procion yellow to prevent dissociation of the two substances and (2) whether this Procion-anesthetic compound would demonstrate the pharmacological actions expected. The advantage of using an intracellular anesthetic is clear since the anesthetic action would be restricted to the cell injected. Such a technique may be useful in demonstrating the relationship between the distance the dye has moved in a cell and the type of synaptic potential which the "unanesthetized" portion of the cell generates following activation of a distributed or remote synapse. In principle it would be possible to remove the different components of synaptic potentials, beginning with the ones near the site of anesthetic injection. This would allow the study of the actual distal synapses as seen after neuronal axotomy (Lux and Winter, 1968; Kuno and Llinás, 1970b). Similar techniques could also be applied for study of dendritic spikes.

Cobalt Ion in the CNS of Vertebrates

Pitman et al. (1972) have recently demonstrated the possibility of intracellular injection of cobalt ions. Although metal ions have been tried several times in the past (see Chapter 1), actual development of a useful technique was not achieved until the recent injection of cobalt ions in the CNS of vertebrates. I would like to report here some results in the CNS of mammals and lower vertebrates following iontophoretic staining of axons from the proximal stump of transected dorsal and ventral roots. While, for the most part, a movement of cobalt ion does not appear to advance much above 1–2 length constants in any given pathway, we did find it possible to demonstrate the internal distribution of fibers in the CNS of frogs and cats following this procedure.

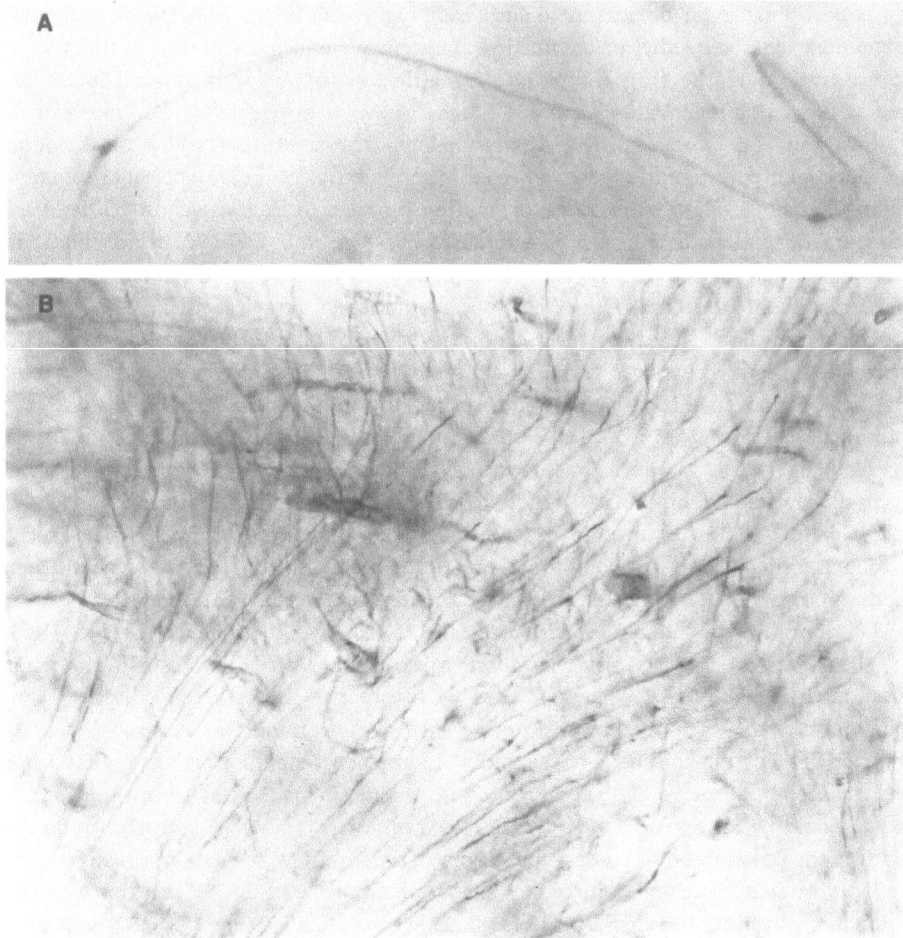

Fig. 4. Cobalt staining of vertebrate myelinated fibers by axonal iontophoresis. A. Single myelinated fiber stained in frog spinal cord following dorsal root axonal iontophoresis at lumbar level. The dye moved as far as one and a half segments above and below the site of root entry. The dark spots in the fiber represent the nodes of Ranvier. B. Cat dorsal roots following two hr of cobalt axonal iontophoresis from severed dorsal root near the site of entry into the cord. Note bifurcation of the fibers and dark staining of nodes.

Among the most interesting findings was the visualization of nodes of Ranvier in the different paths of spinal cord of these two species. One such fiber is illustrated in Fig. 4; note that the fiber itself is rather weakly stained while the nodes are clearly demarcated by the presence of a dark globular bead. On higher magnification, small branches may be observed at nodal level. It has also been observed that following a bifurcation of a fiber both sister branches may be stained (Fig. 4B).

On several occasions, neurons of the dorsal horn have been stained following electrophoretic application of cobalt from the roots. The mechanism for this phenomenon is unknown. It is possible that some neurons may be picking up cobalt from the extracellular fluid. To date cobalt back-filling in the vertebrate system has been very capricious but, on the occasions when it was successful, intriguing results have been obtained.

Acknowledgments. It is a pleasure to acknowledge Drs. R. Baker, D. Frazier, J. T. Hackett, D. E. Hillman, T. Narahashi, C. Nicholson and W. Precht for allowing me to utilize and comment on unpublished results.

The research was supported by U.S. Public Health Service research grant No. NS-09916-01 from the National Institute of Neurological Diseases and Stroke.

References

Barrett, J. N., and W. E. Crill: Specific membrane resistivity of dye-injected cat motoneurons. Brain Res. *28*, 556–561 (1971).

————, and K. Graubard: Fluorescent staining of cat motoneurons *in vivo* with bevelled micropipettes. Brain Res. *18*, 565–568 (1970).

Burke, R. E.: Composite nature of the monosynaptic excitatory postsynaptic potential. J. Neurophysiol. *30*, 1114–1137 (1967).

————: Group Ia synaptic input to fast and slow twitch motor units of cat triceps surae. J. Physiol., Lond. *196*, 605–630 (1968).

————, L. Fedina, and A. Lundberg: Spatial synaptic distribution of recurrent and group Ia inhibitory systems in cat spinal motoneurones. J. Physiol., Lond. *214*, 305–326 (1971).

————, and G. ten Bruggencate: Electrotonic characteristics of alpha motoneurones of varying size. J. Physiol., Lond. *212*, 1–20 (1971).

Dennis, M. J., and H. M. Gerschenfeld: Some physiological properties of identified mammalian glial cells. J. Physiol., Lond. *203*, 211–222 (1969).

Diamond, J.: The activation and distribution of GABA and L-glutamate receptors on goldfish Mauthner neurons: an analysis of dendritic remote inhibition (with appendix by A. F. Huxley). J. Physiol., Lond. *194*, 669–723 (1968).

Eccles, J. C., P. Fatt, and K. Koketsu: Cholinergic and inhibitory synapses in a pathway from motor-axon collaterals to motoneurons. J. Physiol., Lond. *126*, 524–562 (1954).

————, B. Libet, and R. R. Young: The behaviour of chromatolysed motoneurones studied by intracellular recording. J. Physiol., Lond. *143*, 11–40 (1958).

————, R. Llinás, and K. Sasaki: Intracellularly recorded responses of the cerebellar Purkinje cells. Exp. Brain Res. *1*, 161–183 (1966).

Fadiga, E., and J. M. Brookhart: Interactions of excitatory postsynaptic potentials generated at different sites on the frog motoneurons. J. Neurophysiol. *25*, 790–804 (1962).

Frazier, D. T., T. Narahashi, and M. Yamada: The site of action and active form of local anesthetics. II. Experiments with quaternary compounds. J. Pharmac. exp. Ther. *171*, 45–51 (1970).

Granit, R.: The Basis of Motor Control. London: Academic Press, 1970.

————: Mechanisms Regulating the Discharge of Motoneurons. Sherrington Lecture XI. Liverpool: Liverpool University Press, 1972.

Henneman, E., G. Somjen, and D. O. Carpenter: Functional significance of cell size in spinal motoneurons. J. Neurophysiol. *28*, 560–580 (1965a).

———— ———— ————: Excitability and inhibitability of motoneurons of different sizes. J. Neurophysiol. *28*, 599–620 (1965b).

Hild, W., and I. Tasaki: Morphological and physiological properties of neurons and glial cells in tissue culture. J. Neurophysiol. *25*, 277–304 (1962).

Hille, B.: Ionic channels in nerve membranes. In: Progress in Biophysics and Molecular Biology. Ed. J. A. V. Butler and D. Noble. Vol. 21, pp. 1–32. Oxford: Pergamon Press, 1970.

Hillman, D. E.: Light and electron microscopical study of the relationships between the cerebellum and the vestibular organ of the frog. Exp. Brain Res. *9*, 1–15 (1969).

Hultborn, H., E. Jankowska, and S. Lindström: Recurrent inhibition from motor axon collaterals of transmission in the Ia inhibitory pathway to motoneurones. J. Physiol., Lond. *215*, 591–612 (1971).

Iles, J. F. and B. Mulloney: Procion yellow staining of cockroach motor neurones without the use of microelectrodes. Brain Res. *30*, 397–400 (1971).

Jack, J. J. B., S. Miller, R. Porter, and S. J. Redman: The distribution of group Ia synapses on lumbosacral spinal motoneurones in the cat. In: Excitatory Synaptic Mechanisms. Ed. P. Andersen and J. K. S. Jansen. pp. 199–205. Oslo: Universitetsforlaget, 1970.

——— ——— ——— ———: The time course of minimal excitatory post-synaptic potentials evoked in spinal motoneurone by Group Ia afferent fibres. J. Physiol., Lond. *215*, 353–380 (1971).

———, and S. J. Redman: An electrical description of the motoneurone, and its application to the analysis of synaptic potentials. J. Physiol., Lond. *215*, 321–352 (1971).

Jankowska, E., and S. Lindström: Morphological identification of physiologically defined neurones in the cat spinal cord. Brain Res. *20*, 323–326 (1970).

——— ———: Morphological identification of Renshaw cells. Acta physiol. scand. *81*, 428–430 (1971).

Kehoe, J.: Ionic mechanisms of a two-component cholinergic inhibition in *Aplysia* neurones. J. Physiol., Lond. *225*, 85–114 (1972).

Kelly, J. S., K. Krnjevic, and G. K. W. Yim: Unresponsive cells in cerebral cortex. Brain Res. *6*, 767–769 (1967).

Kernell, D.: Input resistance, electrical excitability, and size of ventral horn cells in the cat spinal cord. Science *152*, 1637–1640 (1966).

Kidokoro, Y., K. Kubota, S. Shuto, and R. Sumino: Reflex organization of cat masticatory muscles. J. Neurophysiol. *31*, 695–708 (1968).

Kordaš, M.: The effect of procaine on neuromuscular transmission. J. Physiol., Lond. *209*, 689–699 (1970).

Korn, H., and M. V. L. Bennett: Dendritic and somatic impulse initiation in fish oculomotor neurons during vestibular nystagmus. Brain Res. *27*, 169–175 (1971).

——— ———: Electrotonic coupling between teleost oculomotor neurons: restriction to somatic regions and relation to function of somatic and dendritic sites of impulse initiation. Brain Res. *38*, 433–439 (1972).

Krnjevic, K. and S. Schwartz: Some properties of unresponsive cells in the cerebral cortex. Exp. Brain Res. *3*, 306–319 (1967).

Kuffler, S. W., J. Nicholls, and R. Orkand: Physiological properties of glial cells in the central nervous system of amphibia. J. Neurophysiol. *29*, 768–787 (1966).

Kuno, M., and R. Llinás: Enhancement of synaptic transmission by dendritic potentials in chromatolysed motoneurones of the cat. J. Physiol., Lond. *210*, 807–821 (1970a).

——— ———: Alterations of synaptic action in chromatolysed motoneurones of the cat. J. Physiol., Lond. *210*, 823–838 (1970b).

———, and J. T. Miyahara: Non-linear summation of unit synaptic potentials in spinal motoneurones of the cat. J. Physiol., Lond. *201*, 465–477 (1969).

Llinás, R., and C. Nicholson: Electrophysiological analysis of alligator cerebellum. A study on dendritic spikes: In: Neurobiology of Cerebellar Evolution and Development. Ed. R. Llinás. pp. 431–465. Chicago: Amer. Med. Assn., 1969.

——— ———: Electrophysiological properties of dendritic and somata in alligator Purkinje cells. J. Neurophysiol. *34*, 532–551 (1971).

———, and W. Precht: The inhibitory vestibular efferent system and its relation to the cerebellum in the frog. Exp. Brain Res. *9*, 16–29 (1969).

——— ———, and M. Clarke: Cerebellar Purkinje cell responses to physiological stimulation of the vestibular system in the frog. Exp. Brain Res. *13*, 408–431 (1971).

———, and C. A. Terzuolo: Mechanisms of supra-spinal actions upon spinal cord activities. Reticular inhibitory mechanisms upon flexor motoneurones. J. Neurophysiol. *28*, 413–422 (1965).

Lorente de Nó, R., and G. A. Condouris: Decremental conduction in peripheral nerve. Integration of stimuli in the neuron. Proc. nath. Acad. Sci. U.S.A. *45*, 592–617 (1959).

Lux, H. D.: Eigenschafter eines Neuron-Modells mit Dendriten begrenzter Lange. Pflügers Arch. ges. Physiol. *297*, 238–255 (1967).

——, P. Schubert, and G. W. Kreutzberg: Direct matching of morphological and electrophysiological data in cat spinal motoneurons. In: Excitatory Synaptic Mechanisms. Ed. P. Anderson and J. K. S. Jansen. pp. 189–198. Oslo: Universitetsforlaget, 1970.

——, and P. Winter: Studies on EPSPs in normal and retrograde reacting facial motoneurones. Proc. IUPS 7, 818 (abstract) (1968).

MacIntosh, F. C., R. I. Birks, and P. B. Sastry: Pharmacological inhibition of acetylcholine synthesis. Nature, Lond. *178*, 1181 (1956).

Maeno, T.: Analysis of sodium and potassium conductances in the procaine end-plate potential. J. Physiol., Lond. *183*, 592–606 (1966).

——, C. Edwards, and S. Hashimura: Difference in effects on the end-plate potentials between procaine and lidocaine as revealed by voltage-clamp experiments. J. Neurophysiol. *34*, 32–46 (1971).

Martinez, F. E., W. E. Crill, and T. T. Kennedy: Dendritic origin of climbing fiber responses in cat cerebellar Purkinje cells. Fedn Proc. Fedn Am. Socs exp. Biol. *29*, 454a (1970).

Miller, R. I., and J. E. Dowling: Intracellular responses of the Muller (glial) cells of mudpuppy retina: their relation to b-wave of the electroretinogram. J. Neurophysiol. *33*, 323–341 (1970).

Nelson, P. G., and H. D. Lux: Some electrical measurements of motoneuron parameters. Biophys. J. *10*, 55–73 (1970).

Nicholls, J. G., and S. W. Kuffler: Extracellular space as a pathway for exchange between blood and neurons in the central nervous system of the leech: ionic composition of glial cells and neurons. J. Neurophysiol. *27*, 645–671 (1964).

Payton, B. W.: Histological staining properties of Procion yellow. J. Cell Biol. *45*, 659–662 (1970).

——, M. V. L. Bennett, and G. D. Pappas: Permeability and structure of junctional membranes at an electrotonic synapse. Science *166*, 1641–1643 (1969).

Phillips, C. G.: Intracellular recording from Betz cells in the cat. Q. Jl exp. Physiol. *41*, 58–69 (1956).

Pitman, R. M., C. D. Tweedle, and M. J. Cohen: Branching of central neurons: intracellular cobalt injection for light and electron microscopy. Science *176*, 412–414 (1972).

Precht, W., R. Llinás, and M. Clarke: Physiological responses of frog vestibular fibers to horizontal angular rotation. Exp. Brain Res. *13*, 378–407 (1971).

Purpura, D. P.: Comparative physiology of dendrites. In: The Neurosciences: A Study Program. Ed. G. C. Quarton, T. Melnechuk and F. O. Schmitt. pp. 372–393. New York: Rockefeller Univ. Press, 1967.

Rall, W.: Branching dendritic trees and motoneuron membrane resistivity. Exp. Neurol. *1*, 491–527 (1959).

——: Electrophysiology of a dendritic neuron model. Biophys. J. *2*, 145–167 (1962).

——: Theoretical significance of dendritic trees for neuronal input-output relations. In: Neural Theory and Modelling. Ed. R. F. Reiss. pp. 73–97. Stanford Univ. Press, 1964.

——: Distinguishing theoretical synaptic potentials computed for different soma-dendritic distributions of synaptic inputs. J. Neurophysiol. *30*, 1138–1168 (1967).

——, R. E. Burke, T. G. Smith, P. G. Nelson, and K. Frank: Dendritic location of synapses and possible mechanisms for the monosynaptic EPSP in motoneurons. J. Neurophysiol. *30*, 1169–1193 (1967).

Ramón y Cajal, S.: Histologie du Système Nerveux de L'homme et des Vertébrés. vol 1. Paris: Maloine, 1909.

Ryall, R. W., and M. F. Piercey: Excitation and inhibition of Renshaw cells by impulses in peripheral afferent nerve fibers. J. Neurophysiol. *34*, 242–251 (1971a).

—— ——, and C. Polosa: Intersegmental and intrasegmental distribution of mutual inhibition of Renshaw cells. J. Neurophysiol. *34*, 700–707 (1971b).

Scheibel, M. E., and A. B. Scheibel: Inhibition and the Renshaw cell: a structural critique. Brain Behav. Evol. *4*, 53–93 (1971).

Smith, T. G., R. B. Wuerker, and K. Frank: Membrane impedance changes during synaptic transmission in cat spinal motoneurones. J. Neurophysiol. *30*, 1072–1096 (1967).

Spencer, W. A., and E. R. Kandel: Electrophysiology of hippocampal neurons. IV. Fast prepotentials. J. Neurophysiol. *24*, 272–285 (1961).

Steinbach, A. B.: Alteration by xylocaine (lidocaine) and its derivatives of the time course of the end-plate potential. J. gen. Physiol. *52*, 144–161 (1968a).

——: A kinetic model for the action of xylocaine on receptors for acetylcholine. J. gen. Physiol. *52*, 162–180 (1968b).

Stretton, A. O. W., and E. A. Kravitz: Neuronal geometry: determination with a technique of intracellular dye injection. Science *162*, 132–134 (1968).

Tasaki, I., Y. Tsukahara, S. Ito, M. J. Wayner, and W. Y. Yu: A simple direct and rapid method for filling microelectrodes. Physiol. Behav. *3*, 1009–1010 (1968).

Terzuolo, C. A., and T. Araki: An analysis of intra- versus extracellular potential changes associated with activity of single spinal motoneurons. Ann. N.Y. Acad. Sci. *94*, 547–558 (1961).

——, and R. Llinás: Distribution of synaptic inputs in the spinal motoneurone and its functional significance. In: Nobel Symposium I, Muscular Afferents and Motor Control. Ed. R. Granit. pp. 373–384. Stockholm: Almqvist and Wiksell, 1966.

vanKeulen, L. C. M.: Morphology of Renshaw cells. Pflügers Arch. ges. Physiol. *328*, 235–236 (1971).

Wardell, W. M.: Electrical and pharamacological properties of mammalian neuroglial cells in tissue culture. Proc. R. Soc., Lond B *165*, 326–361 (1966).

Werman, R.: CNS cellular level: membranes. Ann. Rev. Physiol. *34*, 337–374 (1972).

Discussion

Dr. LLINÁS: Dr. Precht, could you elaborate on the axonal iontophoretic method you employ?

Dr. PRECHT (Frankfurt): I think Dr. Llinás has mentioned most of the points regarding the back-filling technique, and Dr. Mulloney has added his experience in invertebrates. It is, generally speaking, very easy to implement this technique. I may add a few technical notes.

Our first step was to produce a pool of Procion yellow and dip the cut end of the nerve. This requires, to be successful in filling the nerve, a strong current for a long period of time. Later we decided to hold the nerve with a suction electrode filled with Procion yellow and pass current between this electrode and an anode located in the CNS. In the case of the frog, the anode was in the body of the cerebellum and we passed between 5 and 15×10^{-6} A for something like 2–3 hr. After the cessation of current, we allowed the frog to survive for another 5 hr maximum and then fixed the brain in formalin, put it on the freezing microtome and sectioned it without dehydration. This allowed us to see very clearly the filled cells and their axons. This technique was utilized for vestibular efferents, as well as for motoneurons in the spinal cord, with similar results.

There is one puzzling point that I wonder about. With intracellular dye injection, you hardly ever see an axon filled. However, with the backfilling technique, the axons show up very nicely. There are still some things we do not understand.

Dr. ROWELL (Berkeley): I would like to ask you about your large pulse technique. The only person I know who has done this sort of thing before is Frank Werblin in the retina, and I wonder if you have the same sort of results.

I believe that Werblin uses 800 V D.C. across his microelectrode. He thinks he breaks open the tip. After that it doesn't matter how soon he loses the cell, he knows that it is filled.

Of course, Werblin does not use Procion, but I wondered if you have found that the electrode tip comes off?

Dr. Llinás: I don't think so, since in many cases, we could penetrate and stain several cells in sequence without changing microelectrodes.

Mr. Merickel (Iowa): Are there any additional techniques to facilitate viewing cells surrounding injected neurons?

Dr. Llinás: Yes, we have used Nomarski optics. I believe Dr. Kater has some technical observations on this.

Dr. Kater (Iowa): Occasionally we would like to see the background structures. It is technically feasible to do both fluorescence, via the epi illumination, and the Normarski interference contrast with a weak white light through a standard bright field condenser.

In that way you can, in optimal situations, see both the background with the Nomarski and the cell you filled with fluorescence.

Dr. Bennett (Albert Einstein): You spoke of cobalt as trying to get out through the nodes when discussing the beaded appearance of cobalt-filled myelinated axons.

I think another explanation is possible. The sulfide is coming in through the node and meeting the cobalt there, so that you get a concentration of precipitate at that point. (Subsequent experiments with Mauthner and giant fibers of the hatchet fish confirm this explanation. Staining occurs initially at nodes and can be much denser at these points independent of axon diameter.)

Dr. Llinás: Yes, there are other possible explanations, such as fiber diameter being slightly larger at the node, and the presence of special membrane structures. There appears to be no gradient in the amount of precipitate between nodal regions.

Dr. S. T. Kitai (Wayne State). Can you control the diffusion of the injected dye in some way so that the dye is localized in the area of injection, for instance, in the dendrite if the electrode happens to be in the dendrite, or in the axon hillock if it is in the axon hillock?

Dr. Llinás: It is possible to use Procion to localize the site of injection; it does require rapid fixation following dye injection. The advantage of using Procion is the resolution you obtain of the neuronal segment penetrated. For instance, in Purkinje cell dendrites the density of spines immediately demonstrates that you are dealing with a Purkinje cell.

Dr. Kravitz (Harvard): On that same point, on the list of Procion dyes (Chapter 2) there are many dyes that come out of electrodes well but do not spread well through cytoplasm. Some of these might be very suitable to just mark the position of the electrode.

Another compound that might be useful for that purpose is mercurichrome which is highly fluorescent, and comes out of electrodes well; it also doesn't diffuse well, and would give you a very highly localized fluorescence (see Chapter 20).

Dr. Kaneko (Keio): Relative to that point, I would like to ask Dr. Kravitz if it is possible to mix two compounds like mercurichrome and Procion yellow, and if it disturbs the fluorescence of Procion yellow. Then we can identify the site of penetration and the type of cell at the same time.

Dr. Kravitz: We never tried it.

CHAPTER 16

Computer Analysis of Neuronal Structure

D. R. Reddy, W. J. Davis, R. B. Ohlander, and D. J. Bihary

A fundamental goal of neuroscientists is to correlate neuronal morphology with neuronal function, both at the level of single nerve cells and that of neuronal networks. This goal has seldom been obtained to a satisfactory degree, partially because quantitative methods for describing neuronal structure are lacking. In this paper we will describe SYNAPS, Symbolic Neuronal Analysis Programming System, which is being developed at Carnegie-Mellon University for the computer reconstruction and analysis of nerve cells and networks. In particular, we will describe programs that perform data gathering, analysis, generation, and representation of three-dimensional models, display of the models from various viewing angles, and analysis of neural structure. The word "model" throughout this paper means a computer-generated representation of a tangible, physical structure, and not a theory as in the case of models of behavior.

The Relationship between Structure and Function of Neurons

All of the complex integrative operations of the nervous system result more or less directly from neuronal structure, i.e., from the geometrical characteristics of single neurons and from the patterns of interconnection between them. On the cellular level, the integrative capacities of single nerve cells are closely related to geometrical features such as size (Henneman, 1957; Henneman, Somjen and Carpenter, 1965; Davis, 1971; Hinkle and Camhi, 1972), the detailed shape of individual dendrites (which helps to determine the space constant), and the proximity of synapses to integrative regions such as spike-initiating zones (Rall, 1967; Rall et al., 1967; Rall and Shepherd, 1968; see also Chapter 18). Important integrative properties of neurons may even depend upon the gross morphology of dendritic fields (Ramón-Moliner, 1962, 1968; Globus and Scheibel, 1967; Munagai, 1967). In frog retinal ganglion cells, for example, the shapes of dendritic trees appear to represent the code for the detection of specific visual shapes (Pomeranz and Chung, 1970). Indeed, the highly characteristic shapes of many nerve cells, as expressed by their diagnostic common names (stellate cells, basket cells, mossy fibers, etc.), may signify that the gross anatomy of dendritic fields is of general integrative significance.

On the population level, it is likely that the relative spatial positions of neurons is an important and in some cases major determinant of the integrative properties of the population. One can imagine, for example, neurons responding sequentially to a traveling wave of neuronal activity and/or performing crucial spatio-temporal transformations on the basis of relative position in the neuronal matrix which supports the wave (Beurle, 1956; Verzeano and Negishi, 1960; Verzeano, 1963; Globus and Scheibel, 1967). Such an arrangement could contribute to the production of stereotyped motor output programs (Davis, 1969). Indeed, in the cat cerebellum, Purkinje cells are regularly spaced along "beams" of presynaptic parallel fibers. As a result of this spatial arrangement, Purkinje cells fire in a stereotyped temporal sequence in response to electrical or natural stimulation of afferent pathways (Fox and Barnard, 1957; Freeman, 1969).

Finally, none of the structural parameters discussed above are static. Instead, the structure of the nervous system is in a constant state of flux, as a result of natural, genetically-regulated processes such as ontogeny and growth, or as a result of pathological processes such as the degeneration associated with many diseases of the nervous system. Changes in neuronal structure may even underlie the most sophisticated operations of the nervous system, including learning and other forms of neuronal plasticity (Hilgard, 1964).

The basic approach used to study the structure of the nervous system has been unchanged for over a century. Small pieces of neuronal tissue are treated with chemicals which selectively color certain nerve cells or regions, thereby emphasizing specific features without altering the original structure significantly. This approach reached a zenith in the gifted hands of Ramón y Cajal, whose studies on the vertebrate nervous system near the beginning of this century set a standard which is still unsurpassed (Ramón y Cajal, 1911). Progress has been slower among invertebrates, at least in part because histological stains which are effective in vertebrate nervous tissue are at best capricious when applied to invertebrates. Recently a new histological technique was introduced which is applicable to invertebrate and vertebrate nervous systems alike—that of intracellular stain injection (Thomas and Wilson, 1966; Kato, Fujimori, and Hirata, 1968; Stretton and Kravitz, 1968).

The full potential of intracellular stain injection has been realized in invertebrate nervous systems, where single neurons can be reliably identified from one animal to the next. In the lobster *Homarus americanus,* for example, each of four segmental ganglia in the abdomen controls a pair of locomotory appendages, the swimmerets. Intracellular stimulation and recording has been utilized to map the soma positions of the motoneurons which operate a swimmeret (Davis, 1971; Figs. 1 and 2). Once the position of a neuron is located in this manner, intracellular stain injection may be applied to unravel the geometry and synaptic connections of the neuron.

The details of the dye injection techniques are fully described elsewhere in this volume (Chapter 20). Basically they involve filling a neuron with a marker substance which remains confined within the neuron so that the central and peripheral projections of the cell may be followed in subsequent whole mounts or thin sections of the tissue. The most commonly used marker substance has been the textile dye Procion yellow, which appears brilliant yellow when viewed with the fluorescence microscope (Fig. 3). Recently a useful new procedure has been developed, utilizing injection of cobalt chloride (Pitman et al., 1972a; Chapter 6). This substance has the advantage that it is electron dense and can therefore be detected using the electron microscope.

The kind of neuronal reconstruction which can be achieved using intracellular stain

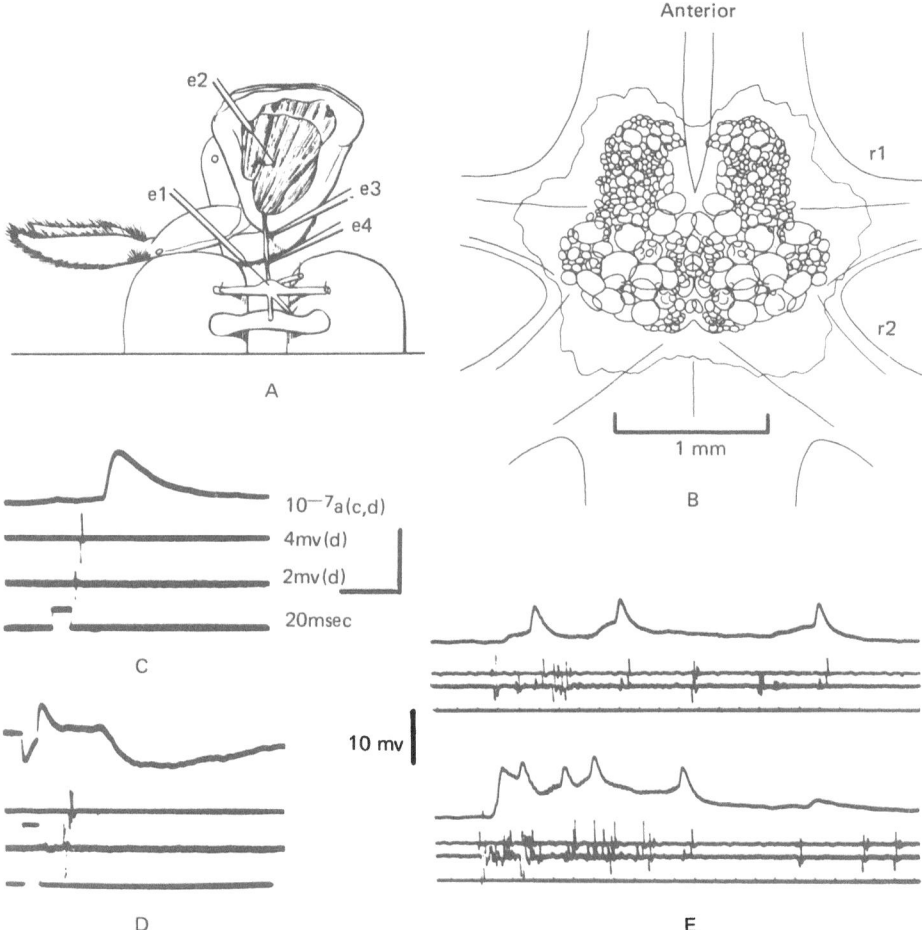

Fig. 1A–E. Identification of the cell bodies (somata) of the swimmeret motoneurons. A. The preparation, consisting of a single swimmeret, its muscles, and its nerve supply (the first abdominal nerve root and the corresponding segmental abdominal ganglion). e 1 and e 2, intracellular microelectrodes for stimulating single motoneuron somata and recording the muscle responses, respectively: e 3 and e 4, suction electrodes for extracellular recording of motoneuron action potentials. B. Ventral view of the desheathed abdominal ganglion, as seen at low power through a binocular microscope. Numerous nerve cell bodies are visible. C and D. Intracellular records of an excitatory (upper trace in C) and inhibitory (upper trace in D) junctional potential in the main power-stroke muscle, caused by stimulating identified somata (lower trace on each record) while recording the corresponding action potentials (middle two traces in each record). E. Simultaneous intracellular/extracellular recordings from an identified motoneuron (from Davis, 1971).

injection is illustrated in Fig. 4 and in the pictures presented by other contributors to this volume. In addition to these reconstructions of neuronal geometry, one can follow the fine branches of the dendrites of injected neurons to their terminals, which are commonly tightly apposed to the branches of other neurons (Figs. 3 and 5). Circumstantial evidence suggests that such connections are functional synapses. In the crayfish

Fig. 2A–F. Soma maps of the left half of the third abdominal ganglion, seen in ventral aspect. A. Somata of flexor and extensor motoneurons (from Otsuka et al., 1967). B–F. Swimmeret motoneuron somata. B. ○, somata of motoneurons to the main powerstroke muscle. ●, main return-stroke muscle. i, common powerstroke inhibitor. C. ○, Abductor muscle of the exopodite, ●, adductor of exopodite. D. ○, Rearward powerstroke muscle, ●, pronator of endopodite. E. ○, Curler of exopodite. ●, curler of endopodite. F. Accessory powerstroke muscle (from Davis, 1971).

abdomen, for example, certain flexor motoneurons can be excited by stimulating the heterolateral but not the homolateral giant fibers. Injection of these same motoneurons has shown independently that they make close contact with heterolateral but not homolateral giant fibers (Kennedy, Selverston, and Remler, 1969). In the lobster swimmeret system, a cross sectional map of all neurons which made contact with a single injected motoneuron has been constructed (Fig. 6A), and this map is similar to a cross sectional map of neurons which are known independently to excite the swimmeret motoneurons (Fig. 6B). Thus, the intracellular stain injection method can be used to determine the geometry of individual nerve cells, and—subject to direct confirmation with the electron microscope—their pattern of interconnection with other nerve cells.

Reconstructions of ganglia achieved to date (Fig. 4) are aesthetically pleasing, and they have provided new information on neuronal geometry. Their utility, however, is

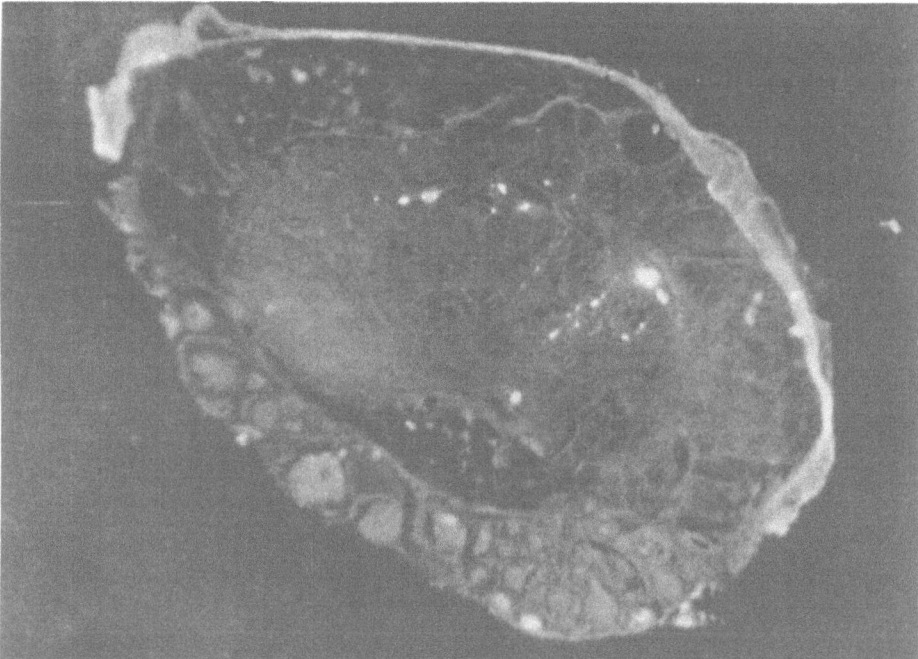

Fig. 3. High power fluorescence photomicrograph of a section containing branches of a power stroke swimmeret motoneuron which has been injected with the dye Procion yellow.

severely limited by the lack of quantification. How many microns in diameter are individual dendrites? How rapidly do dendrites taper and branch, with attendant functional implications? How do particular structural parameters (dendrite size) relate to specific functional characteristics (neuronal threshold)? What are the spatial coordinates of the dendrites relative to other neurons, and how do these relative positions relate to the output patterns of the neurons? How do structural parameters change during learning and development? Satisfactory answers to these questions cannot be obtained until precise methods for accurately describing neuronal structure are developed. In response to this need a number of investigators have begun to apply computer techniques to the problem of reconstructing and qualitatively describing nerve cells (Fox and Barnard, 1957; Mannen, 1964; Ledley, 1964; Levinthal and Ware, 1972; Woolsey et al., 1972; Chapter 17). In this article we describe our initial efforts toward the reconstruction and analysis of neuronal structure, utilizing the specific example of the lobster swimmeret system.

The SYNAPS System

Overview of the System

SYNAPS is the Symbolic Neuronal Analysis Programming System, being developed at Carnegie-Mellon University, for the three-dimensional reconstruction of stain-injected,

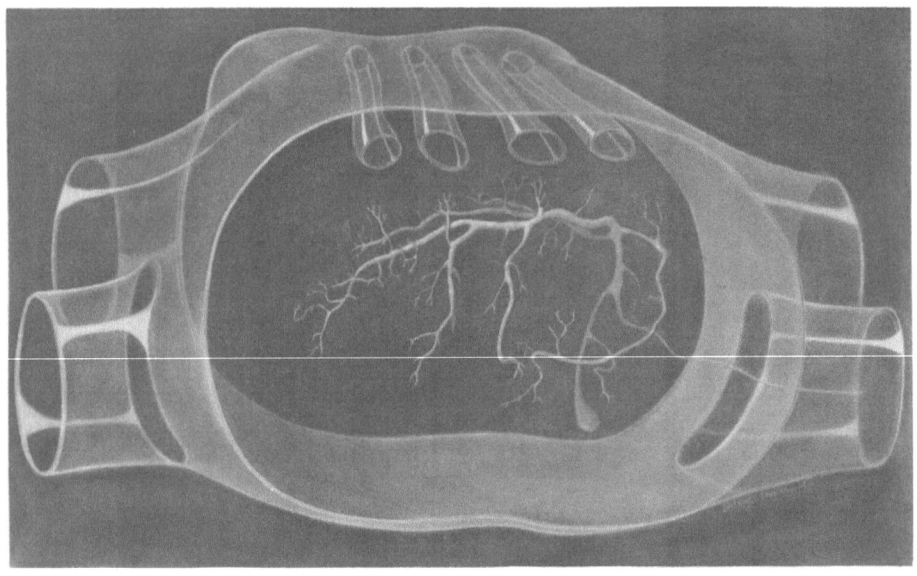

(a)

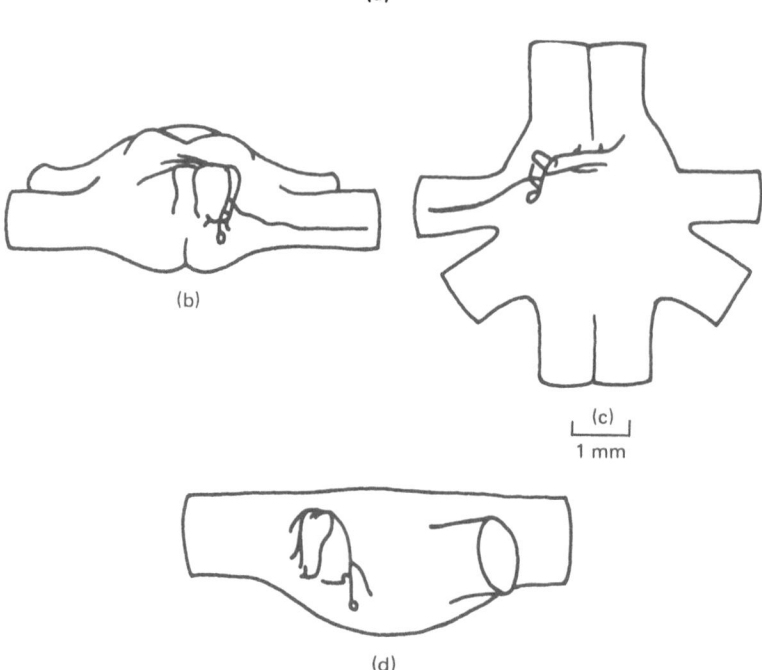

(b)

(c)

1 mm

(d)

Fig. 4A–D. Reconstruction of the third abdominal ganglion and a single powerstroke motoneuron within it, as seen from the anterior. This cell is reconstructed from sections of the type shown in Fig. 3. A. Three-dimensional drawing. B–D. Estimated planar projecttions of the injected neuron. Anterior is toward the top in C and the left in D. (From Davis, 1970).

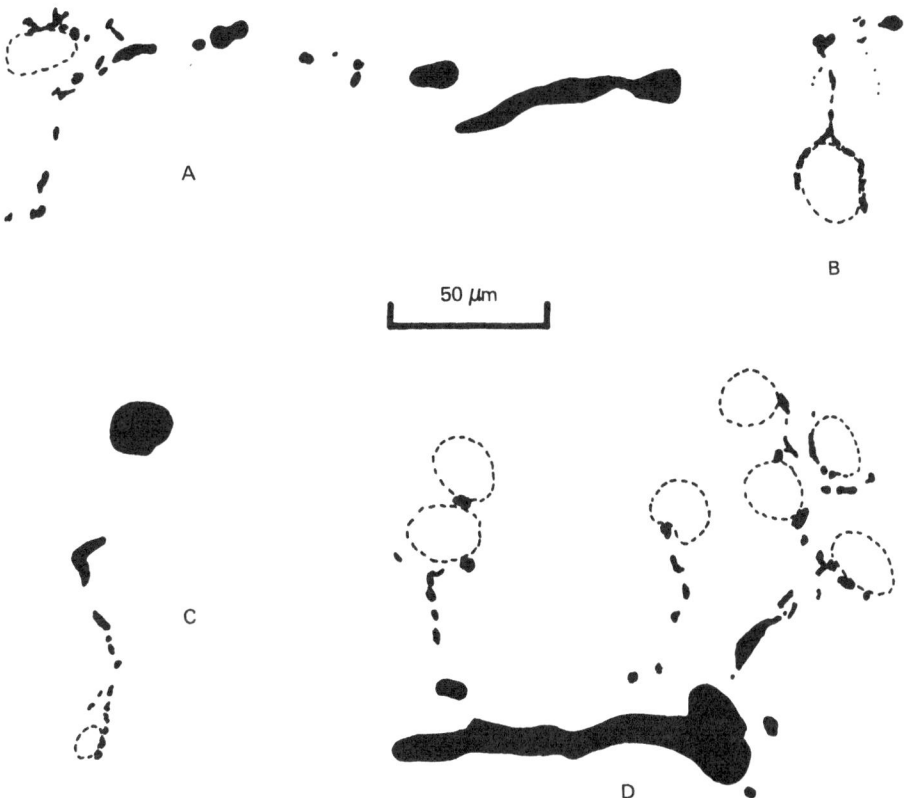

Fig. 5. Typical profiles of an injected powerstroke motoneuron traced from photomicrographs such as the one shown in Fig. 3. (From Davis, 1970).

serially-sectioned neurons on a computer. To achieve the goals implied in our research objectives, this system must satisfy several functional requirements. Specifically, it must: (1) generate digital images of the serial sections of the injected neuron and the ganglion which contains it; (2) process the digitized images to locate the boundary of the ganglion, the dendritic cross sections of the stain-injected neuron, and other structures of interest (giant fibers); (3) align the relevant boundaries from each section with those of adjacent sections; (4) assemble all the sections of the ganglion together to generate a concise three-dimensional model of the ganglion and the relevant neurons within it; (5) display the ganglion and the injected neuron on a cathode ray tube (CRT), or other output devices, at any specified viewing angle; (6) quantitatively describe the structure of identified neurons; and (7) assemble two or more reconstructed neurons from different injection experiments into a single topological map, and perform appropriate quantitative analysis of relative spatial relationships.

The SYNAPS system is envisioned, in its final version, as meeting the above requirements in four operations (Fig. 7); image digitization and processing of information from serial sections, generation of a three-dimensional model from these data, display of the model from any desired viewing angle, and quantitative analysis and comparison

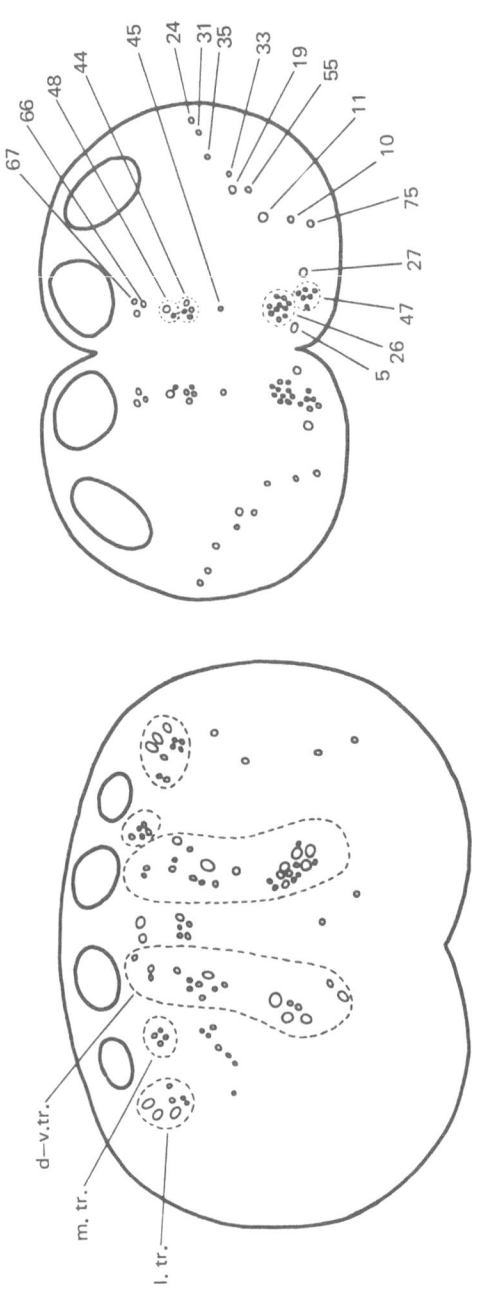

Fig. 6A and B. A. Map of cross sections of all axons passing through the third abdominal ganglion which were contacted (Fig. 5) by the cell illustrated in Fig. 4. The axons are grouped into a bilaterally symmetrical lateral tract (l. tr.), medial tract (m. tr.), and a dorso-ventral tract (d-v. tr.). B. Map of cross section of all axons in the crayfish nerve cord which are known to have functional effects on swimmeret motoneurons (reconstructed from the data of Wiersma and Hughes, 1960 and 1961). Certain populations of axons in the "structural" map (A) are congruent with presumably homologous populations in the "functional" map (B) (from Davis, 1970).

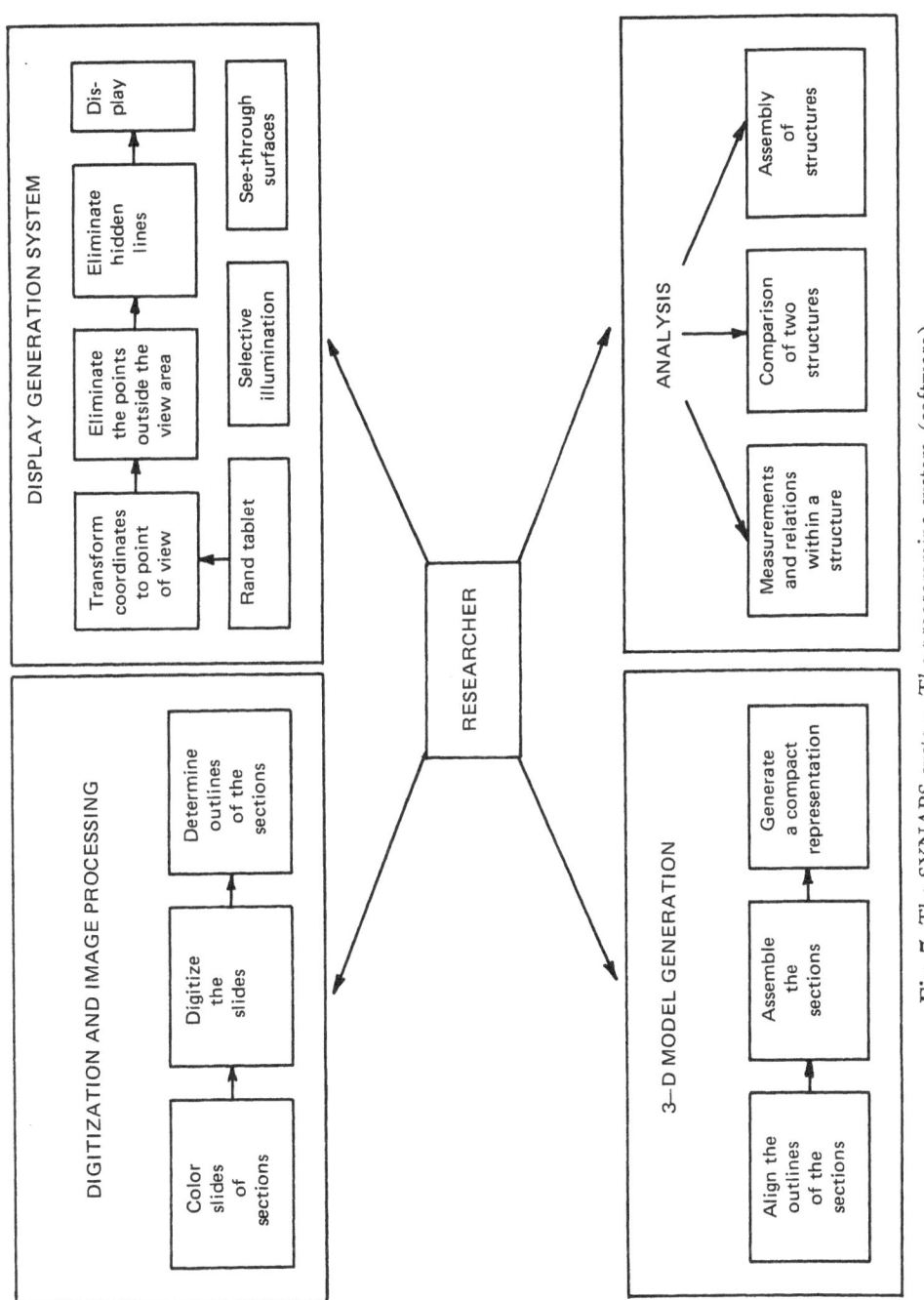

Fig. 7. The SYNAPS system. The programming system (software).

of neuronal structures. The system is highly interactive, i.e., the computer operations are performed in close symbiosis with the researcher. In this fashion, subjective judgements can be introduced—a feature we believe essential in dealing with intracellular stain injection (see also Chapter 17).

The Research Computer System (Hardware)

A computer system capable of performing the above operations must be a large scale, on-line interactive system with several special purpose devices. It must be large scale in order to handle the vast amounts of raw data and extensive processing. It must be an interactive system to allow close supervision by the researcher. In addition, it must have facilities for input of image data and for visual display of neural structures. The computer system we are presently using (Fig. 8) consists of a large PDP-10 computer (192K words of core storage), an image dissector, a Graf-pen and tablet, a Graphic-II regenerative display system, a hard copy Xerox Graphic Printer, and a storage scope.

Input Devices

The image dissector (Information International, Inc.) is used to digitize 35 mm color photographs of serial sections obtained from stain injection experiments. A Kodak carousel projector with a special lens and filter projects the image from the photograph

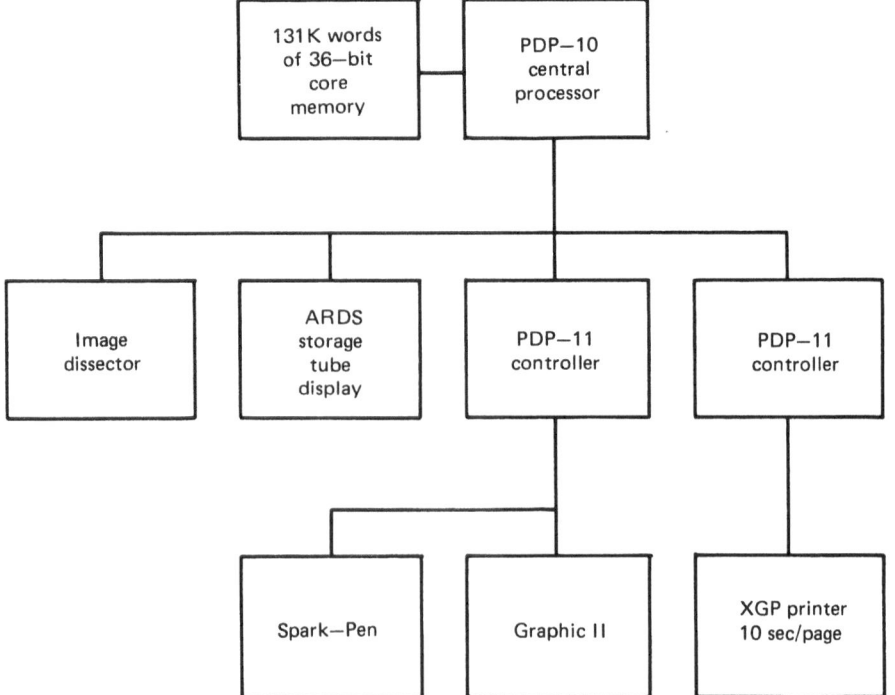

Fig. 8. The SYNAPS system. The computer system (hardware). Conventional input-output devices, such as card reader, line printer, magnetic tape, etc., are not shown.

onto the photo-cathode of the image dissector (Fig. 9A). The image dissector can be used to selectively digitize any point in its field of view, under program control. The outline of the image may also be traced manually using a Graf-pen (Science Accessories 611). In this case the image from the serial sections is projected onto a tablet and traced by the experimenter using the pen (Fig. 9B). The position of the pen is transmitted to the computer and displayed on a CRT to provide immediate feedback to the experimenter.

Output Devices

In order to view the three-dimensional neuronal structure from any desired angle, a display system capable of performing translations and rotations is required. We have several such devices available, each having unique capabilities. The choice is dictated by the particular needs of the experiment.

The storage scope (Tektronix 611) is used for long-term display of a complex, flicker-free image. The main disadvantages of this device are that it is relatively slow (2–3 min for drawing a typical nerve cell of 1000 small, straight-line segments) and modification of any portion of the image necessitates erasing and redrawing the entire image.

The Graphic-II regenerative display (Bell et al., 1971) is used for translation and rotation of simple images in real time. In addition, selected portions of the image can be modified without redrawing the entire image. To avoid "flicker" in the display, however, the complexity of the image is limited to one which can be generated in less than 30 msec.

The Xerox Graphic Printer (Reddy et al., 1972), has the advantages and disadvantages of the storage scope, and in addition yields a permanent copy. Views from many different angles may be produced and examined together later "off-line." All the diagrams presented in this paper were produced using this device.

None of the above devices permit rotation of the complex structures in real time, with the attendant kinetic depth effects. The LDS1 graphics system (Evans-Sutherland Corporation), with its hardware matrix multiplier and clipping divider, displays complex, flicker-free images which can be rotated in real time. This system is expensive to purchase but is available to us on a rental basis over a national computer network (ARPA network).

The Programming System (Software)

Digitization and Analysis of Sections

The purpose of digitzation and analysis of images is to extract the relevant data from each section for use by the assembly program (see next section, Generation of the Three-Dimensional Model). The relevant information usually consists of the boundary of the ganglion, dendritic profiles, and other neuronal "landmarks." The boundaries of the ganglion and of other large structures are approximated by a series of straight lines, and internally represented as a list of lines. Smaller dendritic profiles are approximated by circles, and internally represented by the center location and the radius of

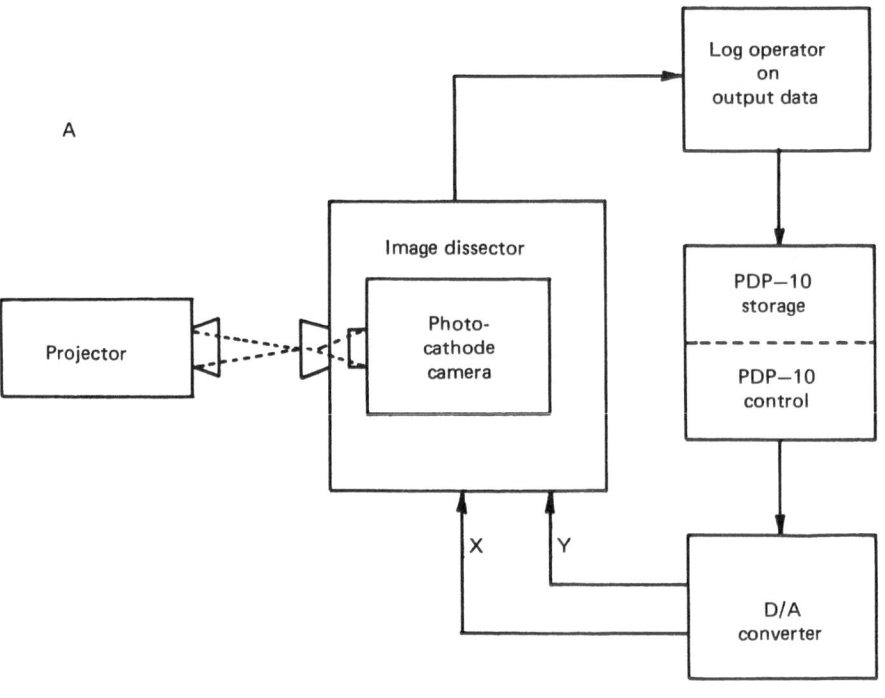

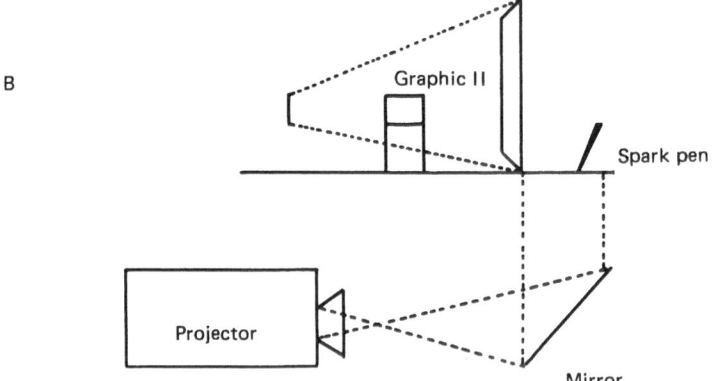

Fig. 9A and B. A. The Image Dissector. The X and Y coordinates of the point to be digitized are converted to analogue inputs providing proportional currents to the deflection coils of the image dissector. The deflected electron beam from the selected point passes through a narrow aperture. Thus measuring light at a selected point on the photo cathode consists of integrating the signal which is proportional to the photo electrons emitted from the photo cathode. The time taken for the integrator to reach a particular threshold is measured. This time is inversely proportional to the average signal amplitude during the integrating period. The log value of this time is calculated and transmitted to the computer for subsequent image analysis. B. The Graf-Pen. The device is based on spark chamber technology. It consists of a stylus which has, at its tip, a repeating spark discharge. Each time a spark is generated a counter is started. Linear microphones placed along the X and Y axis are used to detect the arrival of the sound from the spark discharge. When the sound arrives the counter is stopped. The value of the count is proportional to the distance of the stylus from the microphone.

each circle. The relevant information can be extracted from the section either manually or automatically.

The Manual Method

In the manual method, the experimenter subjectively decides what information is relevant and traces the outline using the Graf-pen (Fig. 9B). Every few milliseconds the location of the pen is transmitted to the computer. A line-fitting program operates on this coordinate data to fit a series of straight lines based on least square error criteria. The experimenter then manually types in whether the outline just traced is a ganglion boundary, a giant fiber, or a dendritic profile. Data prepared in this form can now be used by the assembly program directly (see next section, Generation of the Three-Dimensional Model).

The Automatic Method

Tracing of the outlines by the experimenter is tedious and error-prone. Errors can result from incorrect tracing of the data and from inaccurate judgments. The goal of the automatic method is to obtain dendritic profiles and other boundaries with minimal effort. This is achieved by the use of an "intelligent" image analysis program which operates on the digital representation of the slide from the image dissector (Fig. 9A). The input to this program is a matrix of light values representing the original section, and the output from the program is the location and shape of the dendritic profiles and other landmarks. This approach raises several new problems. Some of these problems are associated with ensuring that the matrix of light values is a faithful representation of the original image, while others are related to image analysis.

Image Digitization and its Associated Problems

The first problem involves the preliminaries associated with the digitization process. The projected slide of a single section of a ganglion is positioned, illuminated, and focussed with reference to the image dissector. Each of these preliminary steps is under program control. Proper illumination of the image, for example, is accomplished by a program which analyzes the entire range of light intensities from all digitized points in the projected image and recommends increasing or decreasing the lens aperture to obtain the maximum dynamic range. A focussing program scans the image, locates a boundary, and recommends adjusting the focus of the projector to minimize blur (to maximize the rate change in digitized light intensities across the boundary). Following these steps, a digitization program scans the section to generate a matrix of light intensities (up to 1000×1000 points). To insure that the image is accurately digitized some form of computer output of the image is required. Conventional output is woefully inadequate for such a purpose. We have developed several alternate output techniques over the last few years. Fig. 10 shows our most accurate form of digital reconstruction using the Xerox Graphic Printer.

The second problem in digitization arises from attempts at selective, high-resolution digitization of specific regions of a section. Such magnification may be desired by the user, or alternatively may be needed by the assembly program. Such a need may arise, for example, when a dendritic profile is absent where one is expected from information in adjacent sections, or when an apparent profile is present where one is not expected.

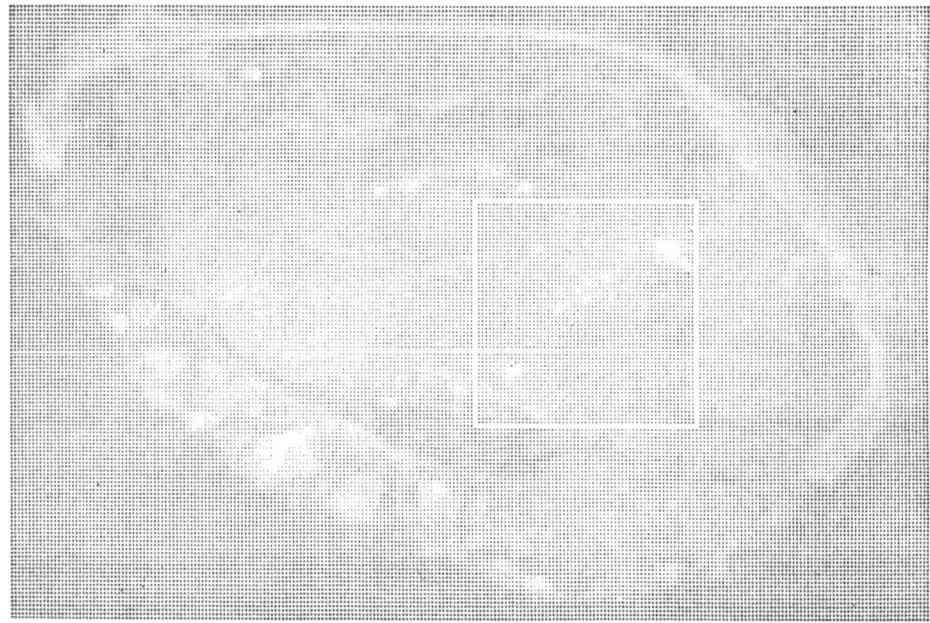

Fig. 10. A digital reconstruction of Fig. 3 using the Xerox Graphic Printer; data is normalized to 16 levels of grey scale.

Selective magnification can be achieved under computer control by an automatic pan, tilt, and zoom mechanism. Such techniques are not yet available on our system but are under development. At present we manually perform the pan, tilt, and zoom operations. Fig. 11 shows a selective magnification of the marked area of the original image.

The third problem in image digitization results from excessive demands on storage and bandwidth and cannot be performed conveniently on slower and smaller computers. The data obtained from a single stain injection experiment can generate more information than can be accommodated even by the larger billion-bit mass memories. For example, using the image dissector, a single section can result in 10^7 bits of data (1000×1000 matrix of light values with 10 bit accuracy per sample). If a single ganglion yields 250 sections one has to store and later process 2.5×10^8 bits of data. Assembling a composite neuronal network of only 10 neurons would involve 2.5×10^{10} bits of data! To ease this severe data storage requirement, we developed image compression programs which reduce the storage by an order of magnitude, using special encoding techniques. Both compression and subsequent re-expansion for further processing require special programs and necessitate an increase in computer processing time (so called time-space tradeoff).

Image Analysis and Associated Problems

The image processing program operates on the raw data to locate the boundaries of the ganglion and dendritic profiles. This analysis is based on the fact that different

Fig. 11. Selective magnification of the image delimited in Fig. 10 using the Xerox Graphic Printer. Note that several dendritic profiles missing in Fig. 10 are clearly visible in this magnified image.

regions of interest on the original slide are signified by discontinuities in light values in the digitized raw data. The discontinuities in the light values are located (and regions delimited) by an edge following program. This program uses a difference operator (analogous to differentiation) which indicates the direction and intensity of the discontinuity at a given point in the image. Low values of intensity indicate the absence of an edge. When an edge is present, the direction of the discontinuity can be used to predict where the next point on the boundary of that region may be found. Fig. 12A shows the results obtained after using a boundary detection operation, followed by edge detection using a difference operation. Note that many undesired regions appear in the output. By applying a thresholding operation to the subsection given in Fig. 11 we obtain the results of Fig. 12B. This operation involves ignoring all the variability in light values over a given threshold thereby eliminating undesired regions. Fig. 12C shows the results of thresholding and edge detection on the subsection to locate dendritic profiles. Dendritic profiles obtained by thresholding and edge detection on the entire image are shown in Fig. 12D.

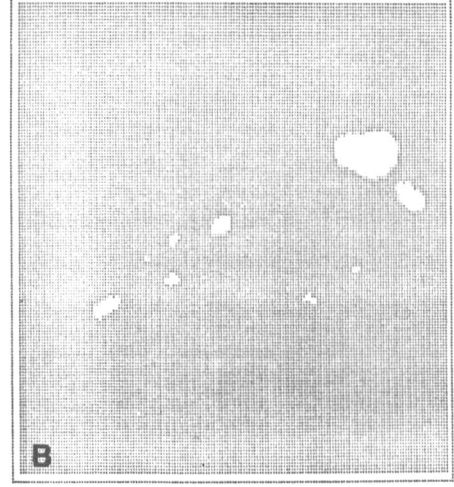

Fig. 12A–C.

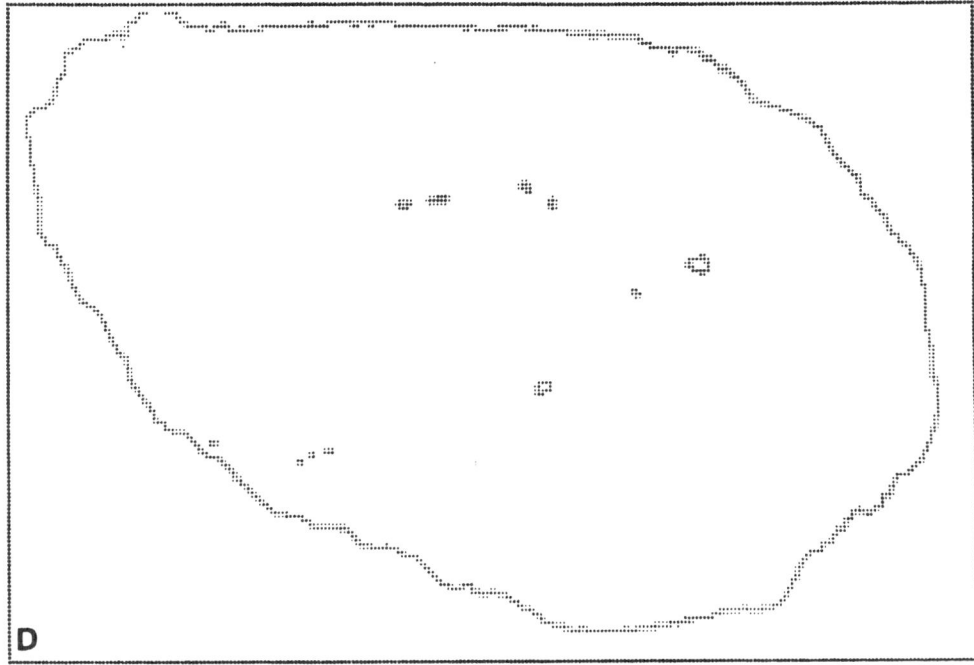

Fig. 12A–D. Results of image analysis programs operating on digital data shown in Fig. 10. A. Output after shell boundary detection and edge detection operations on the original image; note that many undesired regions ("noise") also appear in the output. B. Elimination of some undesired regions by a threshold operation on the subsection given in Fig. 11 which involves ignoring all the variability in light values over a given threshold. C. Thresholding and edge detection in the subsection to locate profiles. D. Dendritic profiles obtained by thresholding and edge detection on the entire image in Fig. 10.

Although there is an increasing body of literature on image processing (Miller and Shaw, 1968; Rosenfeld, 1969; Barlow et al., 1972), analysis of naturally occurring scenes poses several previously unidentified and unsolved problems (Montanari and Reddy, 1971). The main difficulty is with noise which interferes with the analysis. Noise results from several sources: intensity differences caused by variable light transmission from one region to the next in a section, artifacts such as tissue or dust particles and unanticipated folds in the tissue, photographic distortions, uneven lighting of the microscopic field, undesired leakage of stain from the injected neuron, etc. This interference makes it necessary that the experimenter look at the results of image analysis and modify boundaries using the interactive image editor. Thus the automatic method is in reality a man-machine system with machine doing most of the work.

A variety of improvements in the above procedures are under development. Among the most important are programs which utilize information from adjacent sections for predictive purposes. The assumption that adjacent sections are more similar than different permits the programs to hypothesize regions where dendritic profiles may be found (thereby reducing processing time). Moreover, errors resulting from noise will be reduced by this procedure, since noise is unlikely to occur in corresponding regions of adjacent sections.

Generation of the Three-Dimensional Model

The boundary and profile information from the image analysis is assembled to form a single three-dimensional structure. This is achieved by combining a set of "line-lists" into a single, more complex data structure. In order to assemble the serial sections, a preliminary alignment operation is necessary owing to variations of the image position in the photographic field.

Alignment of Sections

The alignment of two adjacent sections requires the identification of at least two sets of corresponding points of both sections. This can be accomplished manually by visual observations of two superimposed outlines on a graphic display, or automatically by aligning corresponding landmarks in both sections by means of correlation procedures. The two sets of corresponding points are used to calculate a 2×2 rotation matrix and a translation vector, both of which define the amount by which one section must be moved to perfectly align the corresponding points. These transformations are then applied to all the points in the line-list of the section to be moved. This process is repeated between every pair of adjacent sections, resulting in a set of aligned sections.

The alignment described above only corrects for misalignments within the plane of the sections. Corrections to the coordinates are sometimes also required along the axis perpendicular to the plane. Although the serial sections through the ganglion are always parallel to each other, they may not be exactly perpendicular to the long axis of the ganglion. In such cases, the deviation from a precisely transverse angle is calculated from the information about the Z-axis (longitudinal) displacement between the anterior or dorsal margins of obvious, bilaterally-symmetrical structures (the first nerve roots, identifiable daughter cells, etc.). On this basis the necessary rotation of each section can be performed before assembly.

Assembly of Sections

Corresponding dendritic profiles in several adjacent sections are located and combined to form a single, straight-line dendritic branch. This identification of the corresponding dendritic profiles is achieved by locating overlapping regions in adjacent sections. When a profile lacks an overlapping match in the preceding or following section, a search is made to locate an unmatched profile in the immediate neighborhood. If no match is found the profile is discarded. Matched profiles are internally represented by variable-diameter, straight-line tubes. Profiles of small somata assembled from several sections are approximated by spheres. Profiles of large somata and the external surface of the ganglion are approximated by either triangular or irregular strip surfaces.

The result of the assembly operation is an internal data structure in which information about the dendritic structure is abstracted into a three-dimensional tree structure with all the branches located and identified, and with all linear proportions preserved. This representation is not only compact and concise, but it is also arranged in a form suitable for subsequent display and analysis.

Graphic Display of the Model

The graphic display provisions of the SYNAPS system permit the experimenter to view the reconstructed neuronal structure from different angles of view. This facility is not only necessary to inspect and modify the reconstruction but also to formulate intuitions about structural-functional relationships. The angle and position of view is specified by typing in the relevant values. We are at present designing a joystick-like analogue input which will permit the experimenter to "fly" through and around the neuronal structure to obtain different views of it. The effectiveness of the joystick increases when one is able to perform rotation in real time.

A major problem in display generation is that we are dealing with nonmathematical surfaces and objects. An unselective display of all the subparts of the model usually results in a cluttered image (Fig. 13). Algorithms which generate transparent surfaces and/or eliminate hidden surfaces are available, but they are usually complex (Watkins, 1970) and cannot, at present, be implemented in real time without special-purpose hardware. One can simply eliminate the outer shell and other structures from the model (Fig. 14), but this results in the loss of crucial points of reference. Fig. 15 illustrates our present compromise, i.e., elimination of giant fibers and use of a simple dotted-line representation of the shell, simulating a see-through surface. The nerve structure may be displayed as straight lines (Fig. 15 A and B) or as tubes. Fig. 16 shows the same neuron reconstructed in Fig. 15, but as viewed from several different angles.

Often the experimenter desires a close-up view of a subpart of a picture. SYNAPS is capable of giving a magnified, selective display of any portion of the neuron. Close-up views are much easier to provide using special-purpose hardware such as the clipping divider (Sproull and Sutherland, 1968) available on the LDS1-graphic system.

Analysis of Reconstructed Neurons

Our eventual goal is to assemble a three-dimensional model of an archetypical ganglion containing select, identified neurons. Postulated relationships between topological and functional properties of the reconstructed networks (Davis, 1969), if present, should appear especially conspicuous in such selective composite reconstructions.

Qualitative Studies

The qualitative, three-dimensional reconstructions achieved to date (Figs. 15 and 16) are a necessary first step towards the above objective, and moreover they provide the basis for significant substudies. For example, with the computer techniques described above we can now qualitatively and accurately assess general morphological features such as the size and shape of dendritic fields. These features are in turn expected to bear interesting and testable relationships with the functional properties of the neurons.

Combination of several identified neurons into a single ganglion, on the other hand, presents somewhat more complex problems. For example, only one neuron is normally injected in each experiment to avoid ambiguities in the reconstruction. Therefore, data from different ganglia will have to be combined to obtain a topological map of a given network. Homologous ganglia appear qualitatively similar from one animal to the next, but significant quantitative differences in relative proportion may exist. In

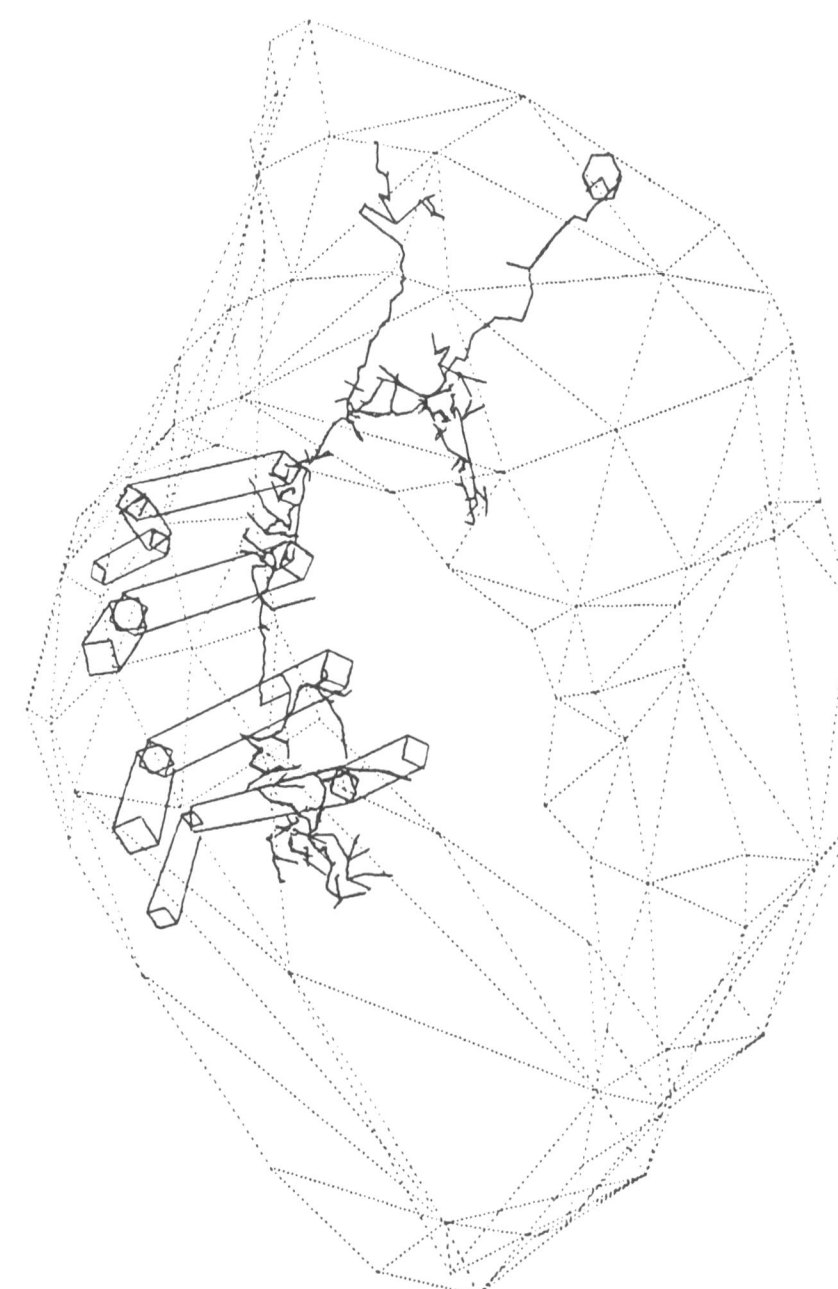

Fig. 13. Cluttered image resulting from display of all the subparts of the model.

Fig. 14. Display of the dendritic structure with the ganglion and giant fibers eliminated. This results in the loss of points of reference.

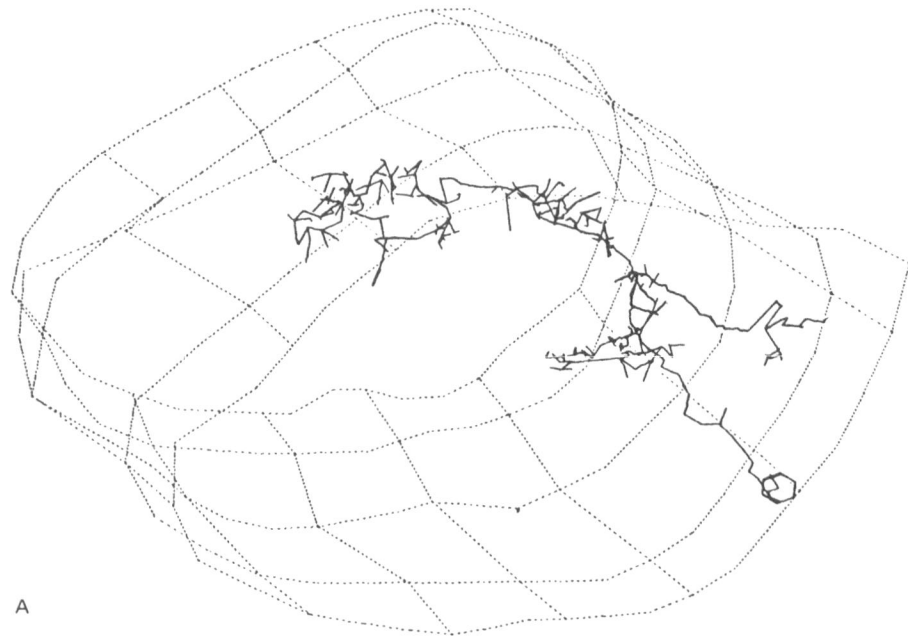

A

B

Fig. 15A and B. Different forms of display. A. Shell represented by irregular patches with giant fibers eliminated. B. Shell represented by strips.

this case normalization of different ganglia may be necessary, and the required transformations may not be linear.

Quantitative Studies

As detailed at the beginning of the chapter, the development of a quantitative language for describing neuronal structure is an important eventual goal of this work. With the programs developed to date, we can extract the following quantitative data from single injected neurons: the number of dendritic branches, dendrite diameter, length, volume, and, with reasonable simplifying assumption, surface area. These measures can be determined either for the entire dendritic field of a neuron, or for select regions of special interest. Additional parameters of interest include dendrite taper, branch pattern, overall shape and distribution, spatial position relative to other dendritic fields, and pattern of connection with other neurons. Some of these measures seem straight-forward and relevant to neuronal function, while others are only vaguely defined at present. In many instances we expect to resort to statistical descriptions, especially with regard to the shape of dendritic fields and their spatial positions. Many of the problems are unforeseeable, and the solutions will simply have to evolve with the research.

Summary and Conclusions

This paper describes research to date on SYNAPS, (Symbolic Neuronal Analysis Programming System), a computer-based system for analysis of the geometry of single nerve cells and the structure of neuronal networks. In its present, unfinished form, the system is useful only in the hands of trained computer scientists, and its capabilities are limited to: (1) acceptance of "edited" data on neuronal structure, consisting of tracings from serial sections of stain-filled branches of an injected neuron; (2) automatic alignment and assembly of the three-dimensional structure of the injected neuron and its surrounding ganglion; (3) rotation of the three-dimensional model so that the reconstructed neuron may be viewed from any angle and its spatial relationships with other structures qualitatively assessed; (4) selective, high-resolution display of desired regions of the injected neuron; and (5) extraction of quantitative data on the number of dendritic branches of an injected neuron, dendrite diameter, length, volume, and surface area.

In its finished form, SYNAPS will be usable by biologists untrained in the computer sciences for studying neuronal structure in any nervous system. Planned capabilities include: (1) automated input of "raw" data (although experience with intracellular stain injection suggests a need for extensive man-machine interaction); (2) quantitative description of the morphology of single neurons; and (3) quantitative analysis of the spatial relationships of many neurons in a network. Completion of SYNAPS will require solving numerous practical and theoretical problems in the computer sciences, but our experience to date suggests that a reasonably complete system can be implemented within five years. Completion of the system will open the way to rigorous studies of neuronal structure and its relations to neuronal function.

Acknowledgments. This research is principally supported by the National Science Foundation under contract No. GJ32784 and in part by NIH Grant NS-09050.

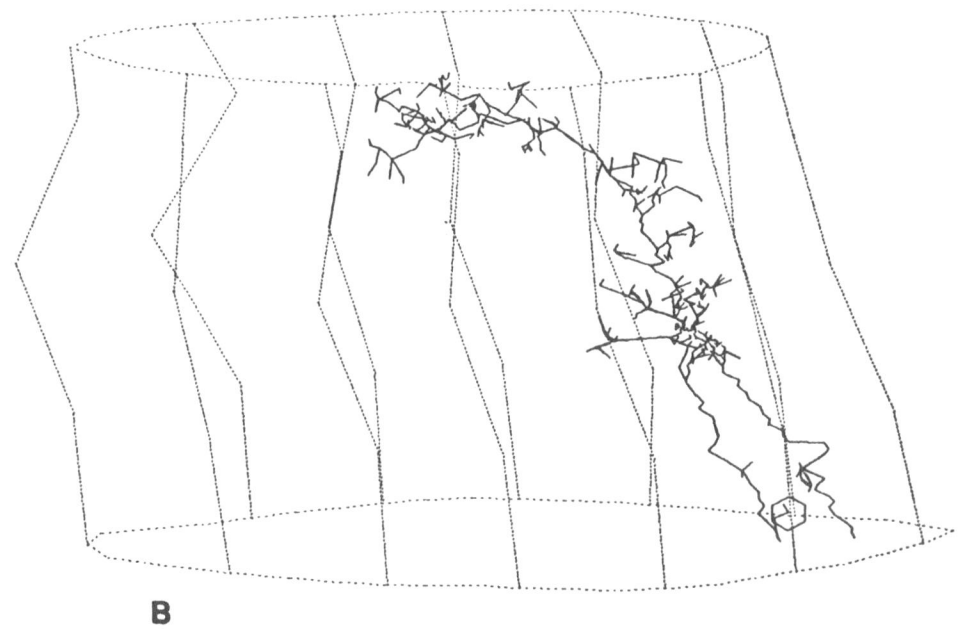

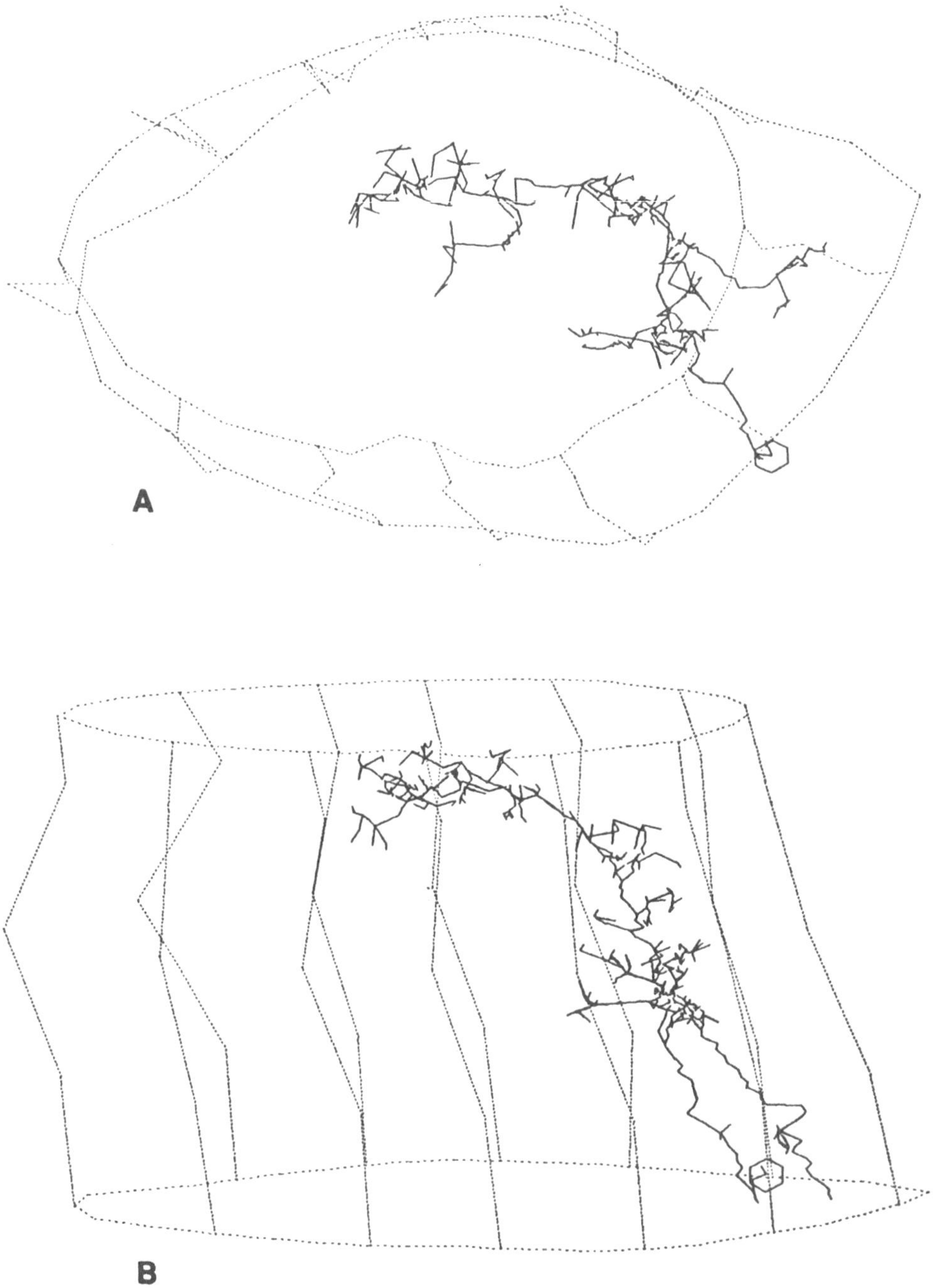

A

B

Fig. 16A and B.

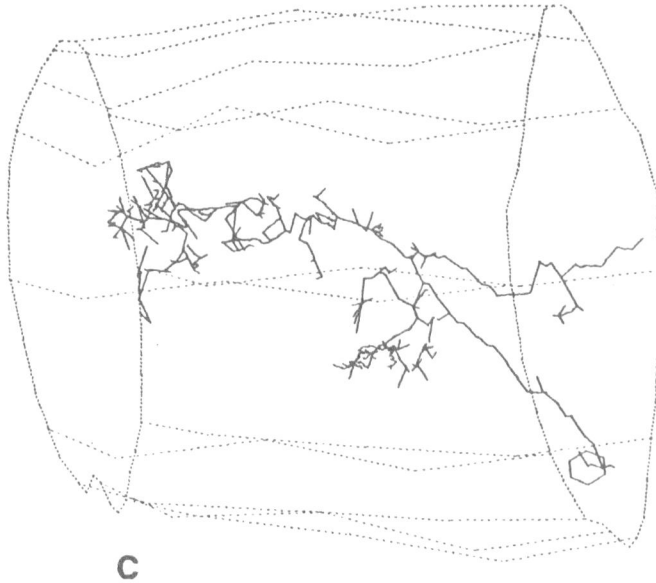

C

Fig. 16A–C. Display of the model from different angles of view. A. Front view—angles of rotation about the x, y, z axis are 0, 0, 0. B. Topview—angles of rotation are 0, 80, 0. C. Side view—angles of rotation are 0, 0, 80 (x-axis out from paper, y-axis pointing right, and z-axis pointing upward). Note that SYNAPS is capable of displaying the model from any desired angle.

References

Barlow, H. B., R. Narasimnan, and A. Rosenfeld: Visual pattern analysis in machines and animals. Science *177*, 567–575 (1972).

Bell, C. G., D. R. Reddy, C. Pierson, and B. Rosen: A high performance programmed remote display terminal. Boston, Massachusetts, Proc. of 1971 IEEE Computer Society Conf., pp. 47–48.

Beurle, R. L.: Properties of a mass of cells capable of regenerating pulses. Phil. Trans. R. Soc., Lond. B *240*, 55–94 (1956).

Davis, W. J.: The neural control of swimmeret beating in the lobster. J. exp. Biol. *50*, 99–117 (1969).

———: Motoneuron morphology and synaptic contacts: determination by intracellular dye injection. Science *168*, 1358–1360 (1970).

———: Functional significance of motoneuron size and soma position in swimmeret system of the lobster. J. Neurophysiol. *34*, 274–288 (1971).

Fox, C. A., and J. W. Barnard: A quantitative study of the Purkinje cell dendritic branchlets and their relation to afferent fibers. J. Anat. *91*, 299–313 (1957).

Freeman, J. A.: The cerebellum as a timing device: an experimental study in the frog. In: Neurobiology of Cerebellar Evolution and Development. Ed. R. Llinás, pp. 397–420. Chicago: American Medical Association, 1969.

Globus, A., and A. B. Scheibel: Pattern and field in cortical structure: the rabbit. J. comp. Neurol. *131*, 155–172 (1967).

Henneman, E.: Relation between size of neurons and their susceptibility to discharge. Science *126*, 1345–1347 (1957).

———, G. Somjen, and D. O. Carpenter: Excitability and inhibitability of motoneurons of different sizes. J. Neurophysiol. *28*, 599–620 (1965).

Hilgard, E. R.: Theories of Learning and Instruction. Chicago: University of Chicago Press, 1964.

Hinkle, M., and J. M. Camhi: Locust motoneurons: bursting activity correlated with axon diameter. Science *175*, 553–556 (1972).

Hughes, G. M., and C. A. G. Wiersma: Neuronal pathways and synaptic connexions in the abdominal cord of the crayfish. J. exp. Biol. *37*, 291–307 (1960).

Kato, M., B. Fujimori, and Y. Hirata: An electron microscopic study of intracellularly stained neurons. Brain Res. *9*, 390–393 (1968).

Kennedy, D., A. I. Selverston, and M. P. Remler: Analysis of restricted neural networks. Science *164*, 1488–1496 (1969).

Ledley, R. S.: High-speed automatic analysis of biomedical pictures. Science *146*, 216–222 (1964).

Levinthal, C. and R. Ware: Three-dimensional reconstruction from serial sections. Nature *236*, 207–210 (1972).

Mannen, H. Arborizations dendritiques etude topographique et quantitative dans le noyau vestibulaire du chat. Arch. ital. Biol. *103*, 197–219 (1964).

Miller, W. F., and A. C. Shaw: Linguistic methods in picture processing: a survey. Proc. 1968 Fall Joint Computer Conference, 279–290 (1968).

Montanari, U., and D. R. Reddy: Computer processing of natural scenes: some unsolved problems. Proc. of the AGARD Symposium on Artificial Intelligence, Rome (1971).

Mungai, J. M.: Dendritic patterns in the somatic sensory cortex of the cat. J. Anat. *101*, 403–418 (1967).

Otsuka, R., E. A. Kravitz, and D. D. Potter: Physiological and chemical architecture of a lobster ganglion with particular reference to gamma-aminobutyrate and glutamate. J. Neurophysiol. *30*, 725–752 (1967).

Pitman, R. M., C. D. Tweedle, and M. J. Cohen: Branching of central neurons: intracellular cobalt injection for light and electron microscopy. Science *176*, 412–414 (1972).

Pomeranz, B., and S. H. Chung: Dendritic-tree anatomy codes form vision physiology in tadpole retina. Science *170*, 983–984 (1970).

Rall, W.: Distinguishing theoretical synaptic potentials computed for different soma-dendritic distributions of synaptic inputs. J. Neurophysiol. *30*, 1138–1168 (1967).

——, and G. M. Shepherd: Theoretical reconstruction of field potentials and dendrodendritic synaptic interactions in olfactory bulb. J. Neurophysiol. *31*, 884–915 (1968).

——, R. E. Burke, T. G. Smith, P. G. Nelson, and K. Frank: Dendritic location of synapses and possible mechanisms for the monosynaptic EPSP in motoneurons. J. Neurophysiol. *30*, 1169–1193 (1967).

Ramón-Moliner, E.: An attempt at classifying nerve cells on the basis of their dendritic patterns. J. comp. Neurol. *119*, 211–227 (1962).

——: The morphology of dendrites. In: The Structure and Function of Nervous Tissue. Ed. G. H. Bourne. Vol. 1, pp. 205–267. New York: Academic Press, 1968.

Ramón y Cajal, S.: Histologie du Système Nerveux de L'homme et des Vertébrés. 2 vol. Paris: Maloine, 1909, 1911.

Reddy, R., B. Broadley, L. Erman, R. Johnson, J. Newcomer, G. Robertson, and J. Wright: XCRIBL: a hardcopy scan line graphics system for document generation. Tech. Report, Carnegie-Mellon Univ., Pittsburgh. (Also published in Information Letters *1*, 246–251 1972).

Rosenfeld, A. Picture Processing by Computer. New York: Academic Press, 1969.

Sproull, R. F., and I. E. Sutherland: A clipping divider. Proc. 1968 Fall Joint Computer Conf. *33*, 765–775 (1968).

Stretton, A. O. W., and E. A. Kravitz: Neuronal geometry: determination with a technique of intracellular dye injection. Science *162*, 132–134 (1968).

Thomas, R. C., and V. J. Wilson: Marking single neurons by staining with intracellular recording microelectrodes. Science *151*, 1538–1539 (1966).

Verzeano, M.: The synchronization of brain waves. Acta Neurol. latinoam. *9*, 297–307 (1963).

————, K. Negishi: Neuronal activity in cortical and thalamic networks. J. gen. Physiol. *43*, 177–195 (1960).

Watkins, G. S.: A real time visible surface algorithm. Computer Science Department, Univ. of Utah, UTECH-csc-70-101 (1970).

Wiersma, C. A. G., and G. M. Hughes: On the functional anatomy of neuronal units in the abdominal cord of the crayfish, *Procambarus clarkii* (Girard). J. comp. Neurol. *116*, 209–228 (1961).

Woolsey, T. A., D. F. Cowan, M. L. Dierker, and C. M. S. Linn: Computer analysis of Golgi impregnated neurons. Washington Univ., St. Louis, presented at the Second Annual Meeting of the Society for Neuroscience, Houston, Texas (1972).

Discussion

DR. KRIPKE (Utah): Have you tried your image processor out on artificially produced sequences of images so that you can compare the results with what you started with? Is your program idempotent: that is, if you process an image and then process the result again, do you get the same result as you would from a single application of the program?

DR. DAVIS (University of California, Santa Cruz): Dr. Reddy and his group began by constructing a tinker-toy model of a neuron. Computer reconstruction programs were developed and initially de-bugged using this model, before applying the same programs to real, and much more complex, neurons.

DR. MACAGNO (Columbia): Your technique is useful when studying a structure that is outstanding within a flat background.

If your contrast level is very good, such as with cobalt or with the fluorescent dye, where you can also do a spectral analysis of the pictures, that is fine.

In our case (Levinthal's laboratory), however, where we do serial section reconstructions from electron micrographs, we have found that if we try to use a digitizer, the amount of information that exists on the micrograph is too large. The programming that you have to do in order to pick out a feature from an electromicrograph is just too extensive, except in some special cases where one can use particular stains, like the PTA stains for synapses, where dark synapses stand out clearly on a light background.

My feeling is that if you want to look at electron micrographs, you have to do it interactively. If you want to study cells in terms of their microstructure, trying to find synaptic ends, or looking at spines and other fine details, you have to do it by looking with your own eyes and telling the computer what it should see.

Let me make just one last statement. It is not very hard to make for yourself a short serial section movie, especially at this level of resolution. I think everybody should make one, even if they do it grossly. Then run the movie back and forth a few times. You will be surprised at the incredible amount of information you can get just by looking sequentially and letting your own brain integrate the images into a continuous structure.

CHAPTER 17

The Use of Intracellular Dye Injections in the Study of Small Neural Networks

Allen I. Selverston

The purpose of this paper is to review briefly some of the literature on the use of Procion dyes in Crustacea, to illustrate some of our work on the stomotogastric ganglion of the lobster, and to describe how computer technology can be used to reconstruct and quantitatively analyze dye-filled cells.

The Use of Procion Dyes in Crustacean Preparations

The lobster *Homarus* was the first preparation used for intracellular injection of Procion yellow (Stretton and Kravitz, 1968 and Chapter 2). Identifiable motoneurons in abdominal ganglia were filled iontophoretically with Procion yellow, and their general geometry as well as the relationship of one cell to the giant fibers was illustrated. This paper was soon followed by one describing the lateral giant interneurons in crayfish (Remler et al., 1968). Dye was injected into the lateral giant axon through a micropipette using a pressure injection technique similar to that described in Chapter 20. Remler et al. also introduced the cleared whole mount as a way of seeing the entire neuron. Injections into the axons of crayfish fast flexor motoneurons revealed their connections with the ventral cord giant fibers (Selverston and Remler, 1972). This method is applicable to any axon which could be stripped from the cord and isolated (Selverston and Kennedy, 1969; Zucker et al., 1971). Injections of giant fibers and sixth ganglion motoneurons of crayfish were shown by Larimer et al. (1971) to be useful in corroborating physiological data.

Several other examples of crustacean cell injections are noteworthy. Paul (1972) has shown in the sand crab *Emerita* that giant stretch receptor neurons have their cell bodies located within the fifth abdominal ganglion, the first unequivocal demonstration of centrally located sensory cell bodies in arthropods. Sandeman (1969) used Procion yellow to facilitate the analysis of the integrative properties of a motoneuron in the crab. Later, Sandeman (1971) filled four identified motoneurons responsible for eye withdrawal with Procion yellow, after determining that ipsilateral pairs were electrotonically connected. The injections revealed points within the neuropil where such contacts might occur.

Barnacle eyes have also been injected by Shaw (1972), who showed that mixing Procion yellow with Procion blue and brilliant red made multiple injections much easier to interpret because each cell fluoresced at slightly different wavelengths. The median ocellus of *Limulus* has been examined anatomically with Procion injections by Jones et al. (1971).

Small Systems Analysis

Our interest is in the use of Procion yellow to study the structural properties of identifiable crustacean neurons. Our premise is that in order to understand the mechanisms underlying behavior, one must be able to describe both anatomically and physiologically those individual neural elements which are involved in patterned motor output. We are still a long way from completely defining the generation of any particular motor output. Significant progress is being made, however, by using a wide variety of invertebrate ganglia as preparations. Such ganglia are capable of generating various kinds of motor output with only a small number of neurons and often without any sensory feedback. One of the major advantages is the ability to return to the same cell from preparation to preparation (see Kennedy et al., 1969). In many cases the location of the cell bodies on the ganglionic surface is constant (Otsuka et al., 1967) and it is possible to investigate the integrative role of a particular cell during a specific behavior pattern (Willows, 1968).

What has been singularly difficult to achieve is a detailed description of the anatomical correlates to the physiological data. Attempts have been made to follow, in sections, the processes of known cells, but this procedure is both difficult and uncertain. The introduction of Procion yellow as an intracellular stain has now made it possible to inject repeatedly and visualize an identifiable cell.

Methods of Dye Injection

Both Procion yellow M4RS and M4RAN give equally good results. An aqueous solution of the dye undergoes photolytic decomposition and forms a viscous gel after several weeks (see discussion after Chapter 3). This process can be retarded by refrigeration of the solution in a light-tight container. I use a 6% (w/v) aqueous solution of Procion yellow which is filtered through a 0.45 μm Millipore filter just prior to use. Glass capillary tubing is cleaned with hot acid before pulling the microelectrodes (see Chapter 20). High impedance electrodes are first filled with distilled water which is then replaced with dye.

When large cell bodies are being filled or the dye is being injected under pressure, larger tip diameters can be used. Empty microelectrodes are filled with filtered dye by connecting an electrode to a syringe via polyethylene tubing. The tip is then broken by rubbing it against the ridges of the palm or across the ground glass part of a microscope slide. When pressure from the syringe is able to force the dye from the barrel into the tip, the electrode is ready for use.

In ganglia with large cells, where the position of the tip can be observed, one can either employ the same electrode for both recording and filling or, once having mapped the position of the cell bodies electrophysiologically, use a fresh Procion-filled electrode for dye injection. The circuit such as shown in Fig. 1 can be used to pass current through the microelectrode. The amount of dye that needs to be injected varies with the size of the cell; the only precaution that one must observe is not to use too high a current density. We have found that currents greater than 10^{-6} A damage the cell and prevent the dye from diffusing freely from the point of injection. We generally use current between 1 and 5×10^{-8} A delivered in 500 msec pulses at 1 Hz for a period of 30–90 min. Our procedure for crustacean cell bodies from 10 to 75 μm in diameter is as follows. Initially cell bodies are identified and mapped using micropipettes filled with 3 M KCl. A fresh pipette is then prepared by filling with filtered Procion and breaking off the tip to give an electrode resistance of 20–50 MΩ. The cell of interest is penetrated with this electrode, the identification of the cell is rechecked, and the dye injected. The cell is observed visually during the next 30–90 min, and the current is turned off when the soma is bright orange in color. The ganglion is then placed in the refrigerator so that the dye can migrate.

We have successfully used the pressure injection technique (Strumwasser, personal communication, also see Chapters 10 and 20) to fill both cell bodies and axons. It is particularly useful in the latter case because a large volume of dye can be injected in a very short time. Current pulses are passed through the electrode tip while the injection is proceeding. Since current passed is proportional to the amount of dye being injected, one can monitor the flow rate and determine immediately if the tip becomes blocked. After sufficient dye has been injected, the microelectrode is removed and the axon tied off with a fine thread proximal to the injection point. The main danger in the use of this technique is excessive pressure. Because hydraulic force is transmitted

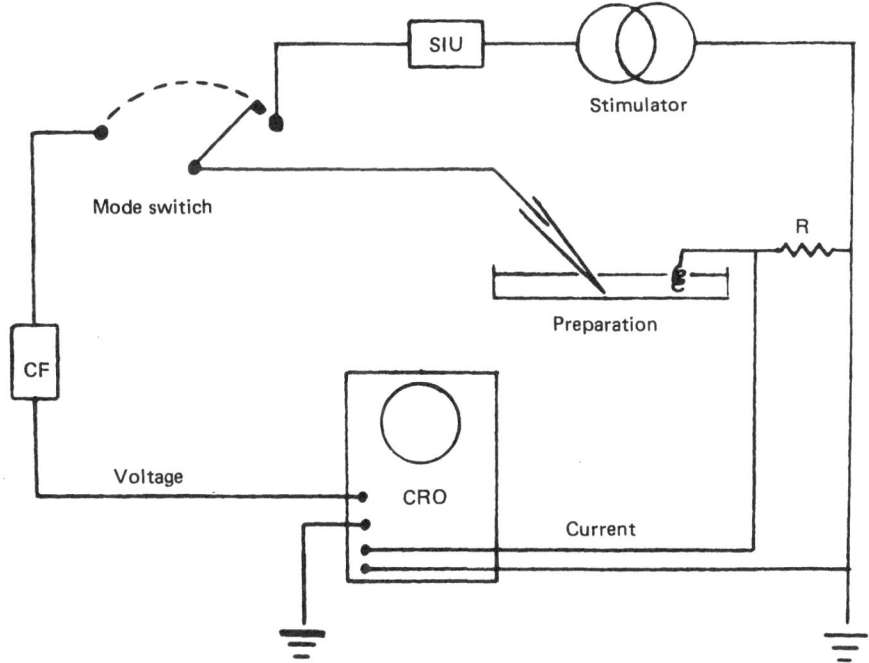

Fig. 1. Electrical circuit used for iontophoresis consists of a switching network which allows the microelectrode to be easily changed from a recording to a stimulating mode. The dropping resistor (R) should have a value of 10^{-5} or 10^{-6} Ω. The SIU is a radio frequency type stimulus isolation unit, and the CF a high input impedance cathode follower or electrometer amplifier.

to all parts of the cell, the dendritic branches can be blown out and the entire ganglion filled with dye. Similarly, electrotonic junctions, particularly large septa, seem to be weak points. Excess pressure will break these septa, allowing the dye to enter the post-synaptic cell. In the crayfish where both pressure injections and iontophoresis have been used to fill fast flexor motoneurons, we could not discern any difference in morphology that might be attributable to the method used.

It seems to be important to keep the cell viable for as long as possible following injection of dye in order to allow migration of the dye. As a rule of thumb, ganglia less than 1 mm in diameter should be left least 12 hr. Ganglia greater than 1 mm in diameter should be left at least 24 hr.

After sufficient time for dye migration has elapsed the ganglion is isolated and covered with fixative, usually Bouin or Susa. We have also had good fixations in buffered glutaraldehyde solutions. The preparation must not be overfixed, especially if it is to be viewed in whole mount. Mercury and picric acid salts must be treated with iodine and lithium carbonate, respectively, in the first alcohol. It is also crucial that all water be removed by adequate alcohol dehydration before clearing. A drying agent should be added to the absolute alcohol to insure the complete absence of water. Clearing can be achieved with methyl benzoate or pure beechwood creosote. A properly cleared ganglion should be perfectly transparent when viewed with transmitted light.

Remler et al. (1968) demonstrated that Procion yellow injected neurons could be well visualized in whole mounts and we routinely use this technique. The tissue is placed in a depression slide containing the clearing agent and is turned and photographed through several angles. When methyl benzoate is used as the clearing agent, the ganglion can be stored in this solution, but only if it is not going to be sectioned since the tissue will become brittle.

For a detailed examination of the injected neuron or neurons, we embed the tissue in paraffin or epon and cut serial sections. Paraffin sections are usually cut at 8–10 μm. The sections are then mounted in a nonfluorescent medium such as Fluoromount. Epon can be sectioned with either glass or steel knives at 5 μm; it does not ribbon so each section must be handled individually.

Whole mounts or sections of Procion yellow-stained material are viewed in a fluorescence microscope having exciter filters between 330 and 500 nm and Zeiss barrier filters which cut off at 470 or 500 nm. Black and white photographs are made using high contrast copy film, which tends to accentuate the filled cell. High speed Ektachrome film (tungsten ASA 125) is used for color slides and prints. Zeiss Luminar macro-lenses are used for whole mount photography. For 35 mm slides in black and white, Kodak direct positive film is used since it produces black neurons against a light background.

Physiology of the Stomatogastric Ganglion

Some of the results from our analysis of the lobster stomatogastric ganglion illustrate the usefulness of stain injection techniques.

In a preliminary account of the physiology of this ganglion, Maynard (1966) described some of the patterned motor output recorded from the isolated preparation in response to various priming stimuli. A more detailed discussion can be found in Dando and Selverston (1972). Basically, the deafferented ganglion is capable of producing two independent classes of rhythmic activity. Each class is responsible for operating the striated musculature of two separate areas of the lobster stomach. The first area is the gastric mill, a dorsally located toothed structure inside the stomach, which macerates food. The second area, the pyloric region, is more caudal and kneads the macerated food from the gastric mill before it is passed to the gut. Of the approximately 30 neurons in the ganglion, 24 are known to be motoneurons. Ten of these innervate the gastric mill and 14 innervate the pyloric region. There are, in addition, two interneurons and four cells of unknown function which may be sensory interneurons.

The ganglion itself is spindle shaped, approximately 1 mm long and 300 μm wide (Fig. 2). It is located within the lumen of the opthalmic artery about half-way between the heart and the supraesophageal ganglion. The nerves which carry the motor axons pass through the arterial wall near the ganglion.

By removing the entire stomach along with the ophthalmic artery from the animal, the ganglion and its motor roots can be kept intact. Each motor axon can be traced to the point where it just contacts an individual stomach muscle. This procedure makes it possible to pin out the ganglion in a Sylgard-lined petri dish, together with all of its identified motor roots and, if required, the stomatogastric nerve which carries most of the input to the ganglion from the central nervous system.

The cell bodies are prepared for intracellular recording by dissecting off the ganglionic sheath and transilluminating. Pin electrodes insulated with petroleum jelly are placed on most of the motor nerves so that the firing patterns of the different motoneurons can be monitored. These electrodes can also be used for antidromic stimulation to identify motoneurons.

Perhaps the most useful feature of the system is that intracellular recordings from the somata reveal much of the subthreshold integrative activity. This is unusual in crustacean cell bodies and makes it possible to determine synaptic pathways by direct intracellular recordings instead of inferring them from the extracellular records.

Fig. 2. The living stomatogastric ganglion as it appears in the dissecting microscope. The ganglion is approximately 1 mm in length and 400 μm at its greatest width. Cell bodies range in size from 20–60 μm in diameter. The photomicrograph is a dorsal view of the desheathed ganglion, using transillumination. The majority of the motoneuron cell bodies are on the dorsal surface.

Unlike many invertebrate ganglia, in which the topography of the somata is relatively constant, cell bodies in the stomatogastric ganglion do not have a predictable position. Cells are identified by correlating spontaneous intracellular spike activity with the extracellular activity of identified axons in the motor roots, or by antidromic spikes following stimulation of identified axons. Although we have described the synaptic pathways for neurons controlling both of the stomatogastric rhythms, we will take as examples some of the units in the pyloric cycle only.

Pyloric Connectivity

The nerves and muscles of the pyloric region are shown in Fig. 3, and the burst pattern of this region is schematized in Fig. 4. These bursts are produced by the units described in Table 1. Some of these bursts are shown in Fig. 5. The phase relationship between PD, AB, LP, and PY cells is relatively constant, while the VD and IC units can show more variability in their phase relationships with the other cells.

Intracellular recording from pairs of pyloric cells has established the following synaptic interactions: (1) The two PD cells are strongly electrotonically coupled to the AB, and these burst together, although their spikes are not synchronous. (2) The PD inhibits the eight PY cells, the LP, the VD and the IC. (3) The LP inhibits the PD-AB group, the PY cells, and the VD. (4) The PY cells inhibit the LP cell only. (5) The VD and the IC cells are reciprocally inhibitory. (6) The PD-AB group is electrotonically connected to the VD cell.

Examples of the physiological data which have been used to elucidate this network are shown in Fig. 6. The anterior burster (AB) and the inferior cardiac (IC) cells (Table 1), while part of the pyloric circuit, are not shown because we do not have good anatomical data from them. Instead, only those cells which will be discussed both anatomically and functionally are shown in the circuit diagram, Fig. 7. It can be seen in Fig. 6A that the two PD cells burst simultaneously. Current injection in one PD neuron causes a change in membrane potential and an increase in burst rate in the other PD and in the AB neuron. Because the coupling coefficient is quite low one might expect a rather broad area of connectivity between the two cells, such as one sees in the crayfish lateral giant septa. An alternative possibility is a small cell-to-cell contact region with an extremely high conductance junction between them. In the second

Table 1. Pyloric System

Neuron	Designation	Number	Efferent nerve	Muscle
Pyloric dilator	PD	2	PDN	cpv 1b
			d-LVN	cpv 2
Anterior burster	AB	1	SGN	?
Lateral pyloric	LP	1	LPN	p2a
			v-LVN	
Pyloric	PY	8	Pyn	p1-13
Ventricular dilator	VD	1	MVN	cv1
Inferior cardiac	IC	1	MVN	cv2

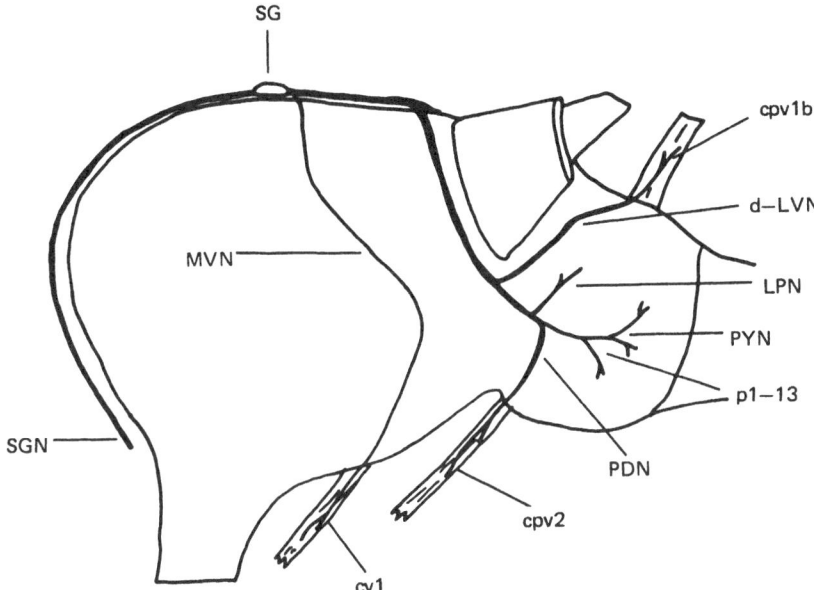

Fig. 3. Diagrammatic side view of the stomach of *Panulirus interruptus* to show arrangement of the nerves and muscles discussed in this paper. SG, stomatogastric ganglion; SGN, stomatogastric nerve; MVN, median ventricular nerve; d- and v-LVN, dorsal and ventral lateral ventricular nerve; PDN, pyloric dilator nerve; LPN, lateral pyloric nerve; PYN, pyloric nerve. Rostral is left, and the pyloric region is the caudal part of the stomach where the LPN, PDN and PYN nerves are shown. Contraction of the cv1, cpv2 and cpv1b muscles tend to dilate the pyloric region, while activity in the pyloric muscles tend to constrict it.

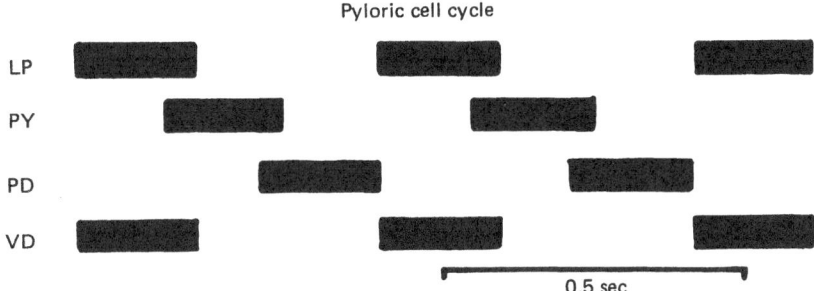

Fig. 4. Block diagram of the contraction sequences for some of the pyloric muscles. The LP and PY units burst between firing of the PD units. Generally the VD unit is out of phase with the PD, but can at times fire with it or be split up so that it fires both before and after the PD unit fires.

trace (Fig. 6B), there is an example of the alternating burst pattern between a PD and a VD cell. The VD burst is seen to be inhibited by the PD burst, and if one looks closely, the IPSPs from the PD's and AB can be seen on the VD intracellular record. The fact that these cells are electrotonically connected is again shown by the passage of DC current pulses between them. Occasionally, the VD burst is changed

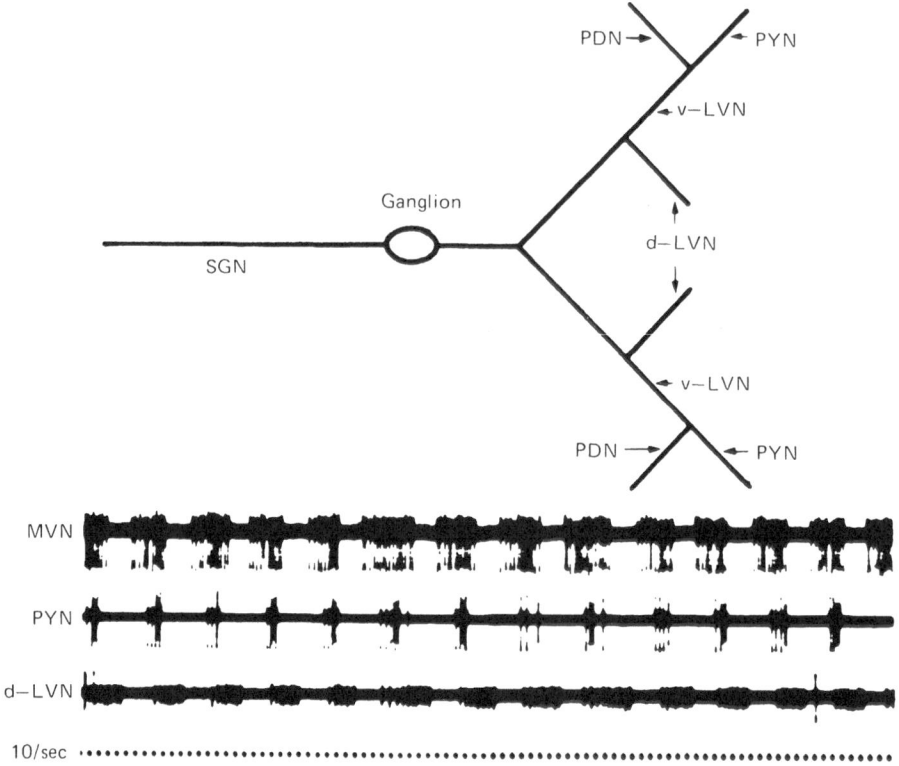

Fig. 5. The upper part of this figure shows how some of the nerve branches would be pinned out. The designations of the nerves are described in the legend of Fig. 3. Extracellular recordings from some of these nerve branches are shown in the lower half of the figure. The position of the MVN nerve relative to the ganglion can be seen in Fig. 2.

(Fig. 8) because its level of excitability rises. When this occurs, the DC shifts in PD and VD parallel each other.

The third trace in Fig. 6C, shows similar alternation, but in the case of PD and PY, the mechanism is due to inhibition of the PY burst by PD activity. This can be demonstrated by direct depolarization of each cell, with inhibitory effects observed only in the PD-PY direction. Finally the PD-LP pattern (Fig. 6D) can be shown to be the effect of reciprocal inhibition between these two cells. It is possible to see the IPSPs both in the LP cell, which result from bursting in the PD-AB cell group, and on the PD cell, which are correlated with firing of the LP cell.

The circuit for these particular cells of the pyloric cycle is illustrated in Fig. 7. The synaptic interactions between these motoneurons is inhibitory (Maynard, 1966), although the cells themselves are excitatory at their neuromuscular junctions (see Maynard and Atwood, 1969). In addition, there are extensive electrotonic connections among many of the cells: among all eight of the pyloric cells, between the two PDs themselves, between the PDs and the AB, and between the PD and VD.

The monosynaptic nature of the connections of the pyloric cells is sometimes difficult

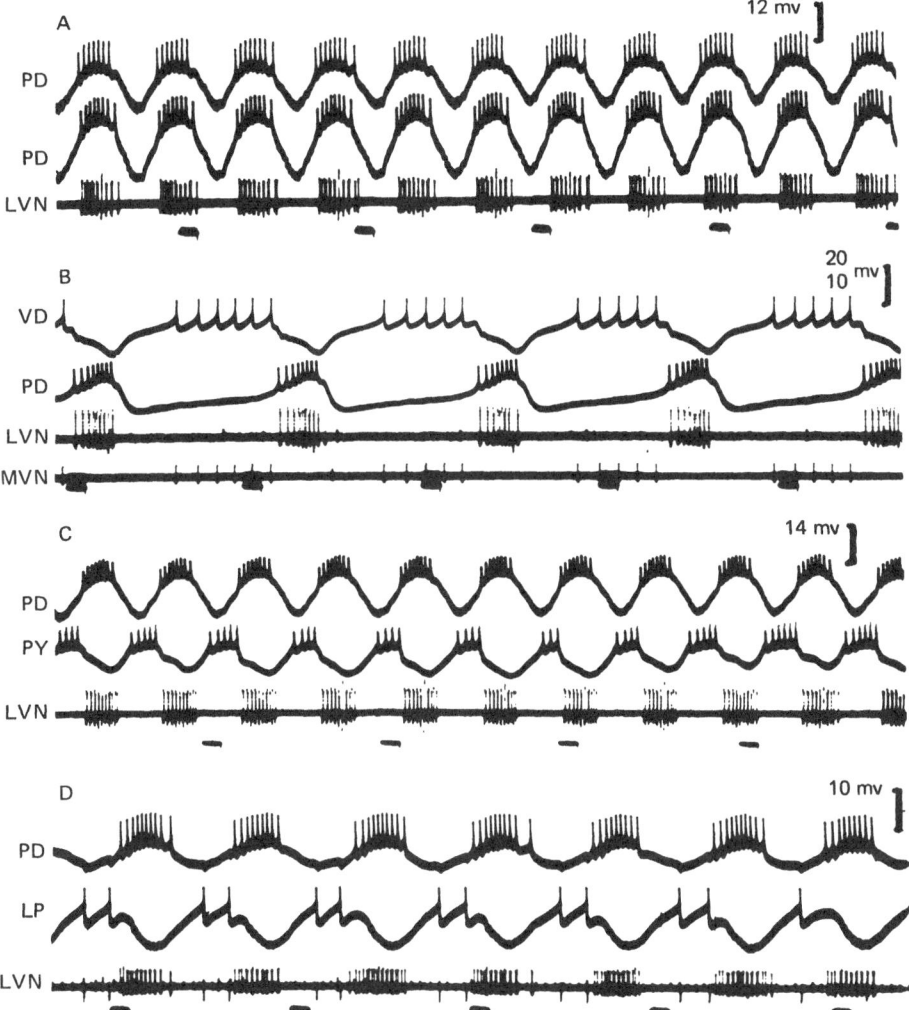

Fig. 6. Intracellular somata recordings from pairs of stomatogastric motoneurons and extracellular recordings from the nerves containing the axons of those motoneurons. Trace A shows the sequential firing of the two PD cells. Note that the spikes occur on the depolarizing phase of oscillating DC potentials and that the spikes are not locked in 1:1. The occasional large spike in the extracellular trace is from the axon of another motoneuron. Trace B shows the alternation between a PD and VD cell. A more detailed description of these two cells is given in Fig. 8. Trace C shows alternation between a PD and PY cell. Only the PD cell is being recorded from extracellularly. Trace D shows the relationship between a PD and LP cell. Axonal spikes from both units can be seen in the LVN recording. Inhibitory postsynaptic potentials can be seen in each cell during activity in the other. Calibration: horizontal, 1 sec.

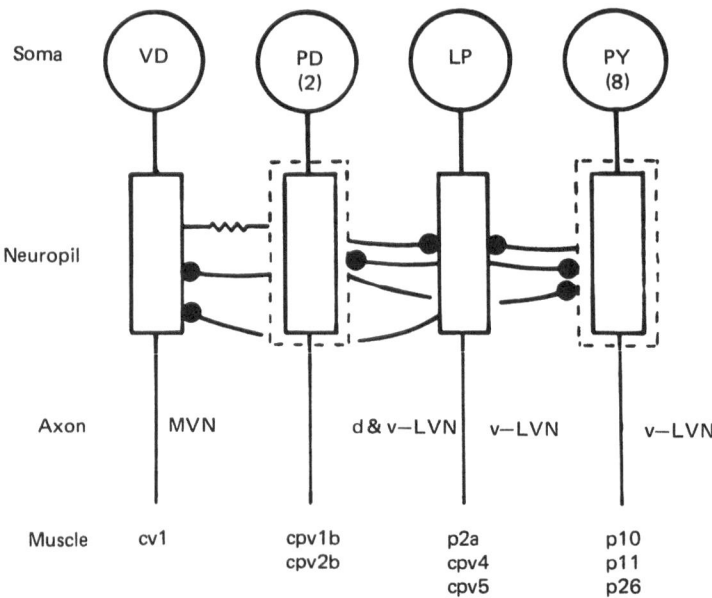

Fig. 7. A partial circuit for the pyloric cycle motoneurons. Only the IC and AB cells have been omitted, the connections shown for the other cells are thought to be complete. The dashed lines around the neuropil region of the PD and PY cells indicate that the two PDs and the eight PY cells are electrotonically coupled. Black dots indicate inhibitory chemical synapses; the sawtooth, an electrotonic connection between the PDs and the VD. Nerves which innervate muscles and from which axonal activity is recorded are also indicated.

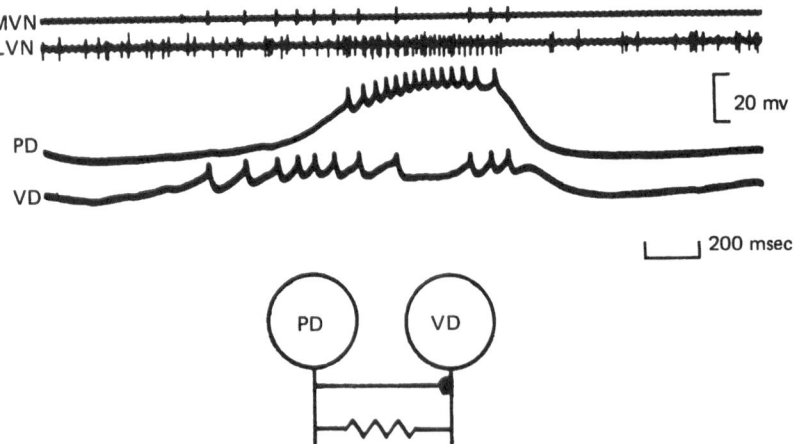

Fig. 8. This recording from a PD and a VD cell shows one common variation in the VD burst, namely a change of the VD sequence by inhibition from the PD. The parallel DC shifts following PD and VD activity are apparent.

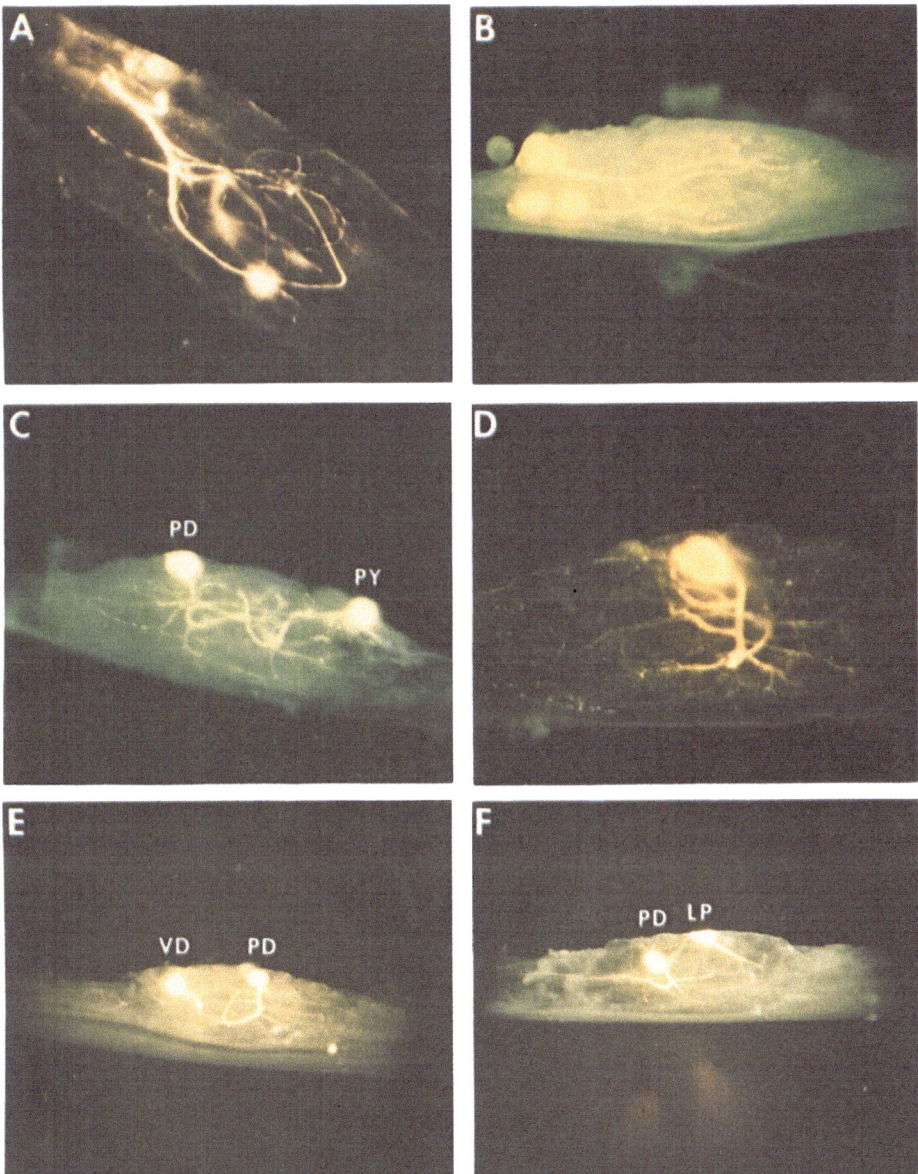

Fig. 9A–F. Fluorescent whole mounts of Procion yellow injected stomatogastric motoneurons. A. A gastric mill cell, one of four functionally identical cells. The soma was lost during the processing. B. Four gastric mill cells are shown in this multiple injection. The large processes are aligned into tracts while the smaller branches are more randomly distributed. C. A PD-PY combination with the PD cell on the left. Graphical reconstruction (Fig. 11) shows that this pair has two points of contact. D. A single PD cell. This is the cell which was used for the computer reconstruction discussed in the text and should be compared with Fig. 18. E. A PD-VD double injection. A reconstructed drawing of this pair is shown in Fig. 13. F. A PD-LP double injection. Note the process of the PD cell to the soma of the LP. The reconstruction of this cell is shown in Fig. 12.

to demonstrate unequivocally because we are not recording from the postsynaptic membrane but from the cell soma, which is some distance away. However, on the basis of cell numbers, there are not enough neurons to provide many interneuronal pathways. Additional evidence for the monosynaptic nature of these connections can be obtained by injecting pairs of cells which one believes are synaptically related.

Structure of Pyloric Cycle Neurons

One of the principle uses of Procion injection techniques is to corroborate structurally the kinds of functional mapping just discussed. In addition to demonstrating graphically points of presumed contact between interacting neurons, the method can also describe the precise geometry of individual cells.

A good example is the PD neuron. The two PD cells, together with the AB neuron, appear to be the pacemakers for the entire pyloric cycle. A whole mount fluorescent photomicrograph of a PD cell is shown in Fig. 9D. The large processes of the cell have a very characteristic shape. The neurite crosses ventrally into the neuropil, turns almost at a right angle to proceed across the ganglion to the other side, then curves slowly in a caudal direction to finally exit the ganglion in the median ventricular nerve. There are many processes which leave this structure, some being quite large themselves (10 μm initial diameters). Most of these large branches divide only once or twice while smaller branches divide extensively. One of the larger branches extends close to the cell body layer and may mediate a particular interneuronal connection. The position of the PD soma varies from ganglion to ganglion. As a result, the structure of its processes undergoes some variation. Fig. 10 shows the outlines of several PD cells

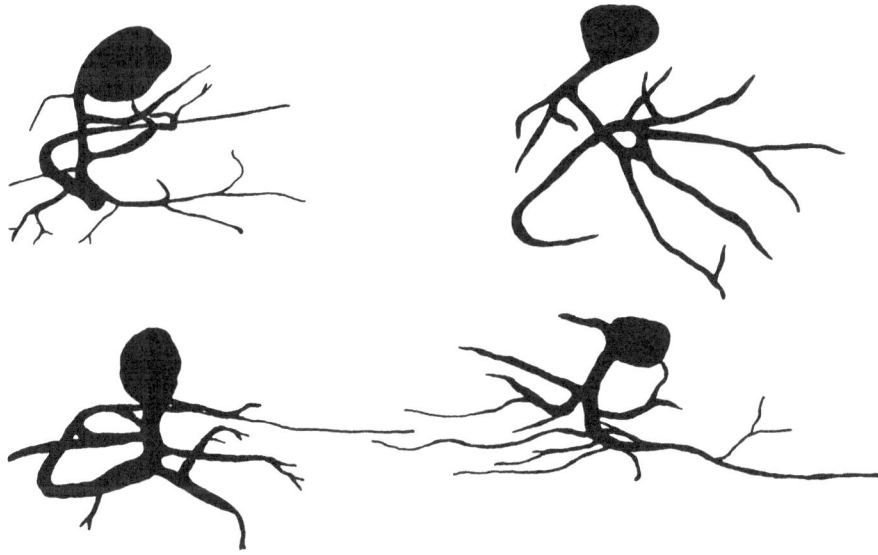

Fig. 10. Drawings made from whole mounts of Procion-filled PD cells to show the variation in neuropilar geometry. A basic structure can be seen for each but there is considerable variability in the secondary and tertiary branches. Positions of the PD cell bodies vary from preparation to preparation.

which were drawn from whole mounts. The structures of the major processes are strikingly similar; with practice this cell can be identified on the basis of its structure alone. The distribution of the secondary processes is quite variable because they have to make specific contacts with neurons whose positions are also variable in location.

Paired Injection

Injection of two or more cells, particularly with different colors of dye, is especially useful in the analysis of cellular interactions. If identifiable neurons are used, one can ask specific questions, such as: (1) Are connections between known cells constant? (2) Do cells contact each other in the same way from preparation to preparation? (3) Is there a difference in structure at the light microscopic level between electrotonic and chemically mediated synaptic connections? (4) Is there a correlation between the size of the synaptic potential recorded at the soma and the distance between the soma and the synapse? (5) Is there more than one point of contact between cells with only one functional synapse? (6) Is there a way of determining which side of a contact is presynaptic and which side is postsynaptic?

The biggest difficulty in interpreting paired injections is knowing if presumed points of contact between cells actually represent synaptic sites. A complete resolution of this question would require an electron microscopic study of the contact points. However, where known synaptic interaction between neurons can be shown electrophysiologically, points of contact between the processes of these cells can be reasonably assumed to be the synaptic sites.

A more difficult problem arises when two neurons have more than one functional interaction. For example, if one refers to the pyloric cycle circuit diagram (Fig. 7), we can define the connections shown in Table 2. Three pairs of cells have two connections and four pairs have one connection between them. The question that arises in the interpretation of paired cell injections with more than one connection is obvious: how do you know which process is presynaptic and which is postsynaptic? Four of the pairs shown have only one connection between them, and we can determine the

Table 2. Synaptic Connections between Pyloric Motoneurons[1]

Cells	Type of connection			Number
PD-PD	PD	~~~	PD	1
PD-VD	PD	~~~	VD	2
	PD	—•	VD	
PD-LP	PD	—•	LP	2
	LP	—•	PD	
PD-PY	PD	—•	PY	1
VD-LP	LP	—•	VD	
LP-PY	PY	—•	LP	1
	LP	—•	PY	
PY-PY	PY	~~~	PY	1

[1] Electrotonic coupling: ~~~
Chemical synapse: —•

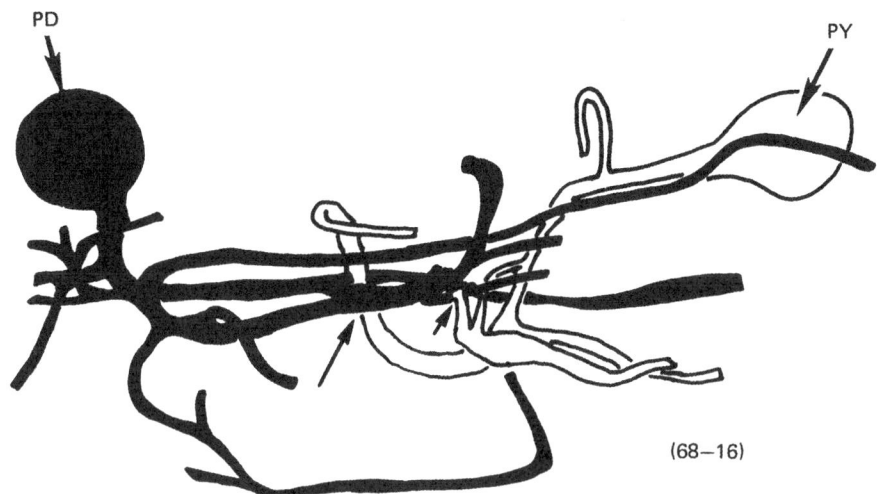

PD   PY

(68—16)

Fig. 11. Drawing of the PD-PY injection shown in Fig. 9D. Arrows indicate the points where the processes of each cell appear to touch in the neuropil.

polarity from physiological data in the case of a chemical synapse. For the three pairs of neurons with more than one connection, the assignment of polarity seems impossible. Before considering some solutions to this ambiguity, it might be useful to look at some of the pyloric cells. Paired injections of a PD-PY, PD-LP, and PD-VD are shown as whole mounts in Fig. 9. One cannot tell by looking at such whole mounts where suspected contacts are unless the ganglion is rolled and photographed from at least two angles, optimally 90° apart.

If the resolution of the finer processes in the whole mount is not satisfactory, the ganglion must be sectioned and photographed; and the injected neuron must be either reconstructed in the form of a plastic model or graphically reconstructed in two views 90° apart, according to the method described by Pantin (1960). Such drawings are shown for the three cells in Fig. 11, 12, and 13. Processes which cross each other when viewed from two angles (arrows) are putative interactive sites. The next question is: how well do these reconstructions correlate with the physiological data?

PD-VD

The VD neuron has not been completely filled, and only one definite point of contact is indicated by the arrow. Since both types of interaction occur, this pair of neurons requires complete filling of both cells and a higher resolution analysis before anything definitive can be said about their structural relationships.

PD-LP

As in the previous example, the PD-LP interaction consists of two connections, both mediated by chemical synapses. Like the VD, the LP did not fill as well as the PD. The graphic reconstruction indicated only one point of contact between the two cells. We can interpret this photomicrograph in at least two ways. First the contact

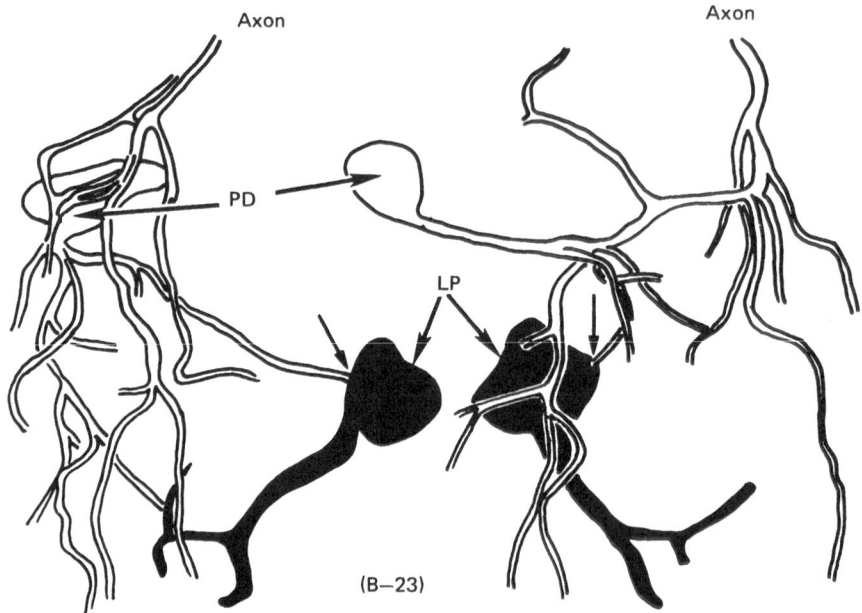

Fig. 12. Drawing of the PD-LP injection shown in Fig. 9F. The drawings were made in two views 90° apart following the method of Pantin (1960). The LP cell is not completely filled but does show one connection indicated by the arrow.

point marked by the arrow may mediate one of the synapses and the site of the reciprocal synapse is missing because of the inadequate LP cell injection. A second, less likely possibility, is that even with complete filling of both cells, still only one point of contact would be visible. In this case we could suggest reciprocal synaptic activity across one anatomical synapse. We have tried several LP-PD injections and have consistently shown only one contact. The one contact always occurs between a secondary PD process and the LP soma. We do not know the polarity of this contact, but the soma membranes of these cells are known to be chemosensitive (Remler and Selverston, unpublished observations): the soma could be the postsynaptic site.

PD-PY

Unlike the previous examples in which there were two functional relationships between the cells, the PD-PY cells have only one inhibitory synapse between them, its direction from PD to PY, so that one would suppose interpretation of a PD-PY injection would be relatively easy. However, it is quite clear from several injections of these pairs that there are two distinct points of contact. While the profiles of the contact points appear to fuse at each site we would suspect that only one of the sites is active, and the other simply a closely apposed unit with no functional contact. At present, the ambiguity remains.

It is apparent that the double injection technique, even when applied to a well defined circuit, does not always clearly establish the anatomical relationships. Showing

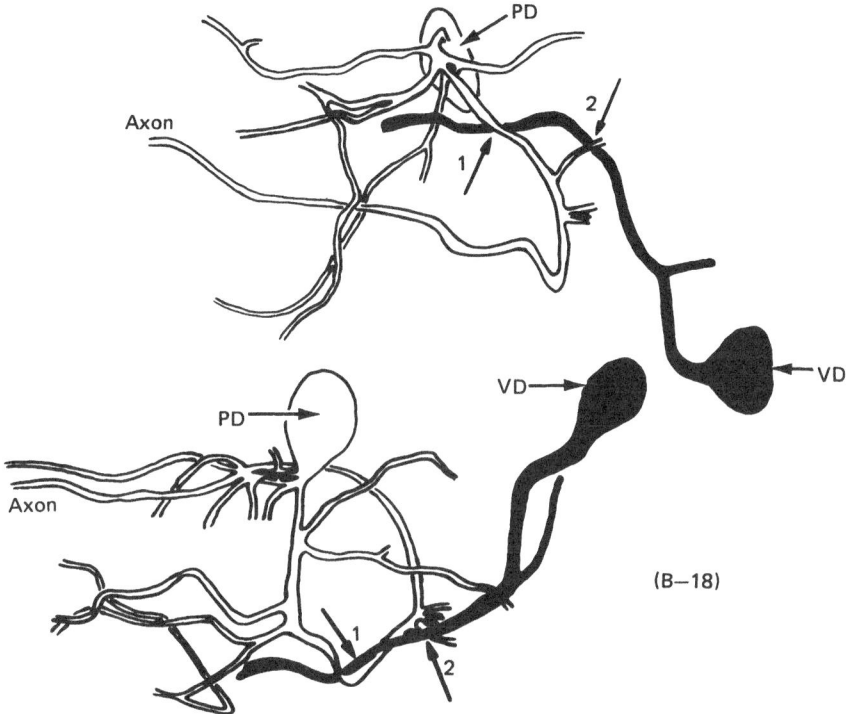

Fig. 13. PD-VD drawing in two views made from the paired injection shown in Fig. 9E. Arrow indicates the suspected point of contact. See text for details.

that two cells make contact is good corroborative evidence; however showing that two cells are definitely not in contact, when one is trying to establish a circuit using physiological analysis, can positively determine that no contact exists.

Sterology

The structure of the individual cells is worth knowing about several reasons: (1) to know how constant in number and positions the branches are; (2) to see if shapes can be modified by use, disuse, or learning; and (3) to see if genetically controlled behavior differences can be explained by structural modifications.

Three-dimensional reconstruction of filled neurons can be realized by many of the available stereological methods. Graphic representations made from sectioned ganglia have already been mentioned. Because the intensity of the Procion fluorescence is so low, it is far easier to generate such drawing from photographs of each section, taken on 35 mm slides, than from the sections themselves. We have found that, at least for such purposes as determining where contacts between cells are located, the graphic representations are quite reliable.

Drawing can also be made by trained medical illustrators if they are given plastic models of reconstructed cells. Such plastic models can be made by projecting each slide onto a clear plastic sheet and tracing out the fluorescent profiles with ink. The sections can then be stacked one on another, and separated to scale with spacers. Alignment of sections is best accomplished by visually aligning the profiles so as to achieve maximum overlap. Incorporating a fiducial mark into each section by embedding threads in parallel with the ganglion has thus far not been successful. Another method of visualizing cells employs whole mounts and stereophotomicrographs, as described in Chapters 6 and 7. It is clear, however, that for handling large numbers of cells, computer graphic analysis is the method of choice.

Computer Analysis

We have explored the use of computer graphics techniques to display neuronal structure with the following objectives: (1) to display the three-dimensional form of a cell by processing the two-dimensional sections or stereo pairs based on whole mount pairs; (2) to give three-dimensional coordinates to the processes of each cell with respect to the the ganglionic boundaries, i.e., to identify the parts of a neuron within the neuropilar space; (3) through rotating the display of a single cell or pair of cells around different axes, to gain insight into the complex geometry of cellular processes and neuropilar structure; (4) to measure anatomical parameters of identifiable neurons such as volume, surface area, number of branch points, distance between synapses and recording sites, distances between branch points; (5) to determine, from repeated measurements of homologues in many ganglia, how consistent the morphology of a given cell is; (6) to correlate structural connectivity with electrophysiologically obtained functional connectivity.

The methodology involved in the computer processing of dye-injected cells can be divided into four categories: (1) Digitizing the structural information contained in the injected neuron; (2) Processing the digitized information; (3) Display of digitized neurons; (4) Measurements of the digitized neuron.

Digitizing

The only way in which a digital computer can deal with graphic information is for the "picture" to be translated into a list of numbers. If we cover a photograph with a hypothetical grid, then we can integrate the gray level of the photograph within each grid square and assign to it a number of "grayness" representing its optical density. The only limiting factor for the number of gray levels and the number of squares in the grid is the storage capacity of our computer. As both the number of grid squares that cover the photograph and the number of possible gray levels increase, the more the digitized picture will resemble the original photograph. The number of bits of information required, however, to adequately resolve a complicated photograph is very large. We are fortunate in our application because there are only two gray levels in our Procion-injected tissue: either the Procion is present or it is not. In addition, it is not necessary to digitize the entire area of the injected process but only the boundary line between the injected process and the rest of the ganglion. Furthermore, the boundary points that we wish to digitize need not be one grid square apart but only close enough to approximate adequately the actual contour. The problem of digitizing these points can be solved in several ways. Because of the low level of fluorescence from Procion, it is not yet possible to read the information directly from the microscope slide (although this may be possible with the cobalt injection technique). We therefore use 35 mm color transparency photographs of the sections. The following are among currently available techniques for digitizing pictures that we have tried:

Rotating Drum Digitizer

This device is in effect a scanning densitometer interfaced with a computer. The fluorescent profiles of cell processes are projected onto a sheet of film and then inked in by hand. The film is fastened to the cylindrical drum of the digitizer. As the drum rotates a small spot of light and a photomultiplier tube mounted on a carriage are moved slowly down the center of the drum. Whenever the light is interrupted by a dark profile on the film, the point on the film is converted to an (x, y) coordinate by an analog to digital converter (ADC) and stored in the computer. At the same time the photomultiplier output is sampled on a grey scale of 256 levels and fed into the computer and stored. A complete description of this device can be found in Xuong (1969).

Fixed Cursor

This extremely accurate device consists of a mechanically movable lens which is part of a 35 mm film projection system. Small areas of each frame are projected onto a large screen imprinted with a fixed cursor so that movements of the lens bring areas of the film over the cursor. The lens also moves a set of moire gratings that by counting cycles of fringes (x, y) coordinates for spots on the film can be stored in the computer.

Computer Driven Cursors

Any computer driven oscilloscope can be used to digitize optical information by use of a "mouse": a hand driven device which can be pushed around on a table in front of the oscilloscope screen. Potentiometers in the "mouse" are mounted at right angles, to produce the voltages proportional to the (x, y) coordinates, which are transmitted to the computer via an ADC. A program can be written which moves a cursor around on the screen as the "mouse" moves. The biological section can be projected on the face of the oscilloscope and the cursor moved around to digitize the profiles.

Data Tablet

In our laboratory, a data tablet (Science Accessories Corp., Southport, Conn.) is currently used for digitizing sections. The tablet is interfaced to a computer which can then drive a graphics display system. Slides are projected from the rear onto the tablet and the stylus generates a brief sonic pulse which is picked up by strip sensors (microphones) on two sides of the tablet. A control unit interprets the information from the sensors as (x, y) coordinates by comparing their time of arrival with the time of generation of the pulse. The coordinates are then transmitted to the computer. The contours traced out are simultaneously displayed on the graphic terminal described below.

Display

For display we use an interactive graphics terminal designed specifically for mini-computers (Vector General, Canoga Park, Calif. U.S.A.). This device is a high-speed computer peripheral using one data channel, one programmed I/O channel and one interrupt channel of the host computer. The complete system shown in Fig. 14 has hardware which allows scaling translation and rotation in three dimensions.

The central processing unit (CPU) that we use a Digital Equipment Corporation (DEC) PDP 11/45 with 24K of core memory, about the minimum amount required for processing one stomatogastric neuron. Other DEC CPUs which can successfully drive the Vector General terminal include the PDP 11/20 and the PDP 11/40. The PDP 11 series processors all have the advantage of speed, program flexibility, and ease of interfacing with the graphics terminal and other peripherals. Since a large amount of data has to be stored, our system also includes a disk.

Computer Aided Reconstruction

The digitization procedures just described are carried out on two-dimensional sections. The role of the computer is to make a three-dimensional reconstruction of the neuron by "stacking up" and aligning the individual sections. This task is complicated by

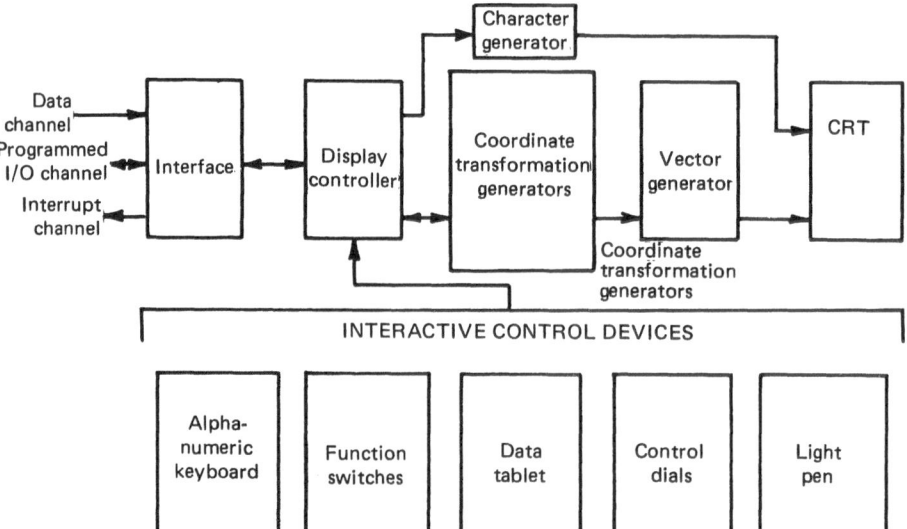

Fig. 14. Vector General interactive graphics terminal system arrangement. The coordinate transformation generators are implemented with hardware. See text for details.

the lack of fiducial marks, i.e., markers having a constant spatial relationship to the whole ganglion for use in alignment. We have used two methods to align sections. In one method, a hardcopy readout of each numbered section is made on tracing paper with a Calcomp plotter (Fig. 15). The paper readouts are placed on a light box two at a time and aligned by eye until all the processes match up. Coordinates for x, y, and Q (rotation) are then punched into the computer; both sections and pairs are displayed. Coordinate values are varied until the displayed sections match the configuration of the paper readouts. The first section and its coordinates are then stored on the disk, and the whole process is repeated, matching a third section with the second

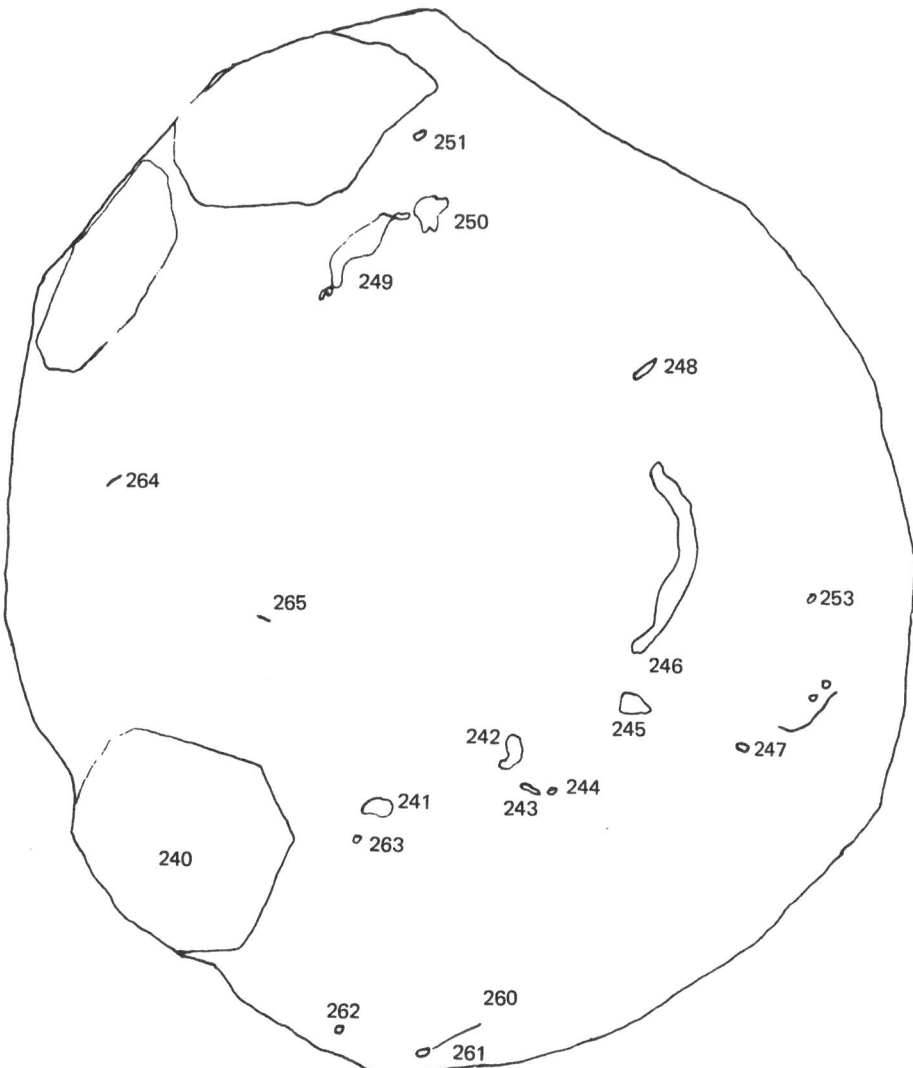

Fig. 15. A Calcomp hard copy readout of one digitized section, in which the profiles have been numbered.

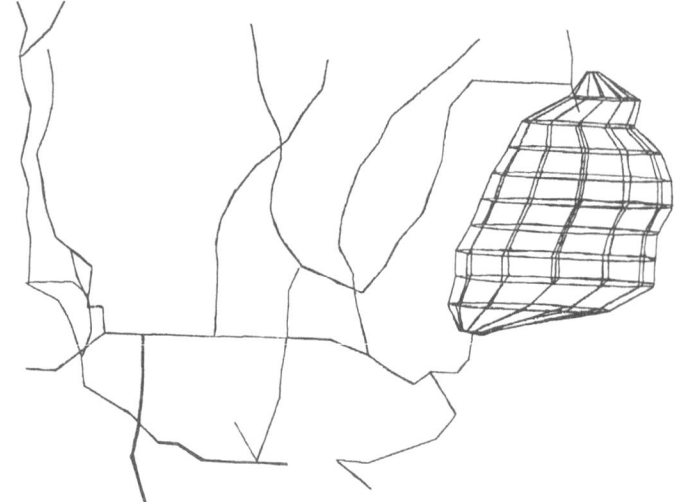

Fig. 16.

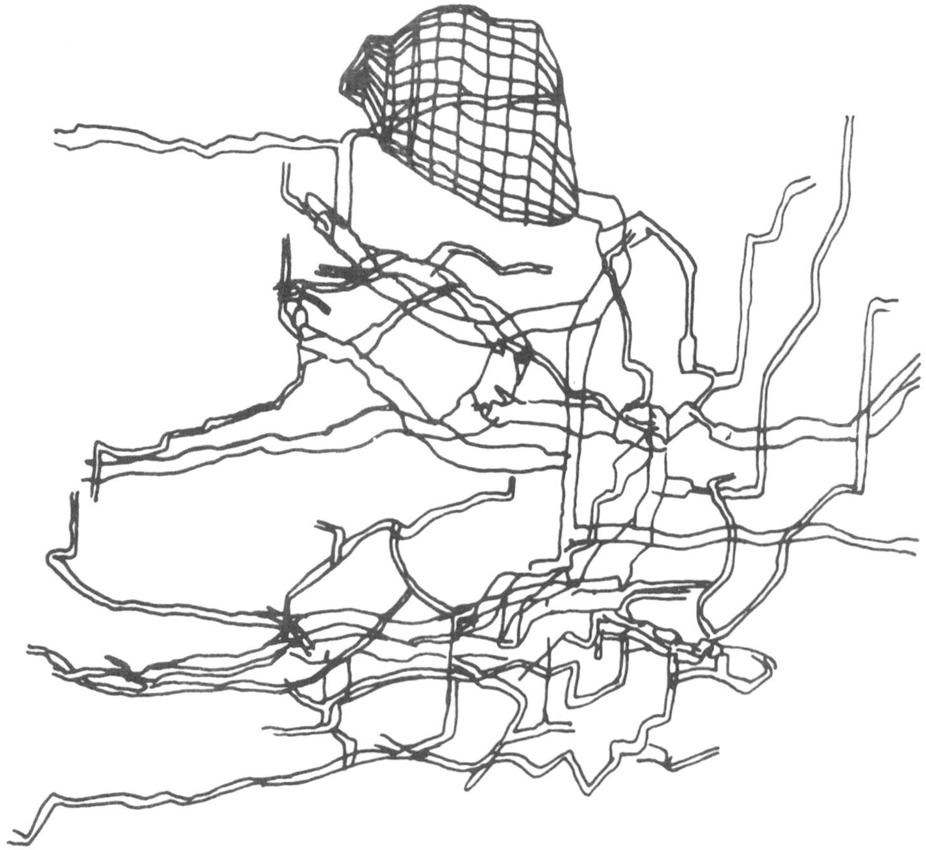

Fig. 17.

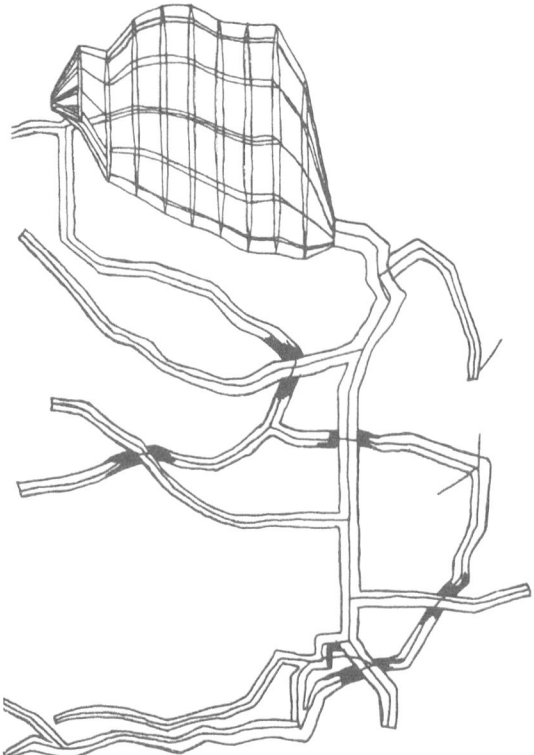

Fig. 18. Drawings of central portion of PD neuron in which the processes have been shaded and points of crossover accentuated by hand.

oscilloscope face. Utilizing a program to connect the points or the radii at each point, and so on until all the sections have been aligned. Hard copies are again made for each section. The profiles are given numbers and the radius of each profile measured as the width of the rectangle of minimum area which fits the inside of each. A "mouse" (see above) is used to obtain the radius. The profiles are connected by manually listing the numbers from one section to another. This may require sensitive operator judgment for some of the smaller profiles since these are difficult to separate from random fluorescent spots. The sets of numbers obtained are stored in sequential lists giving the x, y, and Q coordinates for center point, radius, connectivity, and branching information for each profile. If these points are then connected by straight vectors, one can display stick figure models (Fig. 16) which are useful for analyses such as constancy of branch point position. The filled-out neuron structure is obtained by connecting the endpoints

---

Fig. 16. Stick model as displayed on the oscilloscope by connecting the center of the profiles in each section. Only the center of the neuron is shown. The same neuron in whole mount is shown in Fig. 9D.

Fig. 17. CRT display of the PD neuron after lines are drawn connecting the perpendicular radii at each profile. See text for details.

of radial lines perpendicular to the stick figure (Figs. 17 and 18). These figures can be rotated and are much easier to visualize in stereo pairs of about 5° rotation or in a movie showing a continuously rotating cell. The rotations may be done either using software or the appropriate hardware for the graphic terminal.

The procedure used for reconstructing cells when an interactive terminal is available is somewhat different and a great deal faster. The sections can be aligned quickly by means of knobs which translate and rotate one section with respect to another on the a three-dimensional model can then be displayed almost immediately. In addition, the stick figure can be rotated with the graphics terminal hardware without software computations.

Cell Measurements

The quantification necessary for the three-dimensional reconstruction facilitates other measurements which are extremely useful in determining constancy of position and structure. It also provides information for the analysis of intraneuronal flow of currents with mathematical models. A list of parameters which may be obtained from the stored data include: (1) surface area, (2) volume, (3) number of branch points, (4) number of terminal branches, (5) position of the cell within the ganglion, (6) distances between branch points, (7) distances between suspected synapses and soma, (8) power relations between parent and daughter branch diameters (9) equivalent shapes of branch patterns (in terms of maintaining electrical constancy).

For example, one question that we wished to answer about the cell was whether or not the branch points conformed to the $3/2$ power law used by Rall (1962) to mathematically model that electrotonic properties of motoneurons. Rall has shown that the input impedance of the dendritic tree can be approximated with an equivalent cylinder if $(\Sigma d_i^{3/2})$ is a constant for each level of branching, where d_i is the diameter of a branch at a given level. Lux et al. (1970) have reconstructed spinal motoneurons by first injecting them with a tritiated glycine, allowing the glycine to diffuse, and finally making serial autoradiographs. Starting at the soma the $\Sigma d_i^{3/2}$ relationship decreased in the first 20–50 μm but remained constant thereafter. Barrett and Crill (1971) have filled cat motoneurons with Procion yellow and found that their anatomical measurement indicate a rapid decrease of the $\Sigma d_i^{3/2}$ relationship over the first 100 μm and a continuous reduction over the remainder of the dendritic tree. Both techniques showed clearly that the decreased values were not due to branch point tapering, but to tapering between branch points.

In order to see how close the stomatogastric neurons came to obeying the $3/2$ rule, we wrote a program to measure the branch point diameters of a PD cell. The $\Sigma d_i^{3/2}/D^{3/2}$ at each branch point is 1 if the relationship exists. The mean ratio was found to be 1.67 indicating a substantial deviation. We did not plot the ratios at increasing distances from the soma as did Lux et al. (1970) and also Barrett and Crill (1971) so we cannot say if the deviation has any relationship to distance.

Acknowledgments. I am particularly grateful to Dr. Brian Mulloney who has collaborated with me on many phases of this work, to David King for the drawings of the paired neurons, and to Sheryl Glasser without whom the computer processing could not have been accomplished. This work was aided by a National Institutes of Health grant no. K5 R01 NS-09322 and by an Alfred P. Sloan award in Neurobiology to University of California, San Diego.

References

Barrett, J. N., and W. E. Crill: Specific membrane resistivity of dye-injected cat motoneurons. Brain Res. *28*, 556–561 (1971).

Dando, M. R., and A. I. Selverston: Command fibres from the supraoesophageal ganglion to the stomatogastric ganglion in *Panulirus argus*. J. comp. Physiol. *78*, 138–175 (1972).

Jones, C., J. Nolte, and J. E. Brown: The anatomy of the median ocellus of *Limulus*. Z. Zellforsch. mikrosk. Anat. *118*, 297–309 (1971).

Kennedy, D., A. I. Selverston, and M. P. Remler: Analysis of restricted neural networks. Science *164*, 1488–1496 (1969).

Larimer, J. L., A. C. Eggleston, L. M. Masukawa, and D. Kennedy: The different connections and motor outputs of lateral and medial giant fibres in the crayfish. J. exp. Biol. *54*, 391–402 (1971).

Lux, H. D., G. W. Kreutzberg, and A. Globus: Direct matching of morphological and electrophysiological data in cat spinal motoneurons. In: Excitatory Synaptic Mechanisms. Ed. P. Anderson and J. K. S. Jansen. pp. 189–198. Oslo: Universitetsforlaget, 1970.

Maynard, D. M.: Integration in crustacean ganglia. In: Nervous and Hormonal Mechanisms of Integration. Symposia of the Society for Experimental Biology No. 20. New York: Academic Press, 1966.

———, and H. A. Atwood: Divergent post-synaptic effects produced by single motor neurons of the lobster stomatogastric ganglion. Am. Zool. *9*, 248 (1969).

Otsuka, R., E. A. Kravitz, and D. D. Potter: Physiological and chemical architecture of a lobster ganglion with particular reference to gamma-aminobutyrate and glutamate. J. Neurophysiol. *30*, 725–752 (1967).

Pantin, C. F. A.: Notes on Microscopical Technique for Zoologists. Cambridge: University Press, 1960.

Paul, D. H.: Decremental conduction over "giant" afferent processes in an arthropod. Science *176*, 680–682 (1972).

Rall, W.: Theory of physiological properties of dendrites. Ann. N.Y. Acad. Sci. *96*, 1071–1092 (1962).

Remler, M. P., A. I. Selverston, and D. Kennedy: Lateral giant fibers of crayfish: location of somata by dye injection. Science *162*, 281–283 (1968).

Sandeman, D. C.: Integrative properties of a reflex motoneuron in the brain of the crab *Carcinus maenas*. Z. vergl. Physiol. *64*, 450–464 (1969).

———: The excitation and electrical coupling of four identified motoneurons in the brain of the Australian mud crab, *Scylla serrata*. Z. vergl. Physiol. *72*, 111–130 (1971).

Selverston, A. I., and D. Kennedy: Structure and function of identified nerve cells in the crayfish. Endeavour *28*, 107–113 (1969).

———, and M. P. Remler: Neural geometry and activation of crayfish fast flexor motor neurons. J. Neurophysiol. *35*, 797–814 (1972).

Shaw, S. R.: Decremental conduction of the visual signal in barnacle lateral eye. J. Physiol., Lond. *220*, 143–175 (1972).

Stretton, A. O. W., and E. A. Kravitz: Neuronal geometry: determination with a technique of intracellular dye injection. Science *162*, 132–134 (1968).

Willows, A. O. D.: Behavioral acts elicited by stimulation of single identifiable nerve cells. In: Physiological and Biochemical Aspects of Nervous Integration. Ed. F. D. Carlson. Englewood Cliffs, N.J.: Prentice-Hall, 1968.

Xuong, N.: An automatic scanning densitometer and its application to x-ray crystallography. J. Sci. Inst. *2*, 485–487 (1969).

Zucker, R. S., D. Kennedy, and A. I. Selverston: Neuronal circuit mediating escape responses
    in crayfish. Science *173*, 645–650 (1971).

Discussion

DR. DUBIN (Colorado): Regarding three dimensional reconstruction by computer, what
do you see as the state of the art for the rest of us?

Clearly, there are only a small number of places doing this, at anywhere from $200,000
to $2,000,000. This cost will make it impossible for many people to start their own computer
analysis project.

However, there are a large number of people studying cells that they would like
to analyze in this way. How will these people be able to use the results of your work?

DR. SELVERSTON: (Dr. Selverston commented on a computer-microscope system that
has been developed in St. Louis for the analysis of Golgi-impregnated neurons. The following
remarks describing that system were solicited from Drs. Woolsey, Wann and Cowan and
Mr. Dierker of Washington University, St. Louis, Missouri, and are abstracted from a
fuller report which has been published elsewhere [Wann, Woolsey, Dierker and Cowan,
1973]).

Glaser and Van der Loos (1965) were the first to show that a small computer could
greatly facilitate the extraction of quantitative data from specimens containing neurons
which have been selectively impregnated by the Golgi method. In our system we have
interfaced a small digital computer (Digital Equipment Corp. PDP-12) to a standard
research microscope equipped with a commercially available scanning stage (Zeiss). The
scanning stage contains stepping motors which can drive the stage in 0.5 μm steps in the
x and y directions and we have attached a stepping motor to the fine focus of the microscope
to drive the stage in 0.5 μm steps in the z direction. The computer senses changes in
the position of a joystick (x and y) and a toggle switch (z) which are under the control
of the observer. In response to changes in the positions of these controls the computer
sends pulses to the stepping motors causing the stage to move. Since the computer can
count the number of steps taken in any direction, it can keep track of the relative x,
y and z position of the specimen. Using the joystick and toggle switch the observer can
accurately position and focus any object of interest by centering it in the field of view
with the aid of crosshairs in the ocular of the microscope.

As the observer tracks an impregnated neuron he can signal mutually exclusive topological identifiers to the computer by a set of special controls. Following each signal, the
identifier and its x, y and z coordinates are stored by the computer. The identifiers are:
1) cell body origin (the center of the coordinate system); 2) dendrite (or axon) origin;
3) sample points (taken at various intervals along a process); 4) branch-points; and
5) end-points. The observer· systematically tracks the processes of a cell, recording these
identifiers sequentially. Since the coordinates of a branch-point are stored by the computer,
when the end of one of the branch processes is reached and signaled, the computer can
automatically reposition the microscope stage (in x, y and z) so that the observer can
track the other branch. During tracking, the computer program monitors the number of
branch and end-points entered by the observer, and when the appropriate number of
end-points has been signaled to it, the computer repositions the stage so that the origin
of the next dendrite is centered under the crosshairs and focused. This sequence of events
is repeated until the whole structure has been tracked. The data from a cell thus consist
of a series of topological identifiers and their coordinate triplets.

Another section of the computer program can utilize these data points for a variety
of computations which can be printed out on an associated high speed printer. For example,
the actual three-dimensional length of each dendritic segment is calculated in microns
and the segments are arranged according to their "order" in the dendritic tree (primary,
secondary, etc.). Not only does the system permit the recording and the numerical analysis

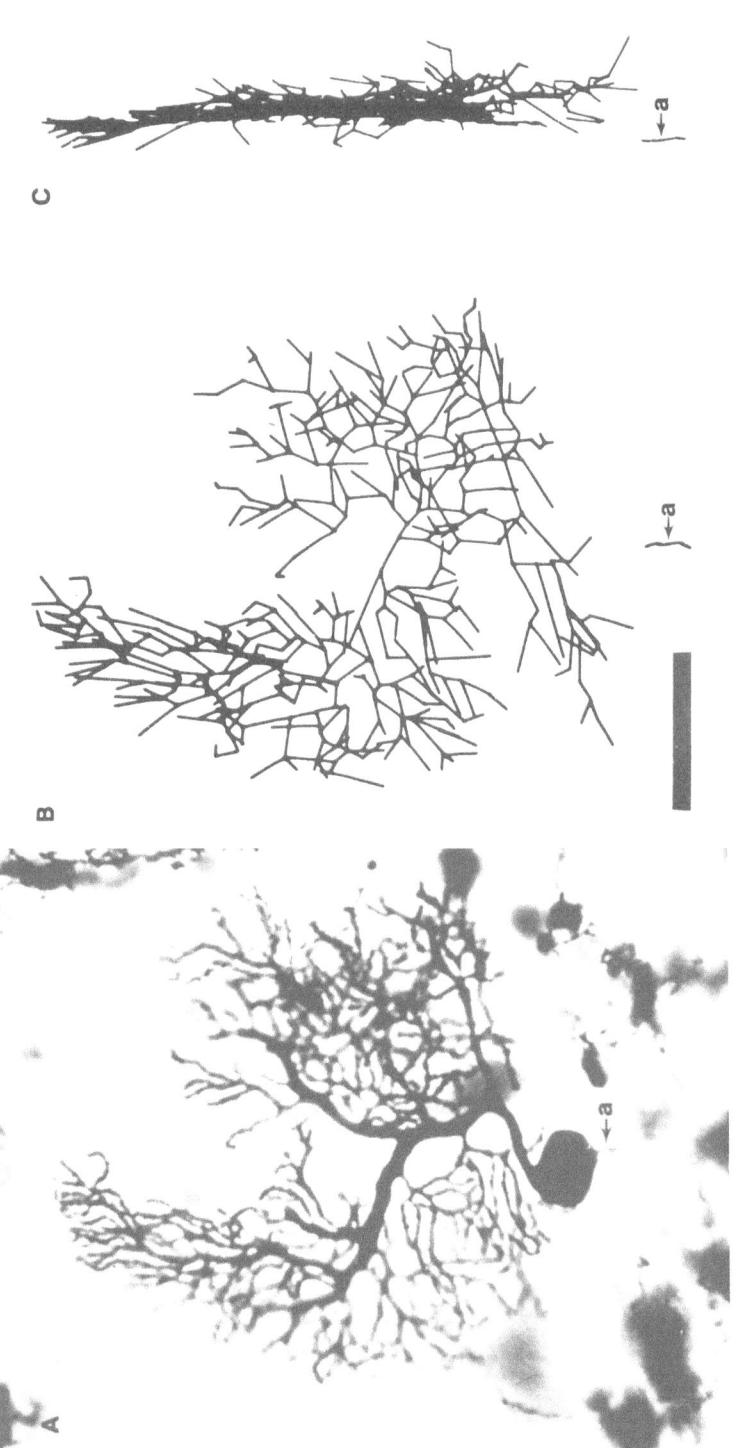

Fig. 19A–C. Illustrations show the display features of the computer-microscope system. A. Photomicrograph of a Purkinje cell from a parasagittal section of the cerebellum of a mouse. Golgi-Cox method, 150µ thick. The number of data points recorded from this cell was 553 and the time required to enter them was about 2 hours. The total three-dimensional length of the dendritic tree is 4,335µ, the number of dendritic segments is 414 and the highest order of bifurcation is the 23rd. B. Photograph of the computer oscilloscope display generated from the data logged from the Purkinje cell shown in A. The cell body is not depicted. Observe the excellent agreement between the appearance of this reconstruction and the actual biological specimen. C. Photograph of the computer generated oscilloscope display from the Purkinje cell shown in A. A coordinate transformation of 90° about the vertical axis has been applied to the data displayed in B. Note the planar distribution of the Purkinje cell dendrite which is viewed as if sectioned parallel to the long axis of a cerebellar folium. All figures are reproduced at the same magnification. Bar = 50µ; a = axon.

of a structure, but it can construct stick figure graphic displays of the cells which can be viewed on an oscilloscope screen. The three-dimensional cell structure is defined by the data, and by applying three-dimensional coordinate transformations it is possible to generate and display a variety of projected views of the structure. For instance, a cell from a specimen cut in the parasagittal plane can be viewed as if it had been sectioned in a horizontal plane (see Fig. 19, B and C). We also have access to facilities at the Computer Systems Laboratory of Washington University which permit these coordinate transformations and displays to be generated in real time and the investigator can visualize important spatial relationships as the structure is rotated dynamically. In addition to these specific examples, a variety of other computation and display features have been incorporated into the system (Wann, Woolsey, Dierker and Cowan, 1973).

A number of tests have been performed to determine the system's accuracy and repeatability. When computer measurements of the total length of a structure were compared to manual measurements, a difference on the order of 1.0% was obtained. When the same cell was carefully tracked twice, the difference in the totals of the two measurements was about 0.1%. The accuracy of the system can be appreciated visually by comparing the stick figure reconstruction displayed on the oscilloscope screen to the appearance of the biological specimen.

Two features of the engineering of this system are particularly advantageous for the biologist. First, the disposition of the various controls necessary to track a specimen is conveniently arranged so that the operator can devote his entire attention to viewing the specimen through the microscope. Second, all of the programs have been designed to be used by the morphologist with a minimum of computer experience and have a high degree of flexibility for each interaction with the various programs, using a simple question and answer format.

With the computer-microscope the time required to log approximately 250 data points, which are needed to adequately sample a simple neuron (cortical stellate cell), is about 40 minutes. This represents less than 5% of the total time required to gather the same data manually. Our facilities are considerably more elaborate than necessary to perform the operations described above. However, a system (computer, microscope and interface) capable of doing the tasks described (with the exception of the on-line dynamic rotation of complex structures) could be assembled for an estimated 1972 cost of $25,000. Although we have applied our computer-microscope system only to individual neurons impregnated by the Golgi method, similar analyses could be performed on single neurons which have been visualized by the intracellular injection of dyes.

Glaser, E. M., Van der Loos, H.: A semi-automatic computer-microscope for the analysis of neuronal morphology. IEEE Trans. BME *12*, 22–31 (1965).

Wann, D. F., Woolsey, T. A., Dierker, M. L., Cowan, W. M.: An on-line digital computer system for the semi-automatic analysis of Golgi-impregnated neurons. IEEE Trans. BME, *20*, 233–248, (1973).

Determination of Neuronal Membrane Properties Using Intracellular Staining Techniques

John N. Barrett

As previous chapters in this book demonstrate, recently developed Procion dye injection techniques make it possible to study both the electrophysiological and the morphological properties of single neurons. This chapter describes how the electrophysiological and anatomical information derived from the use of dye-filled pipettes can be used to calculate passive membrane properties of nerve cells. Knowledge of the membrane parameters enables prediction of the size and relative effectiveness of synaptic potentials originating on the dendritic tree (Rall, 1962, 1967, 1970). In order to obtain accurate data, particular attention must be paid to micropipette recording techniques, and these will be discussed in the first part of this chapter. The experiments reported here were performed on cat motoneurons, but the techniques should be applicable to many other cells.

Intracellular Measurement of Neuronal Input Resistance

Input resistance is measured by recording the steady-state voltage response of the neuron after a small step of current is applied through an intracellular electrode. Ideally, the voltage-recording and current-passing electrodes should be completely separate, with no resistive or capacitive coupling. It is very difficult, however, to record healthy responses from motoneurons impaled by two separate electrodes; therefore, most studies of input resistance employ single- or double-barreled micropipettes.

Single-barreled Electrodes

The easiest method of measuring input resistance is to use a single barrel both to record voltage and to pass current. This technique minimizes damage to the neuronal membrane. However, when voltage is recorded from the same electrode that applies current, a bridge circuit must be used to subtract the current-induced voltage drop across the electrode tip. The bridge can be balanced before penetrating the neuron or, after penetration, by assuming that the very fast components of the voltage response to a current step originate across the electrode tip, and that only the slower components originate across the neuronal membrane (Nelson & Frank, 1967).

Unfortunately, the high and often variable resistance of dye-filled pipettes makes it difficult to balance the bridge accurately. Errors in bridge balance can be reduced by (1) lowering the resistance of the electrode, by mixing the dye solution in 0.15 M KCl and by beveling the electrode tips (see section on *Electrode Beveling Techniques*) and (2) by checking bridge balance when the resistance of the neuronal membrane is shunted by the conductance changes accompanying an intense synaptic barrage (e.g., stimulation of the dorsal roots at 1000 Hz for 50 msec). The bridge balance achieved with such stimulation usually agrees well with the balance obtained by separating the fast and slow components of the voltage response, as described above, but both methods could overestimate the balance slightly and so underestimate neuronal input resistance.

The damage to motoneurons caused by the enlarged tip openings of the beveled pipettes does not appear to be greater than that produced by standard, nonbeveled tips. Resting potentials, action potentials, and time constants of motoneurons impaled with beveled, dye-filled pipettes were similar to those recorded with standard fine pipettes (Barrett & Graubard, 1970). Beveling seems to facilitate cell penetration, and the lower resistance of the beveled pipettes permits more accurate recording of transient voltage responses. When beveled electrodes are filled with standard 3 M KCl or K-acetate solutions, however, diffusion of these extremely hypertonic solutions from the enlarged tip may damage neurons.

Double-barreled Concentric Electrodes

When a motoneuron is impaled with both barrels of a concentric electrode, and the inner electrode tip then advanced 5 to 10 μm ahead of the outer barrel, the resistive (or DC) coupling between the barrels can be reduced well below the neuronal input resistance (Tomita, 1969). Such decoupling allows very accurate measurement of the steady-state input resistance of neurons. Concentric electrodes offer no advantage in measuring transient voltage responses because of the large capacitative coupling between the inner and outer barrels. The distributed nature of this coupling appears to make it impossible to balance the capacitative artifact completely with a cross-compensation circuit.

Double-barreled Electrodes drawn from θ-tubing

Intracellular recording from motoneurons is relatively easy with double-barreled electrodes made from θ-shaped tubing, but the neighboring barrels share a common resistive pathway to ground (Coombs et al., 1955; Nelson and Frank, 1967). Grinding the tip of the θ-tubing electrode at an angle perpendicular to the center partition reduces this shared resistance, and also reduces the resistance of each barrel. The remaining 0.1–0.5 MΩ coupling resistance must be balanced by a bridge circuit. Capacitative coupling between the barrels is less than in concentric electrodes, but the capacitative transient still distorts the first 0.1–0.2 msec of the voltage response to a current step, and thus makes accurate bridge balance difficult. The capacitative coupling between both concentric and θ-tubing barrels is somewhat reduced when only the first millimeter of the electrode tips is filled with electrolyte.

Electrophoretic Dye Injection

Dye injection into mammalian neurons is difficult because the cells cannot be observed during the staining procedure. The next sections therefore describe (1) an apparatus for testing the dye-injecting capabilities of various electrodes, (2) the dye-injection techniques that have proved most successful in our experience, and (3) the effects of dye injection on the electrophysiology and morphology of the neuron.

Testing the Dye-injecting Properties of Electrodes

The proportion of successful neuronal dye injections can be increased by selecting electrodes that pass dye readily. The tip of the outer barrel of a concentric electrode provides an excellent small test chamber with sufficiently small volume to enable the dye efflux to be estimated (see Fig. 1). The electrode to be tested (inner barrel) is filled with dye solution and lowered into the test chamber until its tip is about 20 μm from the 3–5 μm diameter tip of the outer barrel. When current is applied between the inner barrel and the solution bathing the tip of the outer barrel, the dye leaving the inner barrel accumulates rapidly in the small test chamber and is observed with a horizontal microscope. When the current stops, the accumulated dye diffuses away, clearing the test chamber in a few seconds.

Using this test system, I observed that conventional unbroken glass micropipettes (30–100 MΩ) filled with a 5% (w/v) aqueous solution of Procion dye will deliver dye into a very dilute electrolyte solution (0.01 M KCl in water), but that when the concentration of the external solution is increased to 0.15 M KCl, to approximate intracellular fluid, these fine-tipped electrodes usually block and cease to pass dye after only 5–10 min of applied current (10^{-7} A). Dye precipitate builds up as more current is applied. The blocking and precipitating behavior is probably due to the fact that dye ions carry most of the current in the shank of the inner barrel, but near the tip much of the current is carried by the more mobile potassium ions moving in from the test chamber, thus leaving dye to accumulate near the tip. Since the dye solution is nearly saturated this leads to rapid precipitation.

This blocking behavior can be largely eliminated by beveling the tip of the micropipette. Apparently the larger opening of the beveled tip allows faster outward diffusion of dye, hence preventing dye accumulation and precipitation.

Injection Procedure

Because dye injection appears to alter the membrane properties of the motoneuron (see below), dye-injecting current should be applied only after all electrophysiological observations have been made. Dye-injecting electrodes are back-filled with a 5% (w/v) (approximately 0.1 M) solution of Procion yellow or Procion scarlet (MGS) in water or in 0.15 M KCl (see above). Beveled electrodes pass dye most consistently. Dye is injected by applying a cathodal (hyperpolarizing) current of 4–10 $\times$ 10^{-8} A through the dye-filled pipette for 20–30 min via a bridge circuit. Synaptic potentials evoked in cat motoneurons by peripheral nerve or dorsal root stimulation can be used to verify the intracellular position of the electrode during dye injection.

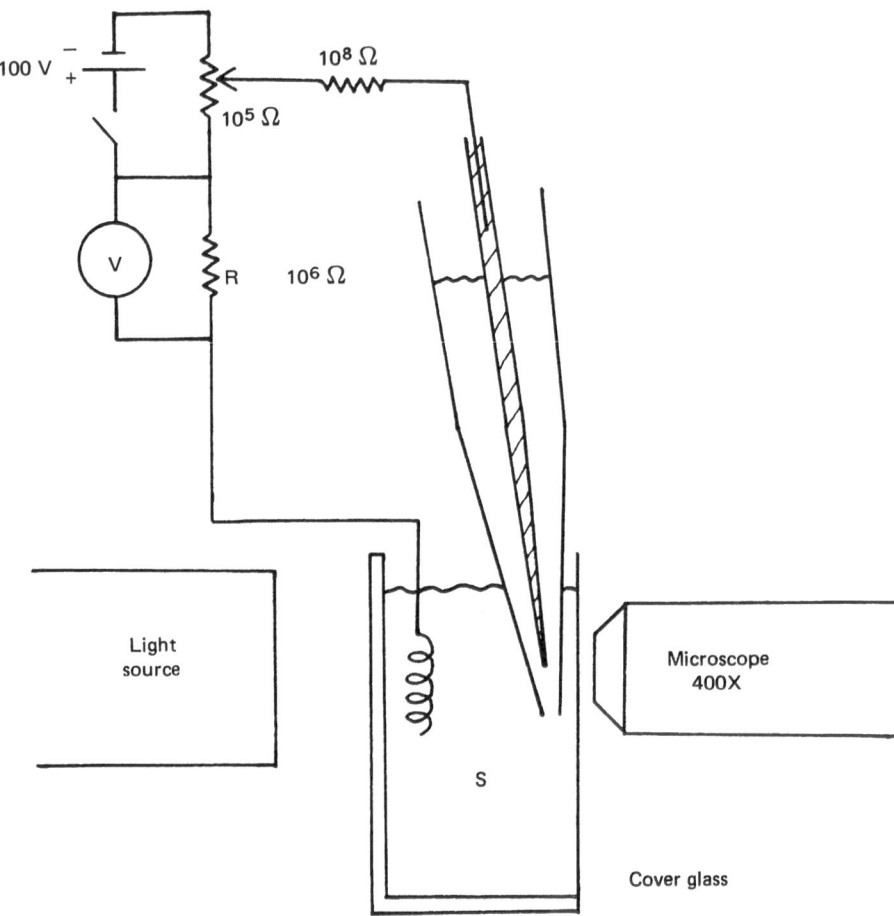

Fig. 1. Microchamber for testing the dye-injection properties of electrodes. When current is passed between the dye-filled inner barrel and the bathing solution (S), dye is injected into the region of the tip of the outer barrel. Current is monitored by measuring the voltage drop across resistor R with a high-impedance voltmeter. Dye accumulating within the tip region of the outer barrel is observed with a compound microscope (400 ×).

Subsequent localization of each stained neuron is aided by stereotaxically placing dyed electrode tracks on the contralateral side of the spinal cord. These tracks are made with electrodes of about 10 μm tip diameter filled with Procion blue or Procion scarlet (Procion yellow tracks cause too great an increase in background fluorescence). These contralateral markers enable each stained neuron to be matched with the appropriate electrical records.

Electrophysiological Effects of Dye Injection

After about 10 min of dye-injecting current, the antidromically evoked action potential of the motoneuron usually shows an increased duration and a smaller amplitude. In some cases the injected neurons display bursting behavior, in which action potentials appear to fire off the delayed depolarization of the previous action potential. The bursts often end in a prolonged depolarization lasting about 1 sec, followed by a rapid repolarization. This aberrant behavior may be caused by the reaction of the Procion dye with neuronal proteins, because no such behavior is observed after passing similar currents from standard electrodes filled with a 1 M K-acetate.

Although some dye does diffuse from the beveled pipette tips, the passive properties and the action potential of motoneurons usually show no observable alteration until dye-injecting current is applied. Neurons impaled for 30 min with a dye-filled pipette, without current injection, are only faintly stained. Thus it is usually unnecessary to apply a bucking current during electrophysiological measurements.

Light microscopic comparison of dye-injected motoneurons with adjacent uninjected neurons suggests that dye injection does not distort the general neuronal morphology, provided that the dye-injecting currents were not excessive.

Electrode Beveling Techniques

Previously reported beveling procedures employed a cylindrical Arkansas stone (1.5 in diameter Arkansas wheel on arbor, L. A. Clark Co., Seattle, Washington, U.S.A.) imbedded with fine glass particles by grinding the stone against the flat surface of a microscope slide (Barrett & Graubard, 1970). More recently, harder grinding materials—ruby, quartz and glass rods painted with fine, dry diamond dust—have been used (see Appendix II).

The electrodes used for passing dye into motoneurons were ground to produce a tip opening 1–3 μm long, keeping the diameter of the extreme point of the electrode smaller than 1 μm.

Histological Procedures

The histological procedures were selected to minimize changes in cellular dimensions, so that measurements of the cell geometry could be used to calculate the specific membrane properties of the neuron.

Fixation

Between 2 and 12 hr after dye injection, animals were perfused through the aorta with a buffered paraformaldehyde solution (dissolve 4 gm of paraformaldehyde in 17 ml of a 2.52% solution of NaOH at 60°C; to the resulting mixture add 83 ml of a 2.24% solution of NaH_2PO_4; adjust the final pH to 7.2 with NaOH) (Westrum & Lund, 1966). After 15 to 20 min perfusion time, the spinal cord was cut into blocks 3–5 mm long and stored for 12 hr at 4°C in the buffered fixative. The buffer osmolarity is important because cells remain osmotically active after aldehyde fixation (Crawford & Barer, 1951). Neurons usually cannot be visualized before or during fixation, and the criteria used to assess the quality of fixation are therefore rather crude. The 450 milliosmole buffer used here caused no detectable alteration in the dimensions of the tissue blocks, and no obvious signs of neuronal swelling or shrinkage. This hypertonic fixative might, however, cause slight neuronal shrinkage, and thus lead to an underestimate of the specific membrane resistance (R_m).

An attempt to eliminate fixation artifacts by rapidly freezing the fresh tissue blocks in liquid propane, followed by freeze substitution, was unsuccessful. The 3 mm blocks could not be frozen rapidly enough to eliminate ice crystal formation and thus possible redistribution of cell water.

Sectioning and Mounting

After storage in the buffered fixative solution, tissue blocks were transferred to a mixture of the same solution in 15% glycerol where they remained for 20 min. Glycerol reduces ice crystal formation when the tissue is subsequently frozen for sectioning. Sections 50–100 μm thick were mounted directly in 95% glycerol for light microscopic observation and measurement. Direct observation of the sections before and after glycerol treatment demonstrated that the glycerol produced no detectable change in neuronal dimensions, even when the tissue had not been treated with 15% glycerol prior to freezing. The glycerol does not clear the tissue as well as other mounting media that more closely match the refractive index of the cell membranes, but the procedures necessary to use these other media cause neuronal shrinkage. For example, the ethanol dehydration necessary for mounting sections in Fluoromount (George Gurr Co., London) produces about a 30% linear reduction in neuronal dimensions, or more than a 50% volume reduction (Fig. 2). Such shrinkage probably caused several previous studies to underestimate R_m. Similarly, a mixture of benzyl alcohol and glycerol also produces better clearing, but at the cost of measurable cell shrinkage. Of course, these latter mounting media can be used in work which does not require quantitative measurement of neuronal dimensions.

Measurements of Neuronal Geometry

Spinal cord sections were viewed with a fluorescence microscope using an excitation wavelength of 460 nm, dark-field illumination, and a Schott 0G1 barrier filter. All the stained neuronal processes on a given section were traced on transparent paper at 600× magnification, using a camera lucida microscope attachment. Tracings from serial sections were fitted together to reconstruct the neuron. Diameters of the neuronal processes were measured at 5 to 100 μm intervals on the original material, using the camera lucida attachment to superimpose the image of a calibrated, back-lighted graticule on the image of the process. The graticule was made by punching an array of 100 μm diameter holes in a sheet of aluminum foil. Back-lighting the sheet made the holes appear as tiny lights superimposed on the neuronal image. The length of each dendritic segment was calculated from the projection of the segment in the plane of the section and the projection in depth (measured with the calibrated focal adjustment of the microscope, and corrected for the refractive index of the mounting medium using Snell's law).

For the calculations described below, the dendritic tree was approximated as a series

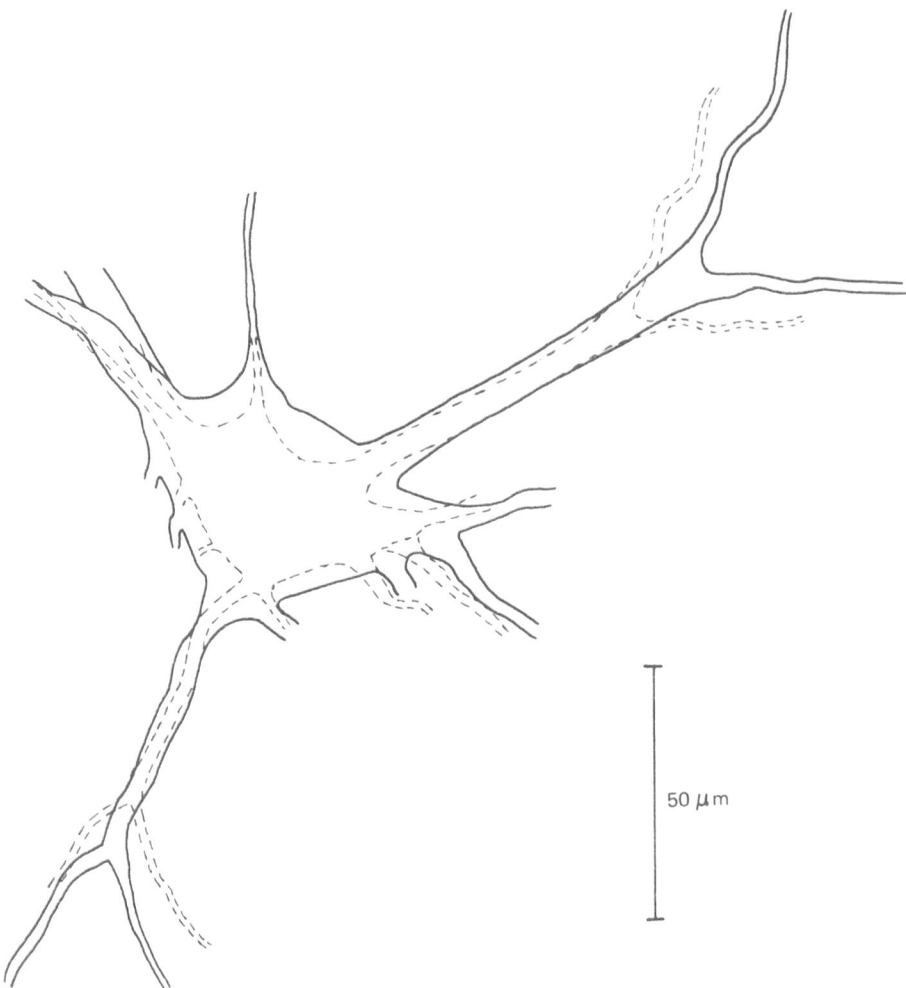

Fig. 2. Shrinkage produced by alcohol dehydration. Continuous lines trace the outline of a motoneuron mounted in the fixative solution; dotted lines show the same motoneuron after alcohol dehydration. Scale 50 μm.

of hundreds of short, interconnecting cylinders, each characterized by its length and diameter and by code numbers indicating its position in the dendritic geometry.

Calculating Specific Membrane Properties of Motoneurons

Specific Membrane Resistivity, R_{m}

If R_{m} is uniform over the membranes of the soma, dendrites and initial axon segment, then for a particular neuronal geometry there is a unique relationship between R_{m} and the steady-state input resistance (R_{N}) measured at the soma. This relation is found by dividing the neuron into hundreds of cylindrical segments, as described above, and using a computer program to apply equation (A9) from the Appendix to each segment. This equation gives the input admittance of a segment in terms of the geometry of the segment (d, the diameter and x_1, the length), the core cytoplasm resistivity R_a, the terminating admittance Y_T, the longest passive time constant of the neuron τ_0 (obtained by the peeling procedure of Rall, 1969), and an assumed value of R_m. Calculations are repeated for many values of R_m. The equation is first applied to the most distal dendritic segments, assuming that they terminate in a negligible admittance ($Y_T = 0$). The input admittances of the more proximal segments are then evaluated iteratively: the calculated input admittance of each segment specifies the terminating admittance of the next more proximal segment. At branch points, the terminating admittance of the parent branch is equal of the sum of the input admittances of the daughter branches. The input admittance that would be seen by an electrode at the soma is equal to the sum of the admittances of all the dendrites and the admittance of the soma membrane. The input impedance at the soma is evaluated as the reciprocal of the admittance. These calculations are first done with the frequency (ω) equal to zero to give the steady state input resistance, R_N, corresponding to different assumed values of R_m. The actual value of R_m for a particular neuron is specified by the intersection of the calculated R_N versus R_m plot with the electrically measured value of R_N. In the example of Fig. 3, this intersection occurs at an R_m of 2480Ω cm². The average R_m value calculated for 10 motoneurons was 1800Ω cm² (Barrett & Crill, 1971). The specific membrane capacity, C_m, is evaluated by dividing τ_0 by R_m.

Somatic Impulse Response, $h(t)$

Using the value of R_m so obtained, the input impedance is then calculated at many different frequencies ($0 \leq \omega\tau_0 \leq 10^5$) and expressed in polar form as an amplitude, $A(\omega)$, and phase shift, $\phi(\omega)$. Numerical integration of equation 1 over the frequency spectrum yields the predicted voltage response of the neuron, $h(t)$, to an impulse of current at the soma:

$$(1) \qquad h(t) = \frac{2}{\pi} \int_0^\infty A(\omega) \cos \phi (\omega) \cos \omega t \, d\omega$$

The passive response of the neuron to any other current input at the soma can be calculated as the convolution of the current input with $h(t)$ (e.g., equation 2).

The predicted impulse response $h(t)$ can be compared to the measured impulse response obtained by differentiating the voltage response of the neuron to a current step (Burke and ten Bruggencate, 1971). This comparison is facilitated by expressing both predicted and measured responses as a series of exponential terms (Rall, 1969; Nelson & Lux, 1970; Burke & ten Bruggencate, 1971).

Is R_m Uniform over Motoneuronal Membranes?

The presence of ion channels and discrete synaptic terminals on the motoneuronal membrane makes it unlikely that each small patch of membrane has identical passive properties. Because the spatial extent of these local nonuniformities is so much smaller

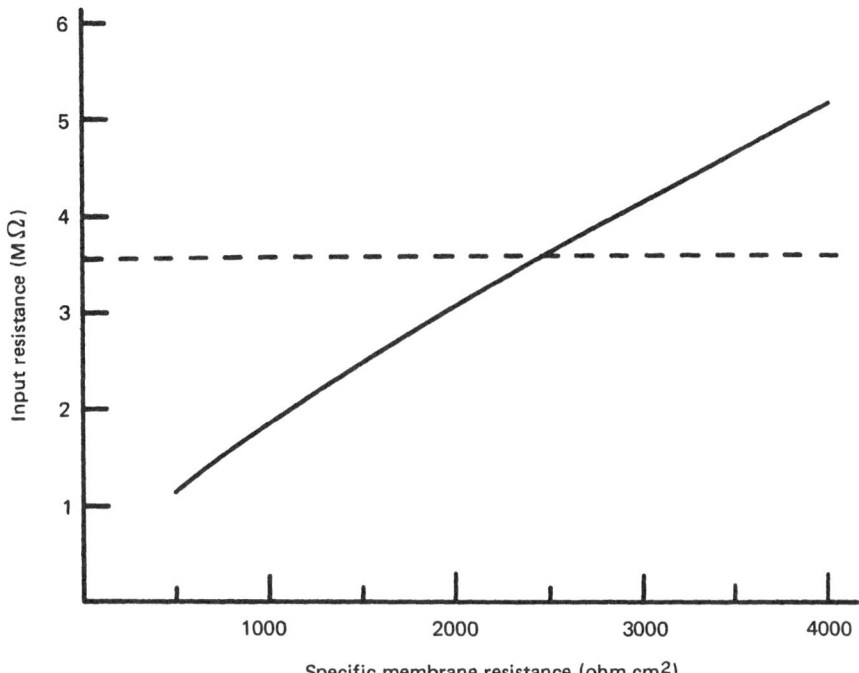

Fig. 3. Continuous curve plots the relationship between input resistance R_N (MΩ) and specific membrane resistance R_m (Ω cm^2), calculated for a Procion dye-injected cat motoneuron using procedures outlined in the text and Appendix. Dashed horizontal line marks the measured value of R_N for this motoneuron, 3.45 MΩ. The predicted curve intersects the dashed line at an R_m of 2480 Ω cm^2. Calculations used an R_a value of 70 Ω cm (Barrett and Crill, 1971), and assumed that the dendrites terminated in closed ends at their most distal visible segments ($Y_T = 0$ for the most distal segments).

than a space constant, they should not cause the passive electrical response of the neuron to differ significantly from that predicted assuming uniform R_m. However, different R_m values for somatic and dendritic membranes would cause significant deviation from uniform behavior. Likewise, differences in R_m values among different dendrites, or between proximal and distal dendrites, could cause nonuniform electrical behavior.

In an attempt to set limits on dendritic R_m, I have made preliminary calculations for the case of different values of R_m in somatic and dendritic membranes. Because the dendrites contribute 80% or more of the total neuronal membrane area, the measured steady-state value of R_N immediately sets a lower limit on dendritic R_m for a particular neuronal geometry. R_m values more than 20% below the value calculated assuming uniform R_m force the predicted R_N below the experimentally-measured value. Unfortunately, measurements of steady-state R_N alone cannot similarly place an upper limit on dendritic R_m.

In theory, analysis of the transient voltage response to an applied current step should yield information about nonuniformities in R_m. Present recording techniques, however, are not accurate enough to permit such detailed analysis. Calculations demonstrate that the dendritic R_m could be as high as 8000Ω cm^2, with a corresponding somatic R_m of

240Ω cm², before the predicted, time-normalized transient responses differ significantly from the measured transients. (The predicted response must be normalized in time because C_m is not known independently of R_m. C_m is adjusted until the predicted τ_0 matches the measured τ_0).

Effects of Anesthesia

The specific membrane parameters calculated here should permit us to predict how synaptic events at various locations on the dendritic tree will contribute to the firing rate of normally-functioning motoneurons. Unfortunately, the neuronal staining involved in determining R_m and C_m requires 10 to 30 min of intracellular dye injection, which is difficult to achieve in a nonparalyzed, nonanesthetized preparation. The low dosage of Nembutal anesthetic used in this study should not directly alter R_N or R_m (Weakly, 1969), but it is likely that anesthesia and paralysis reduce the amount of incoming synaptic activity below that in an alert, active animal. This reduction of synaptic activity could produce a significant increase in the measured R_N and R_m: calculations indicate that an increase in excitatory synaptic activity of 15 quanta per second per bouton (average bouton density: 20 per 100 μm² of motoneuronal membrane, Conradi, 1969) could reduce R_m by as much as 70% (Barrett & Crill, unpublished). Even lower rates of inhibitory activity could produce the same reduction in R_m because quantal inhibitory postsynaptic potentials involve a greater conductance change than quantal excitatory potentials (Kuno & Weakly, 1972).

The effect of anesthesia on R_m might be estimated by measuring the longest average time constant (τ_0) of a neuronal population both with and without anesthesia. Assuming that C_m is unaffected and that the membrane is uniform, changes in τ_0 should directly reflect changes in R_m ($\tau_0 = R_m C_m$).

Calculating Synaptic Potentials

Measured values of R_m, C_m and R_a together with the geometry of the neuron specify all the parameters of a passive, core-conductor model of the cell. This model can be used to predict the amplitude and time course of synaptic potentials, both at the site of the synaptic conductance change and at any other location on the neuronal membrane.

The synaptic potential $V(t)$ at the site of the synaptic conductance change is calculated by convolution of the synaptic current $i(t)$ with the voltage impulse response $h(t)$ at that synaptic site. $i(t)$ is assumed equal to the product of the synaptic conductance change $g(t)$ and the driving potential $(V(t)\text{-}E_{pq})$ (E_{eq} is the equilibrium potential for the synaptic current). $h(t)$, the voltage response to an impulse of current at the synaptic site, is evaluated using the same iterative methods outlined above for calculating the voltage impulse response at the soma. $V(t)$ at the synaptic site is thus given by:

(2) $$V(t) = \int_0^t h(t - \tau)g(\tau)\{V(\tau) - \epsilon_{eq}\}\, d_\tau$$

For the particular case of a unitary (quantal) conductance change, $g(t)$ is evaluated by inserting the somatic impulse response and the measured time course of a unitary postsynaptic potential of somatic origin (Kuno, 1964; Kuno & Miyahara, 1969; Jack et al., 1971) into equation 2, and solving numerically for $g(t)$.

The time course of the synaptic potential at any other site on the neuron is calculated using a Fourier transform technique similar to that used to calculate $h(t)$ at a particular site. This procedure is especially useful for determining the effect of a synaptic conductance change at a dendritic site on the potential at the soma. The dendritic tree is again treated as a series of small, interconnected membrane cylinders. Equation A11 (Appendix I) is applied to each small cylinder to give the voltage transfer function for the cylinder, the relationship between a sinusoidal voltage input into the distal end and the voltage output at the proximal end. At any given frequency ω, this voltage transfer function

$$\frac{\hat{V}_{out}(j\omega)}{\hat{V}_{in}(j\omega)}$$

can be specified by an amplitude ratio $A(\omega)$ and a phase shift $\phi(\omega)$. The calculation is repeated for all segments between the synaptic site and the soma. The product of these voltage transfer functions yields the relationship between a sinusoidal voltage input at the dendritic site and the resulting steady-state sinusoidal voltage at the soma:

$$(3) \qquad \frac{\hat{V}_{soma}(j\omega)}{\hat{V}_{dend}(j\omega)} = \prod_{k=1}^{n} \left(\frac{\hat{V}_{out}(j\omega)}{\hat{V}_{in}(j\omega)} \right)_{k}$$

The subscript k indexes the cylindrical segments and n represents the total number of segments between the particular dendritic site and the soma. The calculations in equation 3 are repeated at many frequencies to give Bode plots of the amplitude and phase shift of the somatic voltage response to the dendritic voltage input. Integrating these data over the frequency spectrum (the inverse Fourier transform, equation 1) gives the response of the soma to an impulse voltage input at the dendritic site. Finally, this somatic voltage response is convolved with the dendritic voltage change calculated using equation 2 to yield the predicted somatic voltage response to a given synaptic conductance change at the particular dendritic site.

Application of these procedures to cat motoneurons demonstrates that even a single quantal conductance change occurring at a high-impedance distal dendritic site will produce a 15 to 25 mV potential change at that site. This voltage is considerably attenuated at the low-impedance soma, but the time integral of the somatic potential produced by a distal synaptic event is still 20% to 50% of the time integral of the somatic potential produced by a quantal event occurring directly on the soma (Barrett & Crill, unpublished). These calculations thus show that, relative to somatic synapses, even the most distant dendritic synapses can have a considerable effect on the somatic potential.

Conclusion

The analytic techniques described here break the dendritic tree into many short, core-conducting, cylindrical segments. The basic iterative calculation procedure applied to these segments was originally suggested by Rall (1959). This detailed model of neuronal geometry is complementary to the simpler procedure of approximating the dendrites by a long equivalent membrane cylinder (Rall, 1959; Lux et al., 1970; Jack & Redman, 1971a, b; Norman, 1972). The detailed approach, made feasible by computer technology, allows a very close approximation to the actual dendritic geometry, and thus provides a check on the accuracy of the equivalent cylinder approximation for a given type of neuron.

The development of dye and isotope injection techniques for studying the electrophysiological and morphological properties of the same neuron has allowed more accurate estimation of the passive properties of the motoneuronal membrane. Several problems deserve further investigation—for example, the effects of dye injection and fixation on neuronal dimensions, possible nonuniformities in R_m, and the value of R_a for dendritic cytoplasm. Further refinements in the passive core-conductor model for motoneurons and other neuron types should help us define more fully the influence of passive membrane properties and neuronal morphology on synaptic function and the shape of the action potential.

Acknowledgments. Much of the experimental work reported here was done in collaboration with Dr. W. E. Crill. Supported by U.S. Public Health Service Grants NS 07987, GM 00739, FR 00374, NS 05934, and NS 08453.

References

Barrett, J. N.: The passive properties of cat motoneurons and their influence on the effectiveness of dendritic synapses. Ph.D. thesis. University of Washington, Seattle, Washington, U.S.A. (1972).
——, and W. E. Crill: Specific membrane resistivity of dye-injected cat motoneurons. Brain Res. *28*, 556–561 (1971).
——, and K. Graubard: Fluorescent staining of cat motoneurons *in vivo* with bevelled micropipettes. Brain Res. *18*, 565–568 (1970).
Burke, R. E., and G. ten Bruggencate: Electrotonic characteristics of alpha motoneurones of varying size. J. Physiol., Lond. *212*, 1–20 (1971).
Crawford, G. N. C., and R. Barer: The action of formaldehyde on living cells as studied by phase contrast microscopy. Q. Jl microsc. Sci. *92*, 403–452 (1951).
Davis, W. J.: Motoneuron morphology and synaptic contacts: determination by intracellular dye injection. Science *168*, 1358–1360 (1970).
Jack, J. J. B., S. Miller, R. Porter, and S. J. Redman: The time course of minimal excitatory post-synaptic potentials evoked in spinal motoneurone by Group Ia afferent fibres. J. Physiol., Lond. *215*, 353–380 (1971).
——, and S. J. Redman: The propagation of transient potentials in some linear cable structures. J. Physiol., Lond. *215*, 283–320 (1971a).
—— ——: An electrical description of the motoneurone, and its application to the analysis of synaptic potentials. J. Physiol., Lond. *215*, 321–352 (1971b).
Jankowska, E., and S. Lindström: Morphological identification of physiological defined neurones in the cat spinal cord. Brain Res. *20*, 323–326 (1970a).
Kaneko, A.: Physiological and morphological identification of horizontal, bipolar and amacrine cells in goldfish retina. J. Physiol., Lond. *207*, 623–633 (1970).
Kuno, M.: Quantal components of excitatory synaptic potentials in spinal motoneurones. J. Physiol., Lond. *175*, 81–99 (1964).
——, and J. T. Miyahara: Non-linear summation of unit synaptic potentials in spinal motoneurones of the cat. J. Physiol., Lond. *201*, 465–477 (1969).

————, and J. N. Weakly: Quantal components of the inhibitory synaptic potential in spinal motoneurones of the cat. J. Physiol., Lond. *224*, 287–303 (1972).

Lux, H. D., P. Schubert, and G. W. Kreutzberg: Direct matching of morphological and electrophysiological data in cat spinal motoneurons. In: Excitatory Synaptic Mechanisms. Ed. P. Anderson and J. K. S. Jansen. pp. 189–198. Oslo: Universitetsforlaget, 1970b.

McMahan, U. J., and D. Purves: An electron-microscopic study of a physiologically identified motoneurone in the leech C.N.S. after injection of the fluorescent dye Procion yellow. J. Physiol., Lond. *222*, 64–66 (1972).

Nelson, P. G., and K. Frank: Anomalous rectification in cat spinal motoneurons and effect of polarizing currents on excitatory postsynaptic potential. J. Neurophysiol. *30*, 1097–1113 (1967).

————, and H. D. Lux: Some electrical measurements of motoneuron parameters. Biophys. J. *10*, 55–73 (1970).

Norman, R. S.: Cable theory for finite length dendritic cylinders with initial and boundary conditions. Biophys. J. *12*, 25–45 (1972).

Rall, W.: Branching dendritic trees and motoneuron membrane resistivity. Exp. Neurol. *1*, 491–527 (1959).

————: Theory of physiological properties of dendrites. Ann. N.Y. Acad. Sci. *96*, 1071–1092.

————: Distinguishing theoretical synaptic potentials computed for different soma-dendritic distributions of synaptic inputs. J. Neurophysiol. *30*, 1138–1168 (1967).

————: Time constants and electrotonic length of membrane cylinders and neurons. Biophys. J. *9*, 1482–1508 (1969).

————: Cable properties of dendrites and effect of synaptic location. In: Excitatory Synaptic Mechanisms, Proceedings of the Fifth International Meeting of Neurobiologists. Ed. P. Andersen and J. K. S. Jansen. Oslo: Universitetsforlaget, 1970.

Remler, M. P., and A. I. Selverston, and D. Kennedy: Lateral giant fibers of crayfish: location of somata by dye injection. Science *162*, 281–283 (1968).

Stretton, A. O. W., and E. A. Kravitz: Neuronal geometry: determination with a technique of intracellular dye injection. Science *162*, 132–134 (1968).

Weakly, J. N.: Effect of barbiturates on 'quantal' synaptic transmission in spinal motoneurones. J. Physiol., Lond. *204*, 63–77 (1969).

Westrum, R. E., and R. D. Lund: Formalin perfusion for correlative light- and electron-microscopical studies of the nervous system. J. Cell Sci. *1*, 229–238 (1966).

APPENDIX I

John Barrett

This section derives the equations used to calculate (1) the input admittance of a cylindrical dendritic segment and (2) the alteration of a sinusoidal voltage signal along the length of the segment.

All equations are derived from the cable equation for a core—conducting membrane cylinder (Rall, 1959):

(A1)
$$\lambda^2 \frac{\partial^2 V(x, t)}{\partial x^2} = \tau \frac{\partial V(x, t)}{\partial t} + V(x, t)$$

where V is voltage, t is time, x is the distance along the cylinder ($0 \leq x \leq x_1$), τ is the membrane time constant ($\tau = R_m C_m$, where R_m, is the specific membrane resistance and C_m is the specific membrane capacitance) λ is the space constant $\left[\lambda = \frac{\sqrt{d}}{2} \sqrt{\frac{R_m}{R_a}}, \text{ where}\right.$

d is the diameter and R_a is the cytoplasmic resistivity, 70 Ω cm (Barrett and Crill, 1971) $\Bigg]$.

It is assumed that within each segment the diameter remains constant while the diameter of adjacent segments can be different.

The Fourier transform of equation (A1) with respect to time is:

(A2)
$$\frac{d^2 \hat{V}(x, j\omega)}{dx^2} = \hat{V}(x, j\omega) \left(\frac{1 + j\omega\tau}{\lambda^2} \right)$$

whose general solution is:

(A3)
$$\hat{V}(x, j\omega) = A_1 e^{\gamma x} + A_2 e^{-\gamma x}$$

where
$$\gamma = \frac{\sqrt{1 + j\omega\tau}}{\lambda}$$

The transform of the axial current $[I(x, t)]$ entering the cylinder is proportional to the derivative with respect to distance of the voltage transform, thus at the point $x = 0$,

(A4)
$$\hat{I}(0, j\omega) = -\frac{1}{R_a} \frac{d\hat{V}(x, j\omega)}{dx} = \frac{\gamma(A_2 - A_1)}{R_a}$$

Dividing the current transform, $\hat{I}(x = 0, j\omega)$, by the voltage transform at $x = 0$, $\hat{V}(x = 0, j\omega)$ yields the Fourier transform of the input admittance at $x = 0$, $Y_{in}(j\omega)$:

(A5)
$$Y_{in}(j\omega) = \frac{\hat{I}(0, j\omega)}{\hat{V}(0, j\omega)}$$

294

Combination of equations (A3), (A4), and (A5) defines $Y_{in}(j\omega)$ in terms of the ratio A_2/A_1:

$$(A6) \qquad Y_{in}(j\omega) = \frac{\gamma}{R_a} \left\{ \frac{A_2/A_1 - 1}{A_2/A_1 + 1} \right\}$$

This ratio is evaluated from the boundary conditions at the terminal end of the segment $(x = x_1)$, as specified by equation (7):

$$(A7) \qquad \hat{I}(x_1, j\omega) = \frac{-1}{R_a} \frac{d\hat{V}(x, j\omega)}{dx} \bigg|_{x_1} = Y_T(j\omega)\hat{V}(x_1, j\omega)$$

where $\hat{I}(x_1, j\omega)$ is the transform of the axial current leaving the cylinder at x_1, and $Y_T(j\omega)$ is the Fourier transform of the terminating admittance. Substituting the expression for $\hat{V}(x, j\omega)$ from equation (A3) into equation (A7) gives an expression for (A_2/A_1):

$$(A8) \qquad \frac{A_2}{A_1} = e^{2\gamma x_1} \left\{ \frac{\gamma + Y_T(j\omega)R_a}{\gamma - Y_T(j\omega)R_a} \right\}$$

Combining equations (A6) and (A8) yields the input admittance looking into the segment at $x = 0$:

$$(A9) \qquad Y_{in}(j\omega) = \frac{\gamma}{R_a} \left\{ \frac{(\exp (2\gamma x_1))(\gamma + R_a Y_T(j\omega)) + R_a Y_T(j\omega) - \gamma}{(\exp (2\gamma x_1))(\gamma + R_a Y_T(j\omega)) - R_a Y_T(j\omega) + \gamma} \right\}$$

Because $Y_T(j\omega)$ is a complex number, this expression is separated into real and imaginary parts for the actual computation of $Y_{in}(j\omega)$ (Barrett, 1972).

The relationship between a sinusoidal voltage input at one end of a membrane cylinder $\hat{V}(0, j\omega)$ and the voltage output at the opposite end $\hat{V}(x_1, j\omega)$ is obtained by evaluating equation (A3) at $x = 0$ and at $x = x_1$:

$$(A10) \qquad \frac{\hat{V}(x_1, j\omega)}{\hat{V}(0, j\omega)} = \frac{e^{\gamma x_1} + A_2/A_1 e^{-\gamma x_1}}{1 + A_2/A_1}$$

Substituting the value of A_2/A_1 given by equation (A8) yields:

$$(A11) \qquad \frac{\hat{V}(x_1, j\omega)}{\hat{V}(0, j\omega)} = \frac{e^{\gamma x_1} \left(\dfrac{2\gamma}{\gamma - Y_T(j\omega)R_a} \right)}{1 + e^{\gamma x_1} \left(\dfrac{\gamma + Y_T(j\omega)R_a}{\gamma - Y_T(j\omega)R_a} \right)}$$

In using this equation, $Y_T(j\omega)$ is evaluated for each segment by iterative application of equation (A9) as described above and in the text.

Discussion

Dr. Dubin (Colorado): Are fine dendrites infiltrated by the dye?

Dr. Barrett: It is difficult to determine whether Procion dye reaches all the most distant dendritic terminals. We compared reconstructions of our dye-injected cat motoneurons with the Golgi-stained ventral horn neurons of Aitken and Bridger (1961, J. Anat. 95:38–53).

The surface area of our largest dye-injected motoneuron was 250,000 μm^2, more than twice the surface areas reported for any of the Golgi-stained neurons. Our reconstructions showed more major trunk dendrites and more extensive dendritic branching than the Golgi preparations, and the total length of the dendritic tree in our largest dye-injected cells was about twice that measured in Aitken and Bridger's largest cells. These differences are probably attributable to cell shrinkage and loss of dendritic branches in the Golgi preparations. However, we still may have missed the finest dendritic terminals, because Aitken and Bridger were able to trace some dendrites a distance of 1400 μm, whereas the most distal dendritic tips we observed were about 1000 μm from the soma.

In well-stained dye-injected motoneurons all the dendritic branches could be traced until they were less than 1 μm in diameter. Branches less than 1.5 μm in diameter constituted about ⅓ of the total dendritic length in the reconstructed motoneurons, and about ⅓ of the dendritic profiles in electron micrographic sections of ventral horn neuropil are under 1.5 μm in diameter (measured from Fig. 18 in A. Peters, et al. (1970). This agreement suggests that we did not miss too many fine collaterals.

DR. KRIPKE: I would like to ask Dr. Barrett if he would comment on why Procion fails in some cases to be released from the tip of a micropipette when you pass current.

DR. BARRETT: We found that application of a 100×10^{-9} A current into standard dye-filled electrodes would eject dye into a dilute solution (0.01 M KCl), but not into a solution approximating mammalian intracellular fluid (0.15 M KCl). This observation suggests that the precipitation of dye in electrode tips may be caused by a sudden decrease in the fraction of current carried by the dye (or the transfer number of the dye) in the region of the electrode tip. Dye ions probably carry most of the iontophoretic current in the shank of the electrode, but near the tip some of the current is carried by potassium ions moving into the electrode tip from the bath solution. The rapid drop in the dye transfer number leaves dye to accumulate in the tip. Precipitation is rapid because the dye solution is already nearly saturated. Less blocking and precipitation occur with broken or bevelled electrodes, probably because their larger tip openings allow faster outward diffusion of dye. There is probably a positive correlation between the area of the electrode tip opening and the current intensity at which precipitation starts to occur.

Peters, A., S. L. Palay, and H. deF Webster: The fine structure of the nervous system: the cells and their processes. New York: Harper and Row, 1970.

Technique for Beveling Glass Microelectrodes

John Barrett and David G. Whitlock

Several investigators (Barrett and Graubard, 1970; Burke and ten Bruggencate, 1971) have found that glass micropipettes whose tips are ground or broken to resemble the tips of hypodermic needles cause less damage during cell penetration, possess better recording properties and pass charged dyes better than standard, nonbeveled micropipettes. We describe an inexpensive grinding apparatus that consistently produces beveled (hypodermic needle-like) electrode tips, even when the largest part of the final tip opening is less than 1 μm in diameter.

The basic requirements for electrode grinding are a fine-grained grinding surface and a means of moving that surface past the electrode tip with minimal wobbling or vibration. Such a grinding surface can be produced using a fine sable brush to paint a spinning quartz rod (3 mm diameter) with a thin layer of dry diamond dust (Schuller, 100,000 mesh, $<.25$ μm diameter, extracted with about five acetone washes).[1] The diamond particles adhere strongly to the rod.

The coated quartz rod (*a* in Fig. 1) is mounted between two stainless steel supports (*b*) and rotated by means of a motor-driven thread belt (*c*). The motor (*d*) is mounted on a separate block to avoid vibrations, and should be capable of slow speeds, between 10–100 rpm. Plastic rings (*e*) glued around the rod on both sides of one support keep the rod from sliding along the supports. A layer of cloth adhesive tape (hatches in Fig. 1) wrapped around the center of the rod prevents the thread belt from slipping around the rod, and plastic guides (*f*) keep the thread belt centered on the tape.

Micropipettes, pulled on standard vertical or horizontal electrode pullers, are lowered onto the rotating glass rod with a micromanipulator (*g*) (e.g., Narashige MP-1 or MM-3). The rod rotates away from the electrode tip (counterclockwise in Fig. 1). The angle between the pipette and the tangent to the rod surface at their point of contact should be about 30°. Optimal grinding requires a very light contact, such that only a few microns of the electrode tip contact the rod surface and the electrode is not visibly bent. This light contact is easily detected using a dissecting microscope (*h*) (25–50$\times$) and spot-light (*i*), both focussed on the upper surface of the rod. The spotlight accentuates the shadow cast by the pipette tip as it approaches the rod. When

---

[1] (The Schuller diamond compound is sold by most lapidary shops. An equivalent (.25 μm particle size) Metadi diamond compound is available from Scientific Products Co., 1210 Leon Place, Evanston, Illinois, or Matheson Scientific.

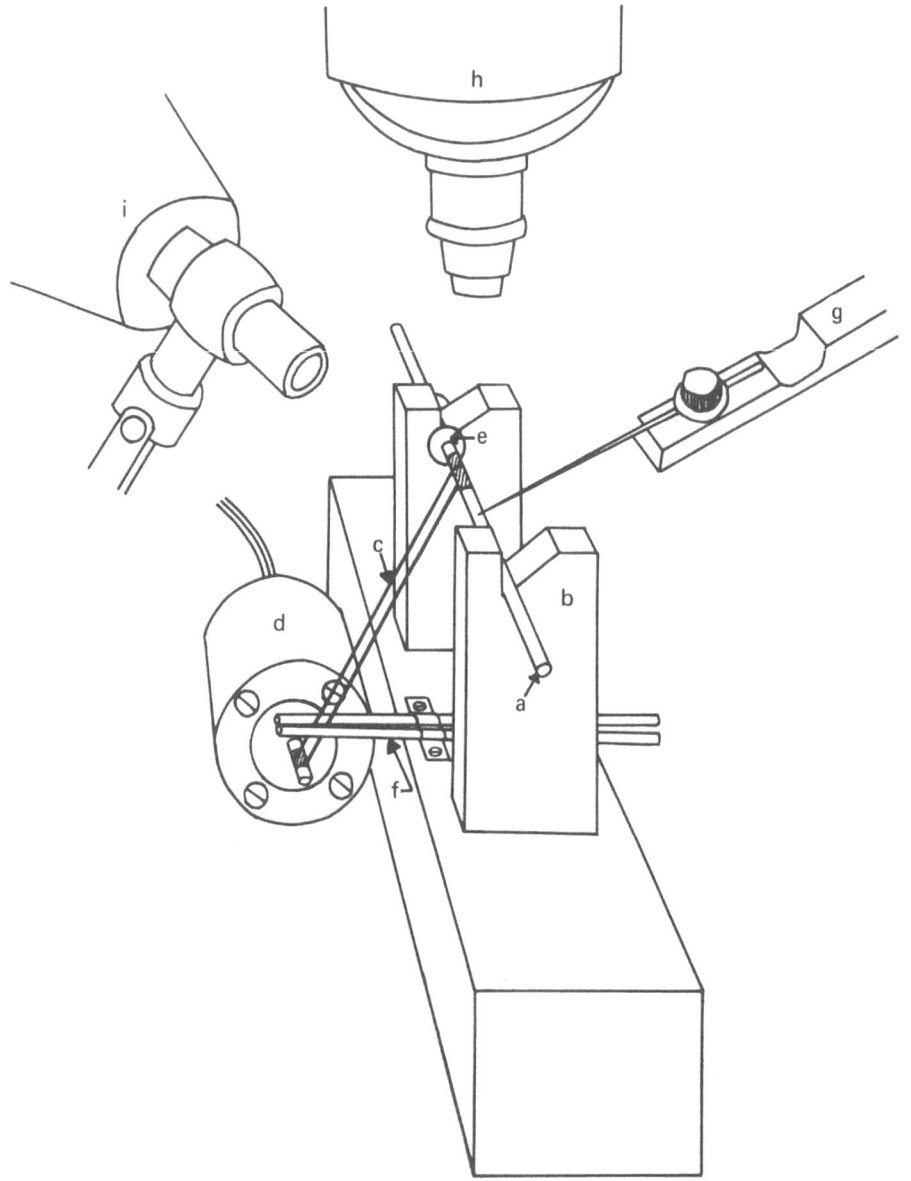

Fig. 1. Electrode grinding apparatus: *a*, quartz, rod; *b*, stainless steel supports; *c*, thread belt; *d*, motor; *e*, guide rings; *f*, thread guides; *g*, electrode holder and manipulator; *h*, microscope (50X); *i*, small spotlight. For details see text.

contact is made, this shadow converges toward the image of the pipette, and the pipette tip picks up refractile diamond particles.

Motor speed (10–100 rpm) and grinding time (10 sec-several minutes) are varied, depending on the desired size of the tip opening. Using this method we have produced beveled tip openings ranging from 3 μm to less than 1 μm, suitable for recording

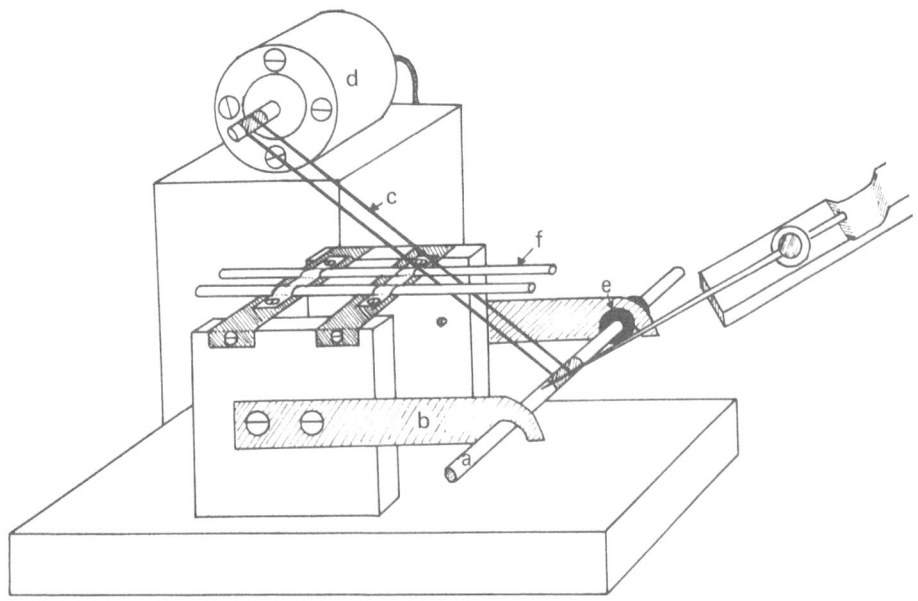

Fig. 2. Grinding apparatus designed to minimize vibration of quartz rod relative to electrode tip. Components are labelled as in Fig. 1; microscope and light source (not shown) should be positioned as in Fig. 1.

intracellularly from such mammalian neurons as motoneurons, dorsal root ganglion cells, and cortical cells. Electrodes may be ground before or after filling. The diamond particles which enter the tip during grinding do not usually interfere with the electrode properties, but if necessary they can be eliminated by applying air pressure to unfilled pipettes during grinding (Lee and Shaw, personal communication).

Asymmetries in the quartz rod cause it to wobble slightly with respect to the pipette tip. In practice this small movement (only a few microns) does not break the flexible pipette tips, even when the glass is stiff and the shank short. Vibrations can be reduced somewhat by grinding the electrode tips near the rod supports (b) or alternatively, the motor and thread belt can be rearranged as shown in Fig. 2, where the upper grinding surface of the rod is held rigidly against the side supports. The electrode contacts the rod a few mm from the metal support, on the same side of the rod as the metal support.

We thank Dr. Robert Lasher for helpful discussions and Kristine McInvaille for drawing the figures. Supported in part by USPHS grants NS08009 and NS08453.

References

Barrett, J. N. and K. Graubard: Fluorescent staining of cat motoneurons *in vivo* with beveled micropipettes. Brain Res. *18*, 565–568 (1970).

Burke, R. E. and G. ten Bruggencate: Electronic characteristics of alpha motoneurones of varying size. J. Physiol., Lond. *212*, 1–20.

CHAPTER 19

Current Practice and Prospects for the Intracellular Stains

Donald Kennedy

If nothing else is clear from this symposium, we have learned that a great many people are using the intracellular stains or plan to do so in the future. To predict what new needs may arise for this technology, and what developments may be necessary to meet them, it may help to summarize the motivations that underlie its contemporary use.

The developers of the Procion yellow method, Stretton and Kravitz (1968 and Chapter 2), were interested in evaluating hypotheses about the constancy of form of identified neurons in lobster ganglia. For that purpose, they needed precise maps of the arrangements of processes; but the processes were hidden from view, whereas the somata were readily accessible and easy to identify by physiological recording. The injection of dye into the functionally labeled part, the soma, allowed an analysis of the previously hidden parts of the cell that were significant for connectivity and its development. Since this kind of arrangement is essentially definitive of ganglionic neuropile in invertebrates, it is not surprising that most of the second-generation applications of the Procion yellow technique employed such preparations.

The questions others asked were not, however, the same as the one posed originally by Stretton and Kravitz. For example, in our own subsequent work on a homologous system—the fast flexor motoneurons of crayfish and the central giant interneurons that excite them—we were primarily concerned with the uniqueness of dendritic architecture in each neuron belonging to a small pool having related function, rather than with the issue of constancy *per se*. Having shown that the branching patterns of each flexor motoneuron in this set were unique, we demonstrated that their forms were associated with differences in their connection patterns with presynaptic giant fibers (Kennedy et al., 1969; Selverston and Kennedy, 1969; Selverston and Remler, 1972). In applying the technique to the lateral giant interneuron, we had a different motive: this element, which provides a highly complex set of motor commands for escape behavior, has been studied only as an axon. As far as physiological analysis was concerned, the soma and dendrites were "lost." Finding them by injecting dye axonally (Remler et al., 1968) enabled us to explain simultaneously why the soma had been difficult to locate and why we were unable to record activity in it. The soma communicates with the dendrites and the main axon via a long and tenuous neurite which is difficult to follow in conventional histological preparations and which attenuates electrical signals sharply. Finding the dendrites made it possible for us to look for the first time with microelectrodes at the active input regions of the cell, instead of blindly probing for them (Zucker, 1972).

The clarification of relationships between soma and axon has turned out to be a more than trivial contribution of the intracellular staining techniques to the comparative cell biology of invertebrate nervous systems. For example, in a number of cases it had seemed that individual axons are syncytial—that is, formed by the fusion of processes from several somata. The segmental giant fibers of earthworms, which had been one of the best advertized cases, were convincingly shown by Mulloney (1970) to have only one soma per segment. Syncytial axons now seem unlikely everywhere save in the third-order giant fibers of cephalopods.

The location of previously occult somata has also yielded an unexpected generality about the evolution of cellular arrangement in bilaterally symmetrical nervous systems: all true interneurons (i.e., cells with processes entirely restricted to the central nervous system) in arthropods have their somata located *contralaterally* to the axon and to the dendritic field if the latter is confined to one side. The number of cases is not large (less than a dozen), but new findings continue to be consistent: since I summarized the matter two years ago (Kennedy, 1971), there have been several additions, including another crayfish tactile interneuron, the caudal photoreceptor interneuron in crayfish (Wilkens and Larimer, 1972), and an insect interneuron (Chapter 9).

A major use category for the intracellular stains—historically the oldest one—involves identification. In many kinds of nervous tissue there are anatomically distinct classes of cells which can be reliably distinguished only if all the processes are visible as, for example, in a Golgi stain. Where the experimenter can also recognize physiological classes, he can test for correspondence between the two if he can inject a migrating dye after recording. The work of Kaneko summarized in Chapter 11 on the vertebrate retina is an especially beautiful example of this kind of use, which is now being extended to the central visual system by Van Essen and Kelley (Chapter 13) and to insect visual systems by Zettler and Järvilehto (1971).

There is, finally, an intermediate application which goes beyond identification but stops short of complete process analysis. In this approach, the aim is to identify that part of a cell in which the recording microelectrode was positioned, and to expose specific features of the nearby morphology in order to account for physiological findings at that locus. The work of Llinás and Nicholson (1971) on alligator Purkinje cell dendrites is an outstanding (and rare) example of this kind of use.

The foregoing account comes fairly close to exhausting the purposes for which Procion yellow was employed during what we may now smugly call its "classical" period. As so often happens, a second generation of innovations expanded the opportunities further.

The first of these was the discovery by Iles and Mulloney (1971) that dye could be introduced into the cut end of an axon. At the time, only Procion stains were available, and it was necessary to pass a current between a pool containing the preparation and one containing the end of the nerve trunk immersed in the dye: the applied current reversed the injury current, which would tend to exclude negatively charged molecules from the axons. With the cobalt method (see below), the applied current can often be dispensed with, since the ions are positively charged. The new method has emancipated the injection technology from its previous dependence upon microelectrode penetration: now small axons may be injected with much less difficulty, and it is also possible to analyze neural structure with intracellular stains at the level of tracts instead of single cells. The potential of this technique for tract analysis in large nervous systems is clearly indicated by the work reported in Chapter 15. It is no overstatement to call these results the vanguard of a new approach to neuroanatomy.

Almost as soon as the Procion technique was announced, various laboratories began searching for ways to do the same thing with a label observable in the electron microscope. The cobalt technique described by Pitman et al. (1972a) represents the first success of this search, but certainly has not ended it. The cobalt method produces excellent whole mount preparations, even though the amplification factor of fluorescence is lost, and the stain seems to display selective affinity for fine processes. Work in our laboratory suggests that cobalt concentrates in fine terminal branches at the junction between giant and motor giant axons in the crayfish (Mittenthal and Wine, 1973), whereas Procion yellow in the same presynaptic giant axons is distributed much more homogeneously, and may not even reach the very fines tips of the processes. This is the only instance to date in which a direct comparison between Procion and cobalt techniques is possible in the same junctional system. We badly need more such cases in order to evaluate the advantages of each method.

At the level of ultrastructure for which the cobalt method was designed, the purpose of intracellular labeling is to look at synaptic distributions on or of particular cells. The aim is identification of processes; but an electron-dense label is not an absolute requirement. As long as processes can be identified in a plastic thick section (or on the block face) as belonging to a particular neuron, it should be possible to find them on the adjacent thin section. The techniques developed by Blackstad (1965, 1970) for carrying out this procedure with Golgi-stained neurons are readily applicable to processes identified only by colored dyes or the examination of serial sections. Blackstad's neurons, unfortunately, were so densely impregnated with silver salts that the identification of post-junctional membrane specializations was impossible. The cobalt technique (see Chapter 6) shares this difficulty to the point that to identify and map synapses seems out of the question with the material produced so far. Some way will have to be found to reduce the density of label, but this will produce other problems. In the meantime, more of us may take the pathway adopted by Purves and McMahan (Chapter 5), which works toward improving the resistance of Procion and other visible dyes to the fixation and staining procedures used for electron microscopy. The alternating-section procedures can then at least be applied to assess synaptic distributions on larger processes.

This summary of the areas in which intracellular staining has made special contributions suggests several likely extensions for the future. First, much remains to be accomplished using the techniques at their present stage of development. Substantial doubt still surrounds the first issue which the new methods attempted to resolve, namely, whether identified central neurons have constant branching patterns in different individuals of a species. The pioneering observation of Stretton and Kravitz (1968) is still the most compelling evidence that they do, but it involved a few reconstructions of one neuron. Repetitions on other systems (e.g., Chapter 17) have suggested much more variability. It seems likely that various neuronal populations will differ in this regard, as they do in others: for example, the epithelially-derived sensory neurons of arthropods apparently lack much of the connection specificity and "identifiability" that characterizes the central neurons (Kennedy, 1973). The reproducibility of branching structure could turn out to be the best single criterion for evaluating this kind of specificity.

The analysis of structure to assist physiology will obviously be a continuing need. The use of the intracellular stains in this way should expand, with or without improvement. It is worth reminding ourselves, however, that dye injection is not the only selective staining technique that can be used in support of cellular electrophysiology. For invertebrate ganglia a superb supra-vital method is available, with remarkable uptake selectivity: Methylene Blue gives results similar to those with Golgi on vertebrate material and the best whole-cell stains made by this method are at least as good as the best Procion yellow or cobalt injections. Methylene Blue has at least two drawbacks: the material is difficult to fix for sectioning, and the experimenter cannot predict in advance which neurons will become stained. The first problem can often be sidestepped, since modern photographic methods allow one to preserve just the right moment of progressive staining. The second problem is actually much less troublesome than it first appears. We now have excellent soma maps for many preparations, and in these cases any cell that takes up the dye can be identified. In such cases, Methylene Blue staining would be a rapid and effective way to work out branch anatomy.

Unfortunately, the art of Methylene Blue staining has not been a careful conserved one. Such Old Masters as Retzius and Zawarzin were not carefully about describing their techniques. The most successful contemporary practitioner, Alexandrowicz, has produced beautiful preparations of peripheral nerve elements (e.g., Alexandrowicz, 1951). These results may inspire emulation, but there is no guarantee that the methods that yield success in the periphery will be best for the quite different elements of the central nervous system. In any event, Methylene Blue is a capricious stain that requires a good deal of experimental manipulation for each new application. In short, even more than most staining techniques it is an art—an art which no modern neurocytologist has attempted seriously to turn into a primary investigative tool. Yet, just to go back over Retzius' drawings of neurons in crayfish ganglia with the background of two decades of single-unit electrophysiology is enough to convince one that a sustained effort with this method would be fruitful.

A development that would enormously expand the usefulness of Methylene Blue would be its conversion from a supravital to a vital dye. At the concentrations presently necessary, Methylene Blue salts kill neurons. Other dyes might be found which would allow the survival of activity, or perhaps a different method of applying Methylene Blue would make it less lethal. The prospect of being able to deal with the entire structure of a living neuron in microelectrode experiments is worth a real effort.

As for extensions of the present *objectives* of intracellular staining technology, I believe it is time to consider seriously the prospects for using injected compounds to study connectivity. Under certain conditions, Procion yellow crosses electrical junctions; Bennett's careful analysis of this transfer process (Chapter 8) has already given us some insight into how this happens in normal cells. But there is also an event, less well understood and doubtless pathological, which sends large amounts of injected dye across junctions in a very short time. It is thus possible on occasions to view the injected cell and several other recognizably different cells into which the dye has passed. If this could ever be promoted from the status of an accident, it would be a powerful tool in network analysis.

The problem of how to visualize parts of identified neurons in the electron microscope is far from solved. The cobalt method obscures enough of the structure required to identity a synapse to give it serious handicaps at its present level of development. The best label ultimately could turn out to be radioactive rather than electron-dense, though the technical obstacles to autoradiographic analysis are formidable. As an interim alternative, improvement of the methods for identifying the same process in adjacent sections ought to reap many of the advantages proposed for the electron-dense stains.

It seems likely, however, that the most rapid development will be made with the iontophoretic methods for introducing dye into the cut ends of axons. The technique has the great advantage that it can be employed on neuronal groups *en masse*, with great power to reveal the cellular sources of functionally identified tracts. It offers greatly reduced equipment and skill requirements, and the rate of information acquisition will grow rapidly as a consequence. The method should yield a really substantial improvement in our knowledge of regional neuroanatomy in complex brains.

It would seem appropriate, in concluding, to turn away from the crystal ball and acknowledge that we are still enmeshed in what we might call the ecology of innovation. In real ecosystems, there is an Eltonian pyramid of numbers in which primary producers outnumber primary consumers, secondary consumers are rarer still, and so forth. New scientific areas seem to feature an Eltonian pyramid that has been stood on its head: the primary producers are rarest, the primary consumers (or first-generation users and modifiers) next, and the numbers then increase through additional levels of consumption. A new innovation, which may be made by any consumer, will start a new pyramid if it is important enough. It is never easy to see where such an initiation point will occur: I have tried to indicate a few possibilities, but they are likely to be wrong. It is much easier to look backward than forward. In doing so it is natural for a greedy and grateful primary consumer to pay tribute to the producers—Stretton and Kravitz—without whom, as the cliché goes, this book would never have been written.

References

Alexandrowicz, J. S.: Muscle receptor organs in the abdomen of *Homarus vulgaris* and *Palinurus vulgaris*. Q. J. microsc. Sci. *92*, 163–199 (1951).

Blackstad, T. W.: Mapping of experimental axon degeneration by electron microscopy of Golgi preparations. Z. Zellforsch. mikrosk. Anat. *67*, 819–934 (1965).

——: Electron microscopy of Golgi preparations for the study of neuronal relations. In: Contemporary Research Methods in Neuroanatomy. Ed. W. J. H. Nauta and S. O. E. Ebbesson. pp. 186–216. New York: Springer-Verlag, 1970.

Iles, J. F., and B. Mulloney: Procion yellow staining of cockroach motor neurones without the use of microelectrodes. Brain Res. *30*, 397–400 (1971).

Kennedy, D.: Crayfish interneurons. Fifteenth Bowditch Lecture. The Physiologist, Wash. *14*, 5–30 (1971).

——: In: The Neurosciences: Third Study Program (in the press).

——, A. I. Selverston, and M. P. Remler: Analysis of restricted neural networks. Science *164*, 1488–1496 (1969).

Llinás, R., and C. Nicholson: Electrophysical properties of dendrites and somata in alligator Purkinje cells. J. Neurophysiol. *34*, 532–551 (1971).

Mittenthal, J. E., and J. J. Wine: Connectivity patterns of crayfish giant interneurons: visualization of synaptic regions with cobalt dye. Science *179*, 182–184 (1973).

———: Structure of the giant fibers of earthworms. Science *168*, 994–996 (1970).

Pitman, R. M., C. D. Tweedle, and M. J. Cohen: Branching of central neurons: intracellular cobalt injection for light and electron microscopy. Science *176*, 412–414 (1972).

Remler, M. P., A. I. Selverston, and D. Kennedy: Lateral giant fibers of crayfish: location of somata by dye injection. Science *162*, 281–283 (1968).

Selverston, A. I., and D. Kennedy: Structure and function of identified nerve cells in the crayfish. Endeavour *28*, 107–113 (1969).

———, and M. P. Remler: Neural geometry and activation of crayfish fast flexor motor neurons. J. Neurophysiol. *35*, 797–814 (1972).

Stretton, A. O. W., and E. A. Kravitz: Neuronal geometry: determination with a technique of intracellular dye injection. Science *162*, 132–134 (1968).

Wilkens, L., and J. L. Larimer: The CNS [sic] photoreceptor of crayfish: Morphology and synaptic activity. J. comp. Physiol. *80*, 389–407 (1972).

Zettler, F., and M. Järvilehto: Histologische Lokalisation der Ableitelektrode. Belichtungspotentiale aus Retina und Lamina bei *Calliphora*. Z. vergl. Physiol. *68*, 202–210 (1970).

Zucker, R. S.: Crayfish escape behavior and central synapses. I. Neural circuit exciting lateral giant fiber. J. Neurophysiol. *35*, 599–620 (1972).

CHAPTER 20

A Guide to Intracellular Staining Techniques

Stanley B. Kater, Charles Nicholson, and William J. Davis

The increasing role that intracellular staining procedures are assuming in neurobiology has been aptly demonstrated in preceding chapters. Successful application of intracellular staining techniques, however, often requires more information than can be included in the "Methods" section of a conventional paper. Such information has usually passed between laboratories by personal communication (e.g., "Notes from the Neurophysiology Underground," see Davis, 1970). In this chapter we collate information into a form that should allow the reader to rapidly acquire expertise in using these procedures.

There are two different methods for filling individual neurons with stain: axonal iontophoresis and microelectrode injection. These differ not only in procedure but also in the kinds of problems for which they are best suited. Axonal iontophoresis is an extremely simple technique for providing information on the microanatomy of groups of neurons whose axons are accessible, and has the advantage that no specialized equipment or training in electrophysiological procedures is required. Microelectrode injection of stain, on the other hand, requires both specialized equipment and previous electrophysiological experience. The latter approach not only provides information on the overall microanatomy of single neurons, but it can also be employed to localize the origin of electrophysiological processes by marking the intracellular location of an electrode.

The most difficult prescription to make is the choice of the appropriate stain for specific applications. Considerable time and effort have gone into the selection of stains for use in microelectrode injection (see Chapter 2), but as yet little comparative work is available on the choice of stains for axonal iontophoresis. Available information suggests that selection of intracellular stain should be based on the following considerations: (1) resolution (see Chapters 2 and 6); (2) toxicity (see last section of this chapter); (3) mobility (see Chapters 2, 6, and last section of this chapter); (4) desired method of injection; and (5) equipment availability (e.g., Procion yellow requires fluoresence microscopy). Some discussion of the comparative features of two widely used intracellular stains, Procion yellow (I.C.I., England, and Wilson Diagnostics Inc., 3 Science Rd., Glenwood, Illinois, U.S.A.) and cobalt(ous) chloride ($CoCl_2$), are found in Chapters 7, 10, and 15 and in the last section of this chapter.

Axonal Iontophoresis

Only limited equipment is required for the application of this method to a variety of neuroanatomical problems. Chapters 7, 10, and 15 provide examples of problems for which axonal iontophoresis is particularly well suited. Briefly, the technique relies on the mobility of stain molecules and the fact that the intracellular space of close-packed axons within a nerve trunk can act as specific channels for this movement.

Later in this section we will present methods for implementing the axonal iontophoretic technique as originally described by Iles and Mulloney (see Chapter 7). Recent results, however, have suggested a much simpler technique which has produced superior results on the limited number of preparations on which it has been tested. N. M. Tyrer, J. Altman, and others at Canberra (personal communication) find it unnecessary to impose an artifical current bias when using cobalt chloride for axonal iontophoresis in arthropod neurons. Employing a configuration like that shown in Fig. 2B, the cut nerve is placed into a pool of 5 to 15% (w/v) cobalt chloride in physiological saline solution. Best results are obtained using hypotonic solutions. The nerve is left for 3 to 10 hr (other preparations require up to 36 hr) and then processed as usual (see Chapter 6). One of the primary assets claimed for this technique is that there is less spread of staining material to extraneural elements. Tyrer and Altman suggest that cobalt moves up the axons by simple diffusion. Another possible mechanism may involve currents generated by the cut nerves as a result of the demarcation, or injury, potential. This current would be in the proper direction to facilitate movement of the positive cobalt ions. Since there is as yet no reported success with injection of the negatively charged Procion yellow in the absence of imposed current, the demarcation potential may well be the driving force for the movement of cobalt.

We strongly suggest that the variation described above be tried first. The results it has provided with insects (*Schistocerca gregaria* and *Acheta domestica*), an annelid (*Hirudo medicinalis*), and molluscs (*Helisoma trivolvis, Pleurobranchea californica*) are far superior to those obtained using an imposed current bias. However, the fluorescent properties of Procion yellow are well suited for certain situations (e.g., in heavily pigmented tissues), and we therefore present the following details for axonial iontophoresis by imposed current.

Equipment

Only basic electrical equipment is required for axonal iontophoresis: (1) a variable current source, (2) a current monitor, and (3) contact electrodes.

A low voltage *current source* is the primary component of this system; it need only be capable of passing 10^{-6} A D.C. across approximately 10 MΩ. A standard 12 V dry lantern battern is the simplest source.

Current flow can be regulated by a fixed and a variable resistor (e.g., a 5 MΩ potentiometer) placed in series in the circuit (see Fig. 1A). In this way one can, in theory, regulate the current by knowing the voltage of the battery and altering the series resistance of the circuit (Ohm's Law: current = voltage/resistance). In practice the current may not remain constant because the electrode resistance can change during the course of

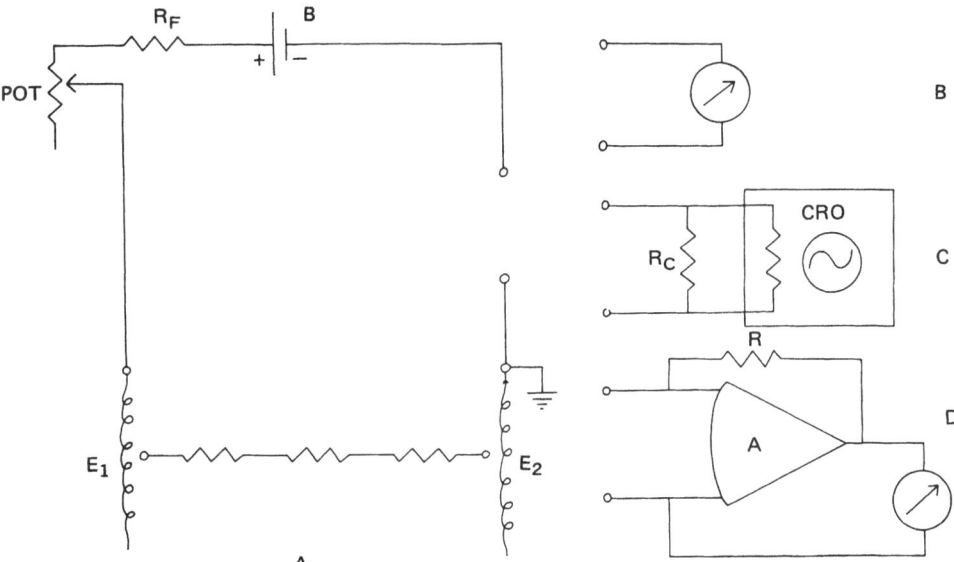

Fig. 1A–D. A diagrammatic representation of the electrical components required for axonal iontophoresis. A. The positive pole of a 12 V battery (B) is connected through a fixed resistor (R_F) to a potentiometer (POT) with a variable resistance of approximately 5 MΩ. From the center tap of the potentiometer a single lead forms a connection with a platinum wire electrode (E_1). An on-off switch can be inserted optionally in a series circuit on either side of the battery. The negative pole of the battery is directly in series with the second electrode (E_2). In this diagram, the circuit is broken between the negative pole and E_2 for inserting one of the current monitoring devices shown at the right (B–D). Resistance symbols between electrodes represent the impedance of the electrodes and preparation. The current monitor in B is a simple ammeter which can measure currents below a microampere. Such an instrument, however, is both expensive and delicate. C. When an oscilloscope with an input impedance greater than 1 MΩ is available, one can monitor current as the voltage drop across a resistor of known value, R_e. This resistor must be of small value so that it does not limit current and so that current flows primarily through it and not the oscilloscope. The circuit in D shows a very accurate alternative method presented by Moore (1971). It employs an operational amplifier (A) and a fixed resistor of known value. The output of this system can be monitored on an oscilloscope or a voltmeter of appropriate sensitivity.

the iontophoresis (see below). Usually the electrode resistance and the preparation resistance combined does not exceed 100 KΩ; therefore the current flow is determined almost entirely by the current-limiting resistors (Fig. 2). The circuit then behaves as a constant current system, and variations in electrode resistance are unimportant.

Current can be monitored throughout the course of the iontophoretic process. Three alternative methods for *monitoring current* are shown in Fig. 1B–D. The simplest method is to insert a microammeter in the circuit. An instrument capable of accurately measuring currents below 10^{-6} A is both expensive and delicate. If an oscilloscope with an amplifier input impedance of about 1 MΩ is available, one may employ one of the alternative circuits shown in Fig. 1C and D.

Electrode wires that actually contact solutions or the preparation should be of special

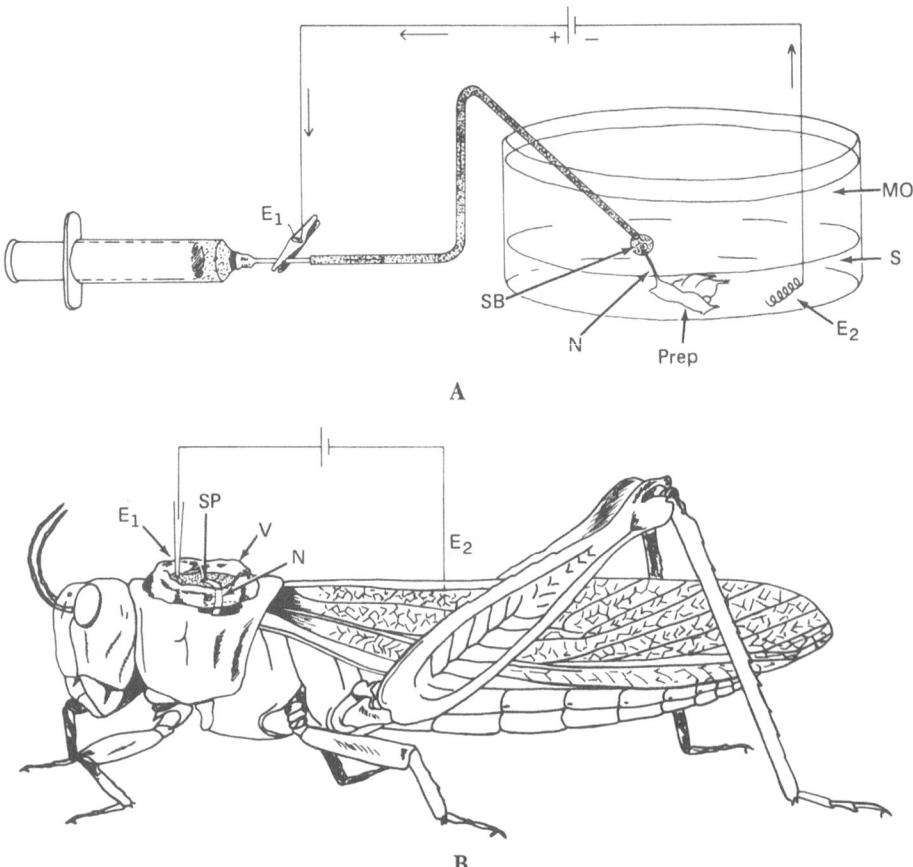

Fig. 2A and B. Two alternative physical configurations for axonal iontophoresis.
A. For work under saline solution. B. For work with 'dry' preparation. (See text for
details.) E₁ and E₂, electrode contacts. MO, mineral oil. N, nerve trunk. PREP, preparation.
S, physiological saline solution. SB, bubble of stain solution. SP, pool of stain solution.
V, 'Vaseline' petroleum jelly well.

construction. Standard metals in contact with salt solutions can 'polarize' while passing
constant current, due to electrolytic action at the metal-saline interfaces (Silver, 1958).
Polarization can greatly increase the resistance of the electrodes and thus retard the
flow of current. Electrode polarization is most pronounced with attempts to pass current
of large magnitude (10^{-6} A). This is mainly a problem when electrodes having a
very small surface area (fine wire) are employed. With axonal iontophoresis one can
normally use a large enough wire to minimize this problem. Polarization can be further
minimized by employing electrodes of 'nonpolarizable' construction. Platinum or silver
wire is effective; however, these should be chlorided for best results. Methods for preparation of silver/silver chloride electrodes are found in a number of technique books
(e.g., Silver, 1958).

A simple suction electrode (Fig. 2A) can be made from a length of small diameter (ca., 1.5 mm O.D.) polyethylene tubing. One end is pushed over a hypodermic needle of appropriate size which is then connected to a syringe. The other end is drawn to a fine tip by pulling it near a small flame. For use, this electrode is filled with stain solution. Care must be taken to avoid air bubbles as they may interrupt current flow. Current is then passed by connecting the appropriate pole of the current source to the hypodermic needle and the opposite pole of the current source to an indifferent electrode (Fig. 2).

Methods

Physical Considerations Dictated by Individual Preparations

An important consideration in axonal iontophoresis is the length of nerve trunk available. The nerve must be long enough to allow effective insulation of the external sheath. Nerve trunks as short as 300 μm can be worked with effectively as long as care is taken in their preparation. At the other extreme, nerve cell bodies have been filled at a distance of 2.0 cm in several different preparations, but the upper limit of nerve trunk length has not yet been established.

Various physical configurations have been employed to fit individual requirements; two are shown in Fig. 2. The configuration shown in Fig. 2A is well suited for isolated neural tissue (e.g., individual ganglia) and certain whole animal preparations that are best dealt with immersed in saline solution. The nerve trunk is exposed and severed in a saline-filled dish. Some saline can then be removed, allowing the nerve trunk to enter a layer of mineral oil previously placed on the saline surface (see Fig. 2A). The mineral oil electrically insulates the nerve trunk exterior from the saline solution. A stain-filled suction electrode (see above) is then introduced into the oil phase and a small bubble of stain solution extruded by pressure on the syringe. The nerve is brought up from the saline phase and its cut end touched to the bubble of stain solution. It is held in the oil phase by the surface tension of the bubble of stain solution attached to the suction electrode.

An alternative arrangement for 'dry' preparations is shown in Fig. 2B. This is the simplest method we have seen for preparations where excessive moisture does not interfere with a suitable insulating seal. The sealing agent, in this case, is white petroleum jelly ("Vaseline," Chesebrough-Ponds Inc., N.Y.). After the nerve trunk is exposed and severed, the nerve and the immediately surrounding area are dried of excess fluid. A 'well' (Fig. 2B) of petroleum jelly is then formed around the nerve by extruding this substance through a fine gauge hypodermic needle. The bottom of the well must form an effective seal around the nerve and the walls must be constructed so they will hold a drop of stain solution. The wire electrode inserted into the well must make good electrical contact with the fluid but should not touch the nerve trunk. The second electrode is located where a good electrical contact with the bulk of the preparation is insured.

Several investigators have added an additional step when using the configuration shown in Fig. 2B. Before placing stain solution in the well, a drop of distilled water is added. After 1–2 min, the distilled water is replaced by stain solution and the current applied. This procedure appears to dilate osmotically the cut ends of neurons, facilitating the passage of stain down individual axons. A treatment of this sort is useful whether or not a current bias is imposed on the preparation.

Parameters of Iontophoretic Injection

It appears that the rate of movement of stain is a function of the diameter of the channels through which it passes. When working with nerve trunks composed of axons whose diameters vary considerably, usually only the larger axons are filled. When there is a higher degree of homogeneity of axon diameters within a nerve, even very small axons (ca., 1 μm) carry stain to their respective cell bodies and neurites. One should note that the presence of stain in the soma is not an assurance of a total fill of that neuron. Stain may enter the soma long before there is any significant invasion in the finer processes of that cell.

When imposing a current bias on the preparation, selection of optimal current values may be an important variable. The exact contribution of this parameter is uncertain, since cobalt can be used with this method without an imposed current bias (see above). Supraoptimal current may result in blockage of the stain in the nerve trunk near the site of iontophoresis, perhaps due to excess heating. In addition, high current can force stain up the extracellular space, thus obscuring any neural elements that might have been filled. While the precise determination of optimal current values for a given nerve will be highly empirical, values between 10^{-6} and 10^{-7} A seem a good starting point when using Procion yellow. The smaller the cross sectional area of a cut nerve, the lower the value of current required. Over a variety of preparations, effective fill times range from 1 to 10 hr.

Microelectrode Injection

In this section we present two different methods for microelectrode injection of stain: iontophoresis and pressure. In addition, we will point out minor variations in technique which can alter the efficiency of staining. Since the technology for the use of microelectrodes in electrophysiological research is beyond the scope of this chapter, we assume some degree of proficiency in this area. For a detailed discussion of microelectrode techniques, see Frank and Becker (1964) and Lavallée et al. (1969).

Iontophoresis

Both Procion dyes and cobalt chloride have sufficient charge and electrophoretic mobility to allow cells to be infused with these compounds by the iontophoretic method. In most situations, the iontophoretic method is preferred over pressure injection if only because the microelectrodes necessary for pressure injection have large tips which often damage the nerve cells upon penetration. Stain-filled microelectrodes used for iontophoresis can have good recording characteristics as long as they are used with impedance-matching amplifiers with sufficient capacitative neutralization capability. The major drawback of Procion-filled electrodes is their rather high resistance (approximately $4\times$ their 3 M

KCl counterparts). Cobalt-filled electrodes, on the other hand, are difficult to use when one wishes to polarize the cell because such electrodes show considerable rectification in their current-passing capability.

Equipment

A laboratory using standard intracellular recording and polarization techniques requires little or no retooling to employ the microelectrode iontophoretic technique. Primary and indifferent metal electrode junctions should be constructed of nonpolarizable elements such as silver/silver chloride (see Silver, 1958) since electrode polarization may cause current blockage.

Methods

Filling Microelectrodes with Stain Solution

Two substances are presently in wide use as intracellular stains: Procion dyes, especially yellow (M4RAN, which is a pure grade of M4RS) and $CoCl_2$ which is available through standard chemical suppliers. Most investigators use a 4% (w/v) solution of Procion yellow, but some make a supersaturated solution of about 10% by gently heating the solution and allowing it to cool slowly. This produces electrodes with lower resistance (NB, a higher current carrying capability) but can cause dye to precipitate and clog the tip when used on preparations kept below room temperature. KCl may be added to Procion solutions; the advantages of this procedure are discussed in the last section of this chapter and in Chapter 18. The initial protocol for iontophoretic injection of $CoCl_2$ (Pitman et al., 1972a) suggested a 1–50 mM solution. We have had good results with a 3M solution of $CoCl_2$, which reduces electrode noise.

All solutions used in microelectrodes should be filtered through a micropore filter with a pore diameter less than that of the electrode tip. For this purpose, we employ 0.22 μm filters in a Swinny filter apparatus attached to a small syringe (Millipore Corp., Bedford, Mass, U.S.A.).

In addition to micropore filtering all solutions used, the glass capillaries can be acid-cleaned. This can be accomplished as follows: (1) cut capillaries into lengths appropriate for electrodes; (2) place about 50 such lengths into a beaker; (3) wash with concentrated H_2SO_4 heated to 80°C; (4) rinse three times with filtered double distilled water heated to 80°C; and (5) dry clean capillaries between two filter papers in a clean petri dish in a dust free oven. When cooled, the capillaries can be made into microelectrodes in a conventional microelectrode-puller.

The easiest method for filling microelectrodes with stain solution involves but slight modification from usual methods (except where glass fibers are used, see below). Micropore filtered, double-distilled water is substituted, for the usual salt (e.g., KCl) solution. Most of this water can then be removed by inserting a 31-gauge needle into the electrode and withdrawing the water from the stem of the electrode by negative pressure on a syringe. A new syringe, filled with stain solution, is then connected to the Swinny filter holder and finally to the fine needle. The needle is reinserted in the back of the electrode and the stem is filled with stain solution. Electrodes are allowed to stand in a hydrated environment for 30 min to 2 hr to allow the stain to diffuse to the electrode tip.

An effective alternative that we have employed for filling microelectrodes is the 'fiber-fill technique' (Tasaki et al., 1968). Before pulling the capillary glass, a few fine strands of fiberglass are inserted into the glass along its full length. Ordinary aquarium filter fiber glass may be used but more consistent results can be obtained with good quality commercial fiberglass. Number 1000 glass fibers may be obtained from Fiber Glass Industries, Inc., Amsterdam, New York, U.S.A. Electrodes are pulled in the usual manner. These electrodes may be filled with stain directly by a filter-coupled syringe and fine needle since the capillarity of the fine fibers will draw the solution directly to the tip of the electrode. Small air bubbles remaining after filling cause no problem as long as electrical continuity exists between the tip and the remainder of the electrode. It is useful to test electrodes for continuity and the ability to pass stain (see Chapters 10, 17, and 18) before attempting to penetrate nerve cells.

Some investigators bevel the electrode tips on a rotating grindstone before filling (see Chapters 13 and 18). This procedure is thought to produce a tip that more easily penetrates cells and also has a greater stain-passing capability (also see last section of this chapter).

Iontophoretic Injection Procedures

Neurons are penetrated as with a standard salt-filled microelectrode; a stable penetration of the neuron is imperative for injections lasting over a long period of time. Once the cell is impaled and the required physiological data are obtained, the stain is passed into the neuron by applying square current pulses of the appropriate sign (hyperpolarizing for Procion yellow, depolarizing for $CoCl_2$). The current passed during iontophoresis should be closely monitored. This allows detection of changes in resistance due to electrode polarization and also provides a reference for future fills. If electrode polarization does cause decreased current passage, the voltage should be raised to maintain a relatively constant value of current. We find that electrode polarization can often be retarded by applying pulses of opposite sign and equal strength for approximately 10 sec every 10 min. For further details on general considerations of microelectrode iontophoresis, consult Curtis (1964).

Unfortunately, the time involved in processing and viewing the results of intracellular staining has discouraged a parametric study of the variables of this technique. The most frequently employed paradigm consists of a repeating series of pulses (of appropriate sign) of about 5×10^{-8} A. Currents an order of magnitude less have been used (see Chapters 6, 10, 11, 12, 13, and 14). A regime of 500 msec rectangular pulses applied at 1 Hz is usually continued for between 5 and 60 min. The major factor determining how long a fill should proceed seems to be how long the investigator remains convinced that the electrode tip is still inside the neuron.

Intracellular staining of certain neurons (especially in vertebrate preparations) is often complicated by the fact that it is difficult to maintain a stable penetration of these cells for the duration prescribed in the preceding paradigm. These cells can be filled using the 'one shot' iontophoretic method. After penetration of the neuron and identification by physiological criteria, the electrode lead is rapidly switched, ideally by relays, from the amplifier input directly to a current source. A single, 1–3 sec pulse of high current (ca., 10^{-6} A) is then passed through the electrode. The electrode blocks rapidly and most often is expelled from the neuron. On inspection. however, a significant area of the neuron is found to contain stain (see Chapter 15).

Pressure

The pressure technique is most appropriate for certain large neuronal elements (e.g., lobster motoneurons and giant fibers, squid giant axons and giant synapse) where the time required for iontophoresis may be prohibitive. Pressure injection could be used for stains which are either uncharged or do not have sufficient electrophoretic mobility for iontophoresis. The major drawback of this method is that it usually requires microelectrode tips that are so large as to damage neurons under approximately 100 μm in diameter. Based on our work with Procion yellow, the results obtained by pressure are in many ways superior to those obtained by iontophoresis. The stain is denser and often migrates considerably farther (see Chapter 10). A disadvantage of pressure, however, is that it occasionally ruptures the fine processes of the neuron, yielding spurious results. Where considerable migration of stain is required for small neurons, we first record from the identified cell with a small, K-acetate-filled electrode to obtain physiological data and then replace this with a stain-filled electrode for pressure injection.

Equipment

In addition to standard electrophysiological equipment, pressure injection requires the following items (see Fig. 3); (1) a tank of compressed air; (2) one regulator, 0–100 lbs; (3) one lever action toggle valve and associated high pressure tubing (e.g., MTV3 Miniature Toggle Valve, Clippard Instruments, Inc., Cincinnati, Ohio, U.S.A.); and (4) pressure injection microelectrode holder (e.g., Model MPH-1, W. P. Instruments, Hamden, Connecticut, U.S.A.) (see also Curtis, 1964, pp. 167–68 and 183–89).

Methods

Filling Microelectrodes with Stain Solution

The considerations discussed in the section on filling microelectrodes for iontophoresis also apply to electrodes used for pressure injection. Contamination, for instance, is still a problem, despite the larger tip size used for pressure injection.

Microelectrodes for pressure injection may be filled with stain as described in the preceding section (taking care to avoid bubbles in the shank); however a faster alternative method is also available. Standard microelectrodes are filled with filtered stain solution by a fine-gauge hypodermic needle inserted into the back of the pipette. The electrode is then mounted in the headstage. Stain is forced down the electrode shank by pressure (ca. 40 lbs). In order to get the stain to exit from the microelectrode tip, it is usually necessary to break the tip by bumping it against a hard surface and then to reapply pressure. This should be done under liquid (water or Ringer's solution) and observed under the high magnification of a dissecting microscope. When the stain begins to flow, the pressure should be released immediately by means of the toggle valve (Fig.

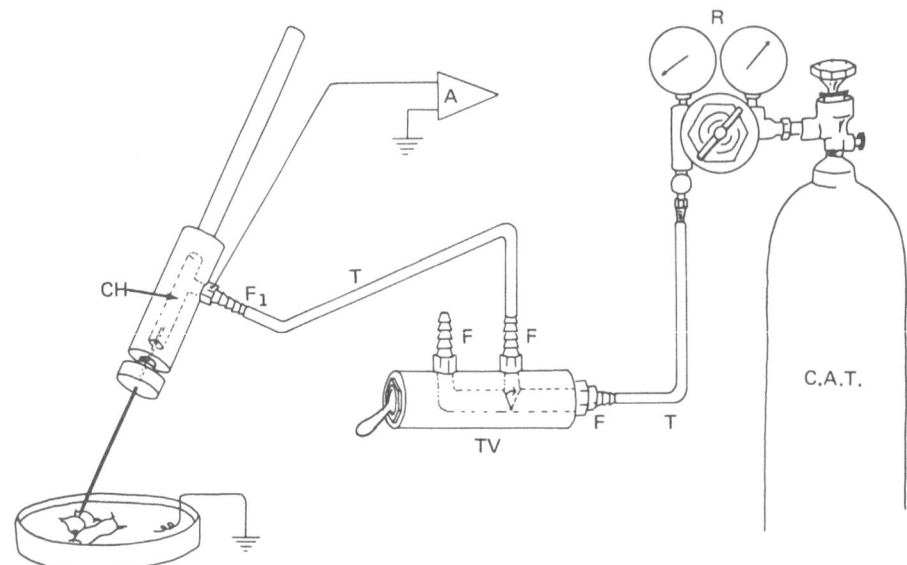

Fig. 3. A diagrammatic representation of a method of pressure injection. A compressed air tank (CAT) is connected to a pressure regulator (R). The output of the regulator is lead to the toggle valve (TV) by high pressure tubing connected to one of the fittings (F) supplied with the valve. Another of these fittings (F_1) is connected into the electrode holder by tapping the holder to appropriate specifications. To this fitting is soldered a lead that runs to the amplifier (A). Stain solution fills the chamber (CH) and provides electrical continuity between the electrode and F_1.

3). It is possible to break off too much of the electrode tip. This is made obvious by the heavy stream of stain that results. Since the commonly employed stain substances are charged, electrode resistance can be measured to further assess the condition of the microelectrode tip. By knowing the approximate resistance of useful KCl-filled electrodes one can determine the acceptability of stain-filled electrodes; e.g., Procion-filled electrodes should have a resistance four times that of KCl-filled electrodes.

The method described above produces tips which are usually too coarse for most intracellular recording applications (D.C. resistance 1–30 MΩ with Procion yellow) but which are adequate for filling neurons which can be identified visually or by intracellular stimulation.

Pressure Injection Procedure

The first requirement of a good fill is a stable penetration of the neuron. In some cases, as with many invertebrate ganglion cells, an acceptable penetration can be accomplished under visual control. Even in these cases it is advisable to monitor the cellular electrical activity as a further criterion of an acceptable penetration. With large neurons that cannot be penetrated under visual control, the electrical characteristics of the cell are the only criteria available for determining whether the penetration is acceptable. Intracellular stimulation, while monitoring the response of postsynaptic elements (e.g., muscles) known to be driven by the neuron, can also provide effective identification.

Once penetrated and identified, most elements are filled under a steady pressure of 1–70 p.s.i. Concurrent weak (c.a., 10^{-8} A, hyperpolarizing) iontophoresis has also been used to help prevent clogging of the electrode tip. The filling process can be completed in as little as 10 sec, or it can require up to 30 min, depending on the pressure employed, the size of the neuron, and the site of the injection (i.e., axon, dendrite, or soma). Excessive pressure can result in leakage of the stain from the cell or complete destruction of the impaled element. Too little pressure, on the other hand, either fails to pass stain from the electrode tip or fills only a small portion of the neuron.

The pressure injection technique is best in situations where one can watch the filling process under a dissecting microscope. As the cell begins to swell visibly, the pressure valve is closed. When the stain begins to migrate away from the locus of injection, the cell may be reinjected. When it is not possible to monitor the passage of stain visually, the tip resistance of the electrode can be monitored continuously as a correlate of stain flow; when stain comes out of the tip the resistance increases markedly.

Post-injection Incubation Period

Many investigators report that irrespective of the method of injection, a post-injection incubation period is useful for allowing stain to migrate throughout the filled neurons. This is a point where critical investigations are sorely missing. First, we have no real understanding of how stains migrate throughout nerve cells. Is it by electrophoresis, osmotic pressure effects, passive diffusion, or active neuronal transport processes? In the absence of any time restrictions, we would recommend the longest incubation time compatible with maintaining a viable preparation. The required time may range from a few minutes in the case of filled retinal cells (Chapters 11 and 12), to many days in the case of *Aplysia* ganglia maintained in artificial culture medium (Chapter 10). Many invertebrate preparations are simply placed in a refrigerator overnight, thus prolonging the viability of the preparation while allowing time for the stain to migrate to the finer processes of the filled neurons.

Preparation of Tissues for Observation of Stain-filled Nerve Cells

Throughout this volume individual preferences are expressed for the optimal preparative procedures. There are, however, three considerations that we would like to stress. First, in selecting a fixative for Procion yellow-filled neurons, one should consider carefully the data on background fluorescences presented by Stretton and Kravitz (Chapter 2). Second, one should assess the amount of tissue distortion that can be tolerated and then consult the protocol presented by Barrett for minimizing tissue distortion during histological procedures (Chapter 18). Finally, whenever possible, whole mount preparations, which minimize the problems inherent in serial reconstruction, should be used (see Chapter 17).

When it is necessary to section material containing stain-filled elements, a number of points must be considered. For Procion-stained material, it is crucial to choose an embedding medium which either has very low intrinsic fluorescence or which can be removed from the sections. In many investigations, paraffin is the embedding medium of choice. It offers two advantages. First, when a large amount of material must be cut to locate a stained element (e.g., vertebrate nervous system), paraffin enables serial sections to be easily cut and managed by virtue of its ribboning property. Second, the background fluorescence of paraffin is sufficiently low to enable major stained profiles to be seen without deparaffinizing. Thus the sections can be quickly scanned until the stained profile is located, the paraffin can then be removed from the required sections and a cover slip can be applied with a nonfluorescent mounting medium (e.g., Entellan, Merck, Darmstadt; Fluoromount, Edward Gurr Ltd., London). The disadvantages of paraffin are that the embedding procedure causes tissue shrinkage (see Chapter 18) and paraffin is not always compatible with the chosen fixative. Fixatives based predominantly on formalin or mercuric chloride usually cut well but when glutaraldehyde is used the paraffin block often becomes excessively friable. Paraffin sections may be cut between about 5 and 200 μm, depending on the hardness of the paraffin selected. Thicker sections, between 50 and 300 μm, can be cut from material embedded in low viscosity celloidin, but this medium fluoresces and should be removed from the section for Procion-stained material (see Barrett and Graubard, 1970). Low viscosity celloidin should be a useful embedding material for cobalt material, since the very low background stain with this technique lends itself to thick sections or whole mounts. Plastic embedding materials have good dimensional stability and yield sections of 5 μm or less. Different plastics vary in their intrinsic fluorescence and Maraglas appears to be the lowest in this respect (see Chapter 2). Frozen sections can minimize tissue shrinkage (Chapter 18) and also enable tissue to be cut soon after the experiment. Preserving serial order with celloidin, plastic, or frozen sections requires care and can be time-consuming. Procion-stained material can be counterstained without losing the Procion; however, the fluorescence may be modified. It is usually unnecessary to stain surrounding neuronal elements since they have an intrinsic green fluorescence which varies with the fixative used (see Chapter 2). It is this tissue fluorescence which often limits the useful thickness at which Procion material may be cut. Basic information on sectioning techniques may be found in standard handbooks, e.g., Carleton and Drury (1957), Jones (1966), and Humason (1967).

Observation and Photography of Stain-filled Nerve Cells

Pitman et al. (Chapter 6) and Mulloney (Chapter 7) present their methods for viewing and photographing cobalt-filled neurons and Van Orden (Chapter 4) provides the fundamentals of fluorescence microscopy. Data on fluorescence excitation emission spectra (Chapters 2 and 4) and on fluorescence degradation as a function of exposure time (Chapter 4) should be consulted before working with Procion yellow. We would also stress the value of Zeiss Luminar lenses for low power observations of both Procion and cobalt-filled neurons.

Problems

There are many unresolved questions which demonstrate appreciable gaps in our knowledge of intracellular staining techniques. Some of these are procedural and result from incomplete analysis of the parameters employed for intracellular staining. Others, like the mechanism of stain migration throughout a neuron, reveal our limited knowledge concerning fundamental questions of cell biology. In an attempt to resolve, or at least reach a concensus on, the points which might affect the efficacy of staining techniques, a series of questions were posed to the contributors of this volume at a meeting held for that purpose. This section represents a distillation of the discussion at that meeting, along with comments made by others at the formal symposium and, to be sure, our own biases.

What is a "Good Fill" and How is it Achieved?

A distinction can be made between staining to identify a neuron and staining to reveal completely neuronal architecture. It seems impossible, in fact, to define a "complete" fill of a neuron (see Discussion of Chapter 18). In invertebrate systems which possess visually identifiable neurons, the most complete picture of a neuron is obtained by a series of fills on the same neuron in different preparations. This procedure has shown that not all neurites fill on any given occasion. In fact, given the reproducibility of neuronal geometry shown in Chapter 2, a composite derived from several preparations would seem to produce the most accurate picture of that neuron. Ramón y Cajal arrived at the same conclusion using the Golgi method.

Often one wishes both to mark the precise locus of recording and to fill the neuron to obtain morphological information. This can be accomplished with Procion yellow by noting the brightest region of the filled neuron, which frequently corresponds to the site of injection and recording. More precise localization has been accomplished by Zettler and Järvilehto (1970). These authors employed a freezing fixation technique that allowed the electrode tip to be broken off and left intimately associated with the actual recording site. The figure presented in a later work employing this method (Zettler and Järvilehto, 1971) clearly shows the close relationship between the region of maximal fluorescence and the electrode tip. Several interesting, but as yet untested, methods of localization have been suggested. Electrodes filled with a mixture of mercurichrome, which fluoresces but does not migrate well, and Procion yellow might be used for this purpose (see Discussion of Chapter 15). Alternatively, selected Procion dyes which migrate poorly (probably those of high molecular weight; see Chapter 3) might be added to Procion yellow solutions.

Factors which May Affect Staining

Electrodes

The larger the opening of the electrode tip the greater the amount of stain that can be passed into a neuron in any given time and the lower the probability of blocking. While the use of microelectrodes with larger tips would facilitate stain injection, too large a tip often causes damage to impaled neurons and produces unreliable physiological data. Barrett (Chapter 18, Appendix II) and Van Essen and Kelly (Chapter 13) have employed a method that seems to partially resolve this problem—beveled electrodes.

A beveled electrode has considerably lower resistance than its conventional counterpart and yet has no increase in tip diameter. Users report an increased facility in penetration of neurons as a result of the sharpness of the tips. A simple method for the production of beveled electrodes is given in Appendix II to Chapter 18. Alternatively, minor modifications of the air-bearing grindstone method published by Vurek et al. (1967) might prove useful. The method applied by A. Gelperin and J. Chang (personal communication) does not rely on a precision grinding element, but rather on a minimum pressure of the electrode on the grinding surface. The grinding surface is a simple motor driven circular plexiglass plate onto which they affix a grinding compound (Alumina Corunda powder type A, 0.3 μm average particle size, available from A. H. Thomas Co., Philadelphia, U.S.A.). The electrode (filled or empty) is mounted on a stiff wire which is then balanced on a razor blade. A counterbalance is added to the opposite end of the wire so that the electrode barely touches the grinding surface. Grinding is performed with the wheel rotating at approximately 80 RPM in the direction that the electrode tip is pointing.

Numerous other modifications have been proposed for standard glass capillary microelectrodes, but there are insufficient data to confirm their usefulness. It is noteworthy that many investigators break the tips of their microelectrodes before use and on occasion this may result in a beveled tip. Glass fiber-filled microelectrodes seem to have a lower resistance than those prepared by boiling methods. Silicone treating glass tubing before pulling may prevent electrode blockage, and scrupulous cleaning of glass and all solutions coming into contact with electrodes may also reduce blockage.

Method of Injection

Axonal Iontophoresis

Our discussions pointed out that we are ignorant of the principles underlying the axonal iontophoretic technique. The fact that cobalt will fill axons over distances up to 2 cm in the absence of an imposed current raises a major question about the mechanism of stain migration. This sort of migration is confirmed by the fact that TEA can migrate from the cut end of an axon into the axoplasm and to the next node of Ranvier (Koppenhöfer and Vogel, 1969). Since Procion, with a negative charge, has not been reported to migrate in the absence of imposed current, it seems possible that the driving force might involve the demarcation currents produced at the cut ends of the axons. As cobalt migrates along this small, but finite, electrical gradient it might then depolarize adjacent regions of the axon, and thus produce a wave of demarcation current which acts as the force for moving the stain further along the axon. In fact, the limitations imposed by the length constants of filled axons obviate electrophoresis as the complete explanation for the distance of migration observed. Another possible mechanism is the possibility that migrating stains increase the membrane resistance, and thus lead to a greater length constant. This, like many other conjectures on this subject, requires experimental analysis.

Microelectrode Injection-Pressure

. A few additional comments should be made concerning pressure injection techniques. The pressure fills reported in Chapter 10 were in somata as small as 50 μm in diameter— clearly in the range of somata of vertebrate neurons, although to our knowledge there have been no attempts to fill vertebrate neurons by the pressure method. The possibility of filling vertebrate neurons with pressure may be important since it is difficult to obtain fills of vertebrate somata which result in stain migration along the axon. Small-tipped electrodes may work for pressure injection if scrupulous cleaning procedures are applied to electrodes and stain solutions. Furthermore, somewhat larger electrodes (and perhaps beveled electrodes) can be used with neurons in the 25 to 75 μm range. Initially, penetration of a neuron will result in depolarization resulting from damage, but the cell can be restored to a "normal" state by a short term treatment with hyper-polarizing current.

Microelectrode Injection-Iontophoresis

The most obvious variable that could influence staining with this method is the amount of current employed to inject the stain into the neuron. Various investigators have successfully employed currents ranging from nanoamperes to microamperes, the very large currents being suggested exclusively for the "one shot" technique (Chapters 10 and 15). In the standard paradigm of alternating current pulses, values range from 1 to 5×10^{-9} A (e.g., A. Kaneko and also R. Murphey) up to 15×10^{-9} A (e.g., A. Selverston and also C. R. S. Kaneko and S. Kater). Logically, more stain is injected if more current is passed into the cell in any given period. Empirically, however, we feel that there may be deleterious results associated with high current values. High current values (tens of nanoamperes) might cause blockage by, for instance, a local heating of the axoplasm near the tip of the electrode, resulting in coagulation.

Stain Migration

We have pointed out above that a simplified view of electrophoretic mobility will probably not account for most results. Nonetheless, Kater and Llinás have applied the one-shot method to 100 μm diameter soma in the snail, *Helisoma* and immediately fixed the preparation. The time from injection to fixation was less than 10 min and yet the axon was filled (with cobalt in this case) at least 500 μm from the injection site. We offer this example to show migration (possibly solely by electrophoresis) in the absence of an incubation period that might allow active neuronal transport processes to carry the stain molecules. It should be pointed out that the nerve trunk containing the axon in this case was cut prior to injection of the current pulse and thus possibly facilitated the movement of stain by creating a more advantageous current path. Most investigators would agree, however, that a post-injection incubation period significantly increases stain migration.

A most interesting difference in stain migration was discerned between axonal ion-tophoresis and microelectrode iontophoresis of stain. In the former method, stain flows from the axons and fills somata quite regularly. Conversely, it appears more difficult to obtain migration from the soma to the axon and neurites of a cell. Stain often stops migrating at sites where there is a significant change in diameter (see Chapter 10), which may be caused by the concomitant increase in axial desistance to current flow. Axonal iontophoresis, with the stain traveling in the other direction, avoids this problem.

Choice of Stain

It is difficult to compare directly the two major compounds, cobalt and Procion yellow, presently available for intracellular staining because of the relatively recent introduction of cobalt. Our discussions did, however, point out a series of advantages, disadvantages, and possible means of modification of technique for both compounds.

Cobalt

One can obtain Golgi-like detail from neurons impregnated with cobalt and viewed as either whole mounts or in thick sections. In insects, where the best comparative information is available, cobalt-filled neurons can give more information about neuronal geometry, at least in whole mounts, than that obtained with Procion. Where rapid information is required, and the level of information contained in whole mounts or very thick sections will suffice, cobalt seems a wise first choice.

Contraindications for the use of cobalt are based, unfortunately, on limited experimental data. The first of these comes from reports of the potential toxicity of cobalt. Murphey and also Rowell, working with insects, and C. R. S. Kaneko and Kater and Waziri (personal communication) on molluscs have observed significant alterations in the electrophysiological characteristics of neurons impaled by electrodes filled with cobalt chloride (50 mM to 3 M). Even when no current is injected into the neuron and the cobalt-filled electrode serves only a monitoring function, one routinely notes a rapid decrease in membrane potential and a concomitant decrease in the amplitude of action potentials. It seems possible that one could overcome leakage from microelectrodes by applying a bucking current in series with the electrode. This would be useful only where such currents would not interfere with the interpretation of synaptic events and would not impose a severe voltage bias on the neuron.

Cobalt also alters potentials recorded from cat spinal motoneurons. Lux and Schubert (1969) demonstrated that intracellular injection of cobalt produced long-lasting changes in the equilibrium potential of IPSPs. Cobalt also produces significant effects when applied extracellularly (see Chapter 1).

While cobalt is electron opaque and thus useful for electron microscopic localization of intracellular injected neurons, there has been some criticism of the quality and fidelity of the information that it provides. Ammonium sulfide, the recommended precipitating agent, causes significant distortion of tissue. This distortion might be great enough to alter neural associations to the point of obviating the usefulness of the method. Another problem is the extreme density of the cobalt reaction product, which can completely obscure cellular organelles. If one is unable to recognize membrane specializations and synaptic vesicles it is difficult to define critically points of synaptic contact.

Procion Yellow

The rationale behind the selection of Procion yellow as an intracellular stain is presented in Chapter 2. In addition to these points we have seen how the use of various colors of Procion dyes can significantly complement physiological data. Kaneko has recorded simultaneously from two retinal cells, followed by stain injection using one electrode filled with Procion yellow and another filled with Procion red (Chapter 11). Shaw (1972), working with the visual system of the barnacle, has added an interesting variation to this technique. One electrode is filled with equal part mixtures of 5% Procion yellow M4RAN and 5% Procion Navy blue H3RS and the other electrode similarly with Procion yellow and Procion Brilliant red H3BNS. This, he felt, allowed him to take advantage of the superior fluorescent properties of Procion yellow, while giving him the added advantage of identification and initial localization of stain filled neurons.

For some time it was felt that the major drawback of Procion dyes was that they were of little use in ultrastructural studies. Their lack of electron opacity may indeed be a drawback, but, as shown in Chapter 5, with diligent application, Procion yellow may be of considerable value for investigations of neuronal fine structure. All factors considered, it appears that Procion dyes remain as the most useful compounds in our intracellular staining repertoire. That is not to say that these compounds are perfectly suited to all our requirements, but rather that even in their present form, they can provide a great deal of reliable information in a variety of experimental conditions. Certainly, from microelectrode work where distortion of the electrical properties of impaled neurons is unacceptable, a comparison of the potential toxic effects of cobalt (see above) to the lack of these effects with Procion (see Chapter 15) would dictate the use of Procion.

Among the initial advantages cited for Procion yellow was the fact that it fluoresced after standard histological treatment of tissue. This feature would theoretically extend the minimum size of a neurite that could be resolved in a stain-filled neuron. In fact, a compound with considerably stronger fluorescence than Procion yellow would be most desirable, since the stain can be in such low concentration in the neurites that resolution is quite poor.

There are several possible technical modifications of Procion yellow that might be useful. The first of these is the addition of salt to the Procion solution. In the textile industry, dye is initially deposited on cloth by the addition of salt. That is, the presence of salt can cause Procion yellow to fall out of solution. This, then, might be one of the bases of electrode blockage. When Procion moves from the electrode into the cell it encounters an environment high in salts. It has been suggested by Stead (personal communication) that the addition of perhaps 0.1 to 0.3 M KCl to Procion solution might decrease the probability of electrode blockage by decreasing the salt difference encountered by Procion molecules flowing from the electrode. It would be important, when using this modification, to carefully filter the stain and salt solution before filing electrodes since some Procion will undoubtedly precipitate during this process. Prior to Stead's suggestion, John Barrett (Chapter 18) had empirically arrived at a similar conclusion and now routinely adds KCl to his Procion solutions.

According to Stead, it should be possible to either modify the existing structure of Procion yellow or synthesize slightly different molecules with modification of the basic Procion formula (Chapter 3) in order to tailor the dye more precisely to our needs. The solubility, for instance, is presently selected as optimal for textile work. Minor structural changes could increase the solubility by several fold (see discussion, Chapter 3). Additionally, Stead has suggested that one might titrate Procion solutions with NH_4OH at 40°C until a stepwise pH change was observed. This would indicate the release of a Cl^- group and thus render the compound less binding. This treatment might be useful for extending the amount of migration Procion would undergo before binding. Reactivity can then be increased after migration (see discussion, Chapter 3). Alternatively, binding is decreased by cold temperature. Cooling Procion-filled preparations might retard binding and thus increase migration, if this treatment did not alter significantly the processes responsible for migration itself. After a cold post-injection incubation period, the preparation could be kept at elevated temperatures (optimum for the dye is about 40°C) to ensure binding.

Prospectus

It is apparent that a new set of words will soon be added to our neuroanatomical vocabulary. New schemes of classification which correlate neuronal structure with neuronal function are within our reach. Several investigators are capable of rendering serial sections of stain-filled neurons into quantified statements of neuronal architecture. The answers that the intracellular stain technology can provide, however, are limited not only by the questions that we pose but also by our willingness to develop more routinely reliable methods. If we remain content with the variability and unknowns that are now a part of these methods, intracellular staining will never realize its full potential as a tool for neurobiologists. We require more precise experimental investigations of the variables inherent in our methods. Such investigations will undoubtedly yield not only technical information on staining but also important data on fundamental questions of neurobiological importance.

References

Carleton, M. M., and R. A. B. Drury: Histological Technique. New York: Oxford University Press, 1957.

Curtis, D. R.: Microelectrophoresis. In: Physical Techniques in Biological Research, Vol. VA. Ed. W. L. Nastuk. pp. 144–192. New York: Academic Press, 1964.

Davis, W. J.: Motoneuron morphology and synaptic contacts: determination by intracellular dye injection. Science 168, 1358–1360 (1970).

Frank, K., and M. C. Becker: Microelectrodes for recording and stimulation. In: Physical Techniques in Biological Research. Vol. VA. Ed. W. L. Nastuk. pp. 23–88. New York: Academic Press, 1964.

Humason, G. L.: Animal Tissue Techniques (second edition). San Francisco: Freeman, 1967.

Jones, R. M.: Basic Microscopic Techniques. Chicago: University of Chicago Press, 1966.

Koppenhofer, E., and W. Vogel: Effects of tetrodotoxin and tetraethylammonium chloride on the inside of the nodal membrane of Xenopus laevis. Pflugers Arch. ges. Physiol. 313, 361 (1969).

Lavellée, M., O. F. Schanne, and N. C. Hebert: Glass Microelectrodes. London: Wiley, 1969.

Lux, H. D., and P. Schubert: Postsynaptic inhibition: intracellular effects of various ions in spinal motoneurons. Science *166*, 625–626 (1969).

Pitman, R. M., C. D. Tweedle, and M. J. Cohen: Branching of central neurons: intracellular cobalt injection for light and electron microscopy. Science *176*, 412–414 (1972).

Shaw, S. R.: Decremental conduction of the visual signal in barnacle lateral eye. J. Physiol., Lond. *220*, 143–175 (1972).

Silver, I. A.: Other electrodes. In: Electronic Apparatus for Biological Research. Ed. P. E. K. Donaldson. pp. 568–584. London: Butterworths Scientific Publications, 1958.

Tasaki, I., Y. Tsukahara, S. Ito, M. J. Wayner, and W. Y. Yu: A simple, direct and rapid method for filling microelectrodes. Physiol. Behav. *3*, 1009–1010 (1968).

Vurek, G. C., C. M. Bennett, R. L. Jamison, and J. L. Troy: An air-driven micropipette sharpener. J. appl. Physiol. *22*, 191–192 (1967).

Zettler, F., and M. Järvilehto: Histologische Lokalisation der Ableitelektrode. Belichtungs Potentiale aus Retina und Lamina bei *Calliphora*. Z. vergl. Physiol. *68*, 202–210 (1970).

——— ———: Decrement-free conduction of graded potentials along the axon of a monopolar neuron. Z. vergl. Physiol. *75*, 402–421 (1971).

Zucker, R. S., D. Kennedy, and A. I. Selverston: Neuronal circuit mediating escape responses in crayfish. Science *173*, 645–650 (1971).

Subject Index